Green Energy and Technology

Climate change, environmental impact and the limited natural resources urge scientific research and novel technical solutions. The monograph series Green Energy and Technology serves as a publishing platform for scientific and technological approaches to "green"—i.e. environmentally friendly and sustainable—technologies. While a focus lies on energy and power supply, it also covers "green" solutions in industrial engineering and engineering design. Green Energy and Technology addresses researchers, advanced students, technical consultants as well as decision makers in industries and politics. Hence, the level of presentation spans from instructional to highly technical.

Indexed in Scopus.

Indexed in Ei Compendex.

Jamal Akhter Siddique · Akil Ahmad ·
Mohammad Jawaid
Editors

Energy Sources

Environmental and Health Prospects

Springer

Editors
Jamal Akhter Siddique
Department of Chemistry
Gandhi Faiz-E-Aam College
Mahatma Jyotiba Phule Rohilkhand University
Shahjahanpur, Uttar Pradesh, India

Akil Ahmad
Department of Chemistry
College of Science and Humanities
Prince Sattam Bin Abdulaziz University
Al-Kharj, Saudi Arabia

Mohammad Jawaid
Department of Chemistry and Petroleum Engineering
United Arab Emirates University
Al Ain, United Arab Emirates

ISSN 1865-3529 ISSN 1865-3537 (electronic)
Green Energy and Technology
ISBN 978-981-95-1208-9 ISBN 978-981-95-1209-6 (eBook)
https://doi.org/10.1007/978-981-95-1209-6

This Springer imprint is published by the registered company Springer Nature Singapore Pte Ltd.
The registered company address is: 152 Beach Road, #21-01/04 Gateway East, Singapore 189721, Singapore

Contents

Introduction, Origin, and Development of Fossil Fuel and Its Importance 1
Babita Saroha, Sukwinder Kaur, Veena Maurya, Naresh Kumar, Neelam Dutt, and Anand Kumar

The Future of Energy and Mobility 17
Abdul Jabbar Khan and Muhammad Sajjad

Composition, Classification, and Properties of Petroleum 27
Rajeev Kumar, Riya Chugh, and Indrani

The Role of Fossil Fuels in a Sustainable Energy System 47
Babita Saroha, Anand Kumar, Naresh Kumar, Deepali Singhal, Rakesh Dhoundiyal, H. V. Pant, Sandeep Negi, Harish Chandra, and Shyam Vir Singh

Coal Energy: Carbon Capture, Utilization, and Storage 67
Shahid Pervez Ansari, Mohammad Ayub Husain Khan, Deepika Varshney, Mohammad Omaish Ansari, and Jamal Akhter Siddique

An Introduction and Industrialization of Crude Oil 81
Rajesh Kumar, Kahkashan Begum, Shahid Parvez Ansari, and Jamal Akhter Siddique

A Fundamental Approach to Renewable Energy: Classification and Development 95
Shaukat Khan

Solar Renewable Energy Systems: Study and Applications with Present and Future Potential 123
Ranjan Dey, Eshita R. Kakodkar, Praseeda P. Nair, and Suvanjan Bhattacharyya

Wind Energy Technology: Application and Impact 147
Riya Chugh, Indrani, and Rajeev Kumar

IoT-Based System for Hybrid Renewable Energy System and Its Application 171
Sashwat Kumar Singh and Debashis Kundu

Introduction and Classification of Fuel Energy 193
Priti Gupta, Shilpa Sharma, Shagufta Jabin, and Ritu Bir

Artificial Intelligence Techniques in Wind Energy Prediction for Sustainable Development 215
Sadiqa Abbas, Shagufta Jabin, Sonal Saluja, Anupama Chadha, and Jyoti Chawla

Transition from Fossil Fuel to Sustainability: Solar Energy—Prospects, Technology Advancement, Policies, Challenges, and Opportunities 243
Deepika Varshney, Rashmi Mishra, Shahid Pervez Ansari, Jamal Akhter Siddique, and Mohammad Omaish Ansari

Introduction, Origin, and Development of Fossil Fuel and Its Importance

Babita Saroha, Sukwinder Kaur, Veena Maurya, Naresh Kumar, Neelam Dutt, and Anand Kumar

Abstract Fossil fuels have played a central role in shaping the modern world. Fossil fuels remain the dominant source of global energy consumption, powering industries, transportation, and households across the world. As conventional energy sources, they are nonrenewable in nature, derived from the ancient remains of once-living organisms buried deep beneath the Earth's surface for millions of years and gradually transformed by intense heat, pressure, and time, and they are destined to be exhausted if reliance continues without sustainable alternatives. Fossil fuels have played an instrumental role in powering the industrial development and economic expansion of human civilization. This chapter provides a comprehensive overview of the origin, classification, development, and global significance of fossil fuels. It explores their geological formation over millions of years, the history of their discovery and utilization, and their classification into coal, oil, and natural gas.

Keywords Fossil fuels · Coal · Petroleum · Natural gas · Conventional

B. Saroha
Department of Zoology, , Roorkee Degree College, Latherdeva Shekh, PO Iqbapur,, Roorkee,, Uttarakhand, India

S. Kaur
GSSS Poohla, Bathinda, Punjab, India

Department of Chemistry, Uttarakhand Open University, Haldwani, Uttarakhand, India

V. Maurya
Department of Zoology, SGRR University, Dehradun, Uttarakhand, India

N. Kumar
Department of Applied Sciences and Humanities, School of Engineering and Technology, H.N.B. Garhwal University, Srinagar Garhwal, Uttarakhand, India

N. Dutt
Faculty of Pharmacy, Babu Banarashi Das Northern India Institute of Technology, Lucknow, Uttar Pradesh, India

A. Kumar (✉)
Department of Chemistry, SGRR (PG) College, Dehradun, Uttarakhand, India
e-mail: anandkciitd17@gmail.com

J. A. Siddique et al. (eds.), *Energy Sources*, Green Energy and Technology,
https://doi.org/10.1007/978-981-95-1209-6_1

1 Introduction

Fossil fuels are a primary energy resource and are called "fossil" fuels because they originate from fossilized biological matter. These fuels are rich in carbon and hydrogen, making them highly combustible and efficient for generating energy. Fossil fuels have played a vital role in the global energy mix for centuries, supplying the majority of the world's electricity and transportation fuels [1]. These fuels are formed through natural processes such as the anaerobic decomposition of dead organisms. The energy stored in fossil fuels originates from ancient photosynthesis [2]. Formed from the fossilized remains of dead plants [3], they are created over millions of years under intense heat and pressure within the Earth's crust [4]. The organisms that make up these fuels are often older than 650 million years [5]. Fossil fuels are carbon-rich and include coal, oil, and natural gas, along with derivatives like kerosene and propane. These fuels range in form from nonvolatile materials like anthracite coal, which contains almost pure carbon, to volatile substances with lower carbon-to-hydrogen ratios, such as methane, and to liquids like petroleum. Methane can be found in hydrocarbon fields either by itself, associated with oil, or in other forms.

Although fossil fuels are technically formed by ongoing natural processes, they are generally considered nonrenewable because their formation takes millions of years, and existing reserves are being consumed much faster than new ones can be replenished [6]. According to data from 2015, fossil fuels made up 86.0% of global primary energy consumption: petroleum accounted for 32.9%, coal for 29.2%, and natural gas for 23.9%. In contrast, nonfossil energy sources contributed much smaller shares: hydroelectric (6.8%), nuclear (4.4%), wind (1.4%), solar (0.45%), and other renewables (0.89%), as shown in Fig. 1 [7].

As global energy consumption continues to rise, the connection between fossil fuels and our lives remains significant. However, it is vital to understand their environmental impact [8]. The combustion of fossil fuels emits approximately 21.3 billion tonnes of carbon dioxide (CO_2) each year. Natural processes can only absorb about half of this amount, leaving an excess of around 10.65 billion tonnes of CO_2 released into the atmosphere annually [9]. It is now widely recognized that CO_2 is a greenhouse gas that contributes to global warming by increasing atmospheric radiative forcing. As a result, there is a growing global movement toward renewable energy generation to reduce greenhouse gas emissions, including those from fossil fuels. Additionally, the interconnection between energy and water has become increasingly evident [10]. Fossil fuel extraction and processing require significant water input, which often negatively impacts water quality.

Fossil fuel use is expected to drop significantly in scenarios aimed at meeting the Paris Agreement goals [11]. However, fossil fuels aren't just used for energy—they are also a major component in making materials, especially plastics. Today, most plastics are still made from fossil-based sources [12]. For a long time, the environmental impact of plastics, including emissions, did not get much attention [12, 11].

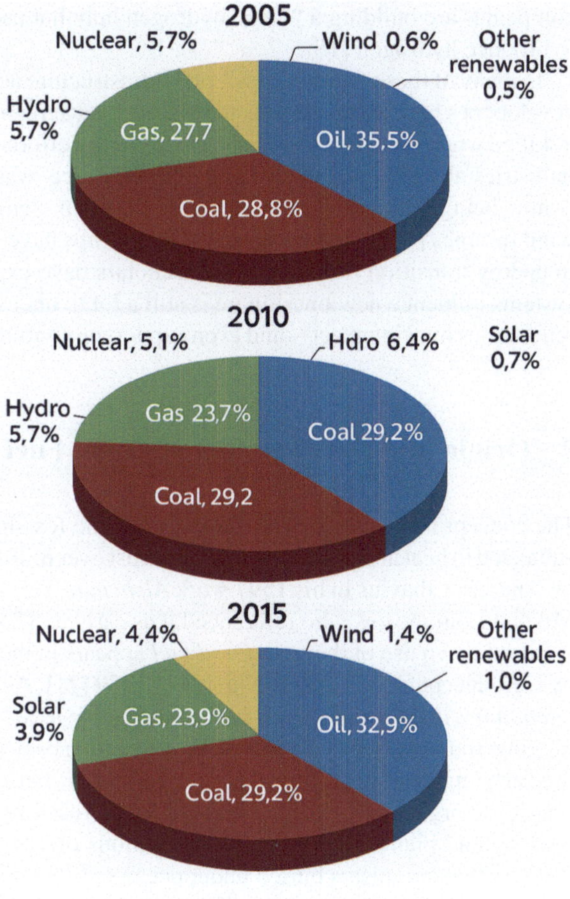

Fig. 1 Global primary energy consumption comparisons. Reprinted from World Energy Council [7]. Survey of world energy resources, London, 2016

In August 2022, U.S. President Joe Biden signed the Inflation Reduction Act (IRA)—the most ambitious climate legislation in American history. The Act renews and expands tax credits for clean energy projects, which is expected to drive significant investment in renewable energy [13]. Notably, it provides a 10% bonus tax credit to projects located in "energy communities"—areas historically dependent on fossil fuel industries, like former coal mining towns [14]. This approach is seen as a way to support both economic recovery and the shift to cleaner energy in places hit hardest by job losses and industrial decline. The Act also expands federal loan programs, sparking renewed interest in repurposing old wells, power plants, and degraded mining lands for clean energy development [15].

Still, the U.S. energy transition is more complex than simply replacing fossil fuels with renewables. In some places, the two are becoming surprisingly interconnected. In West Texas's Permian Basin, for example, solar and wind farms help power the region's booming oil production [16, 17]. In North Dakota's Bakken shale, energy companies are building a "blue" hydrogen hub that uses locally sourced natural gas to produce hydrogen [18].

In many of these regions, fossil fuel infrastructure actually gives renewable energy developers a head start—thanks to existing land agreements, permitting systems, and a skilled workforce. But ironically, clean energy efforts may also help keep extractive industries alive in unexpected ways. For instance, waste from fossil fuel extraction is now being "mined" for lithium and rare earth elements used in solar panels and wind turbines [19]. These complex relationships have not received enough attention in energy transition research. While scholars have explored how traditional energy systems influence new ones, there is still a lot to uncover about how fossil fuels and renewables may intersect—and even support one another—in specific local contexts.

2 Origin and Development of Fossil Fuel

The concept that fossil fuels originate from the fossilized remains of ancient plants subjected to heat and pressure in Earth's crust over millions of years was first proposed by Andreas Libavius in his 1597 work *Alchymia*. This theory was later elaborated by Mikhail Lomonosov, who referenced it as early as 1757 and later by 1763 [20]. The earliest known use of the term *fossil fuel* appears in the English translation of a work by German chemist Caspar Neumann in 1759 [21]. According to the *Oxford English Dictionary*, in the phrase *fossil fuel*, the adjective *fossil* originally meant "obtained by digging; found buried in the earth," a usage dating back to at least 1652. It was not until the early eighteenth century that the noun *fossil* became primarily associated with long-dead organisms. Figure 2 represents the fossil fuel formation: coal forms from ancient land plants compressed over millions of years, while oil and gas originate from marine organisms buried under ocean sediments, later transformed by heat and pressure. Both processes take millions of years and involve deep geological changes.

Petroleum and natural gas are primarily formed from the remains of aquatic phytoplankton and zooplankton that accumulated in large quantities under anoxic conditions millions of years ago. These organic remains became buried under layers of sediment, where they underwent anaerobic decomposition. Over time, increased pressure and temperature transformed the organic matter into a waxy substance called kerogen, commonly found in oil shales. With further heat, this kerogen underwent catagenesis—a process that breaks it down into liquid and gaseous hydrocarbons. Although these transformations are driven by heat, the energy released during combustion originates from ancient photosynthesis [2].

In contrast, terrestrial plants contributed primarily to the formation of coal and methane. Many of the world's major coal fields trace back to the Carboniferous period. Terrestrial plant matter also produces Type III kerogen, a precursor to natural gas. While fossil fuels continue to form through natural geological processes, they are considered nonrenewable resources because their formation takes millions of years, and the rate at which known reserves are being depleted far exceeds the rate of new formation [22, 23].

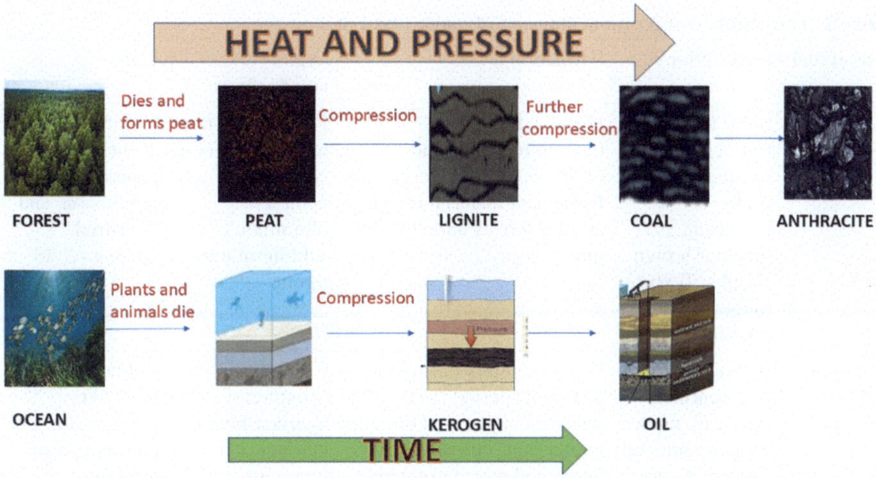

Fig. 2 Formation of fossil fuels. Reprinted from https://histatic.zzdn.net/files/d09/ade0c5638d9f09a2674f63b6671eee47.png

3 Classification of Fossil Fuels

Fossil fuels are classified based on their physical form—solid, liquid, or gas—and their biological origin. As depicted in Fig. 3, the three main types include **coal**, a solid fuel formed from ancient land plants subjected to intense heat and pressure over millions of years; **petroleum (crude oil)**, a liquid fossil fuel derived from the remains of marine organisms like plankton and algae that were buried under sediment and transformed by geological processes; and **natural gas**, a gaseous fuel that forms under similar conditions to oil, often at higher temperatures.

These fuels differ not only in their physical states but also in their **formation processes, chemical compositions, and uses**, as outlined in Table 1. Coal is widely used for electricity generation, petroleum is refined into fuels like gasoline and diesel, and natural gas is used for heating, electricity, and as a feedstock in chemical industries.

Fig. 3 Classification of fossil fuels

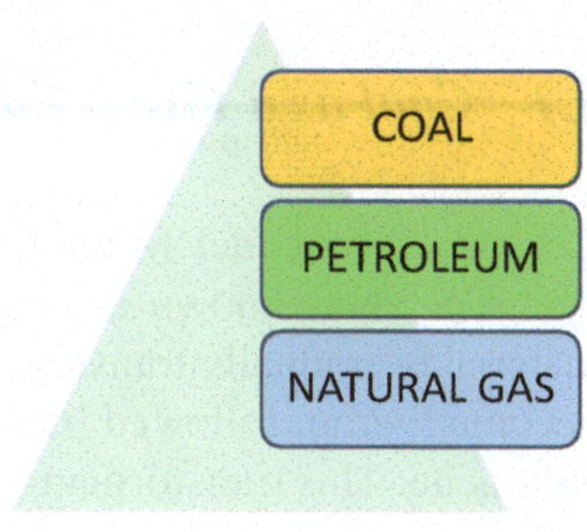

Table 1 Formation, composition, and uses of major fossil fuels

Fossil fuel	Description	Formation	Types/ composition	Uses
Coal	Solid fossil fuel found in sedimentary rocks, appearing as black or brown chunks. Texture ranges from crumbly to hard	Formed during the Carboniferous period (300–360 million years ago) from plant material in swamp forests buried under mud	Four categories based on carbon content: anthracite, bituminous, subbituminous, lignite	Used for electricity generation, steel production, and industrial heating
Petroleum	A liquid hydrocarbon primarily made of hydrogen and carbon. It varies in color, texture, and viscosity	Formed mainly during the Mesozoic period from plankton, algae, and organic matter that settled on sea floors and were buried by debris. Microbial activity and pressure transformed it into kerogen and bitumen. Under the right conditions, kerogen became crude oil	Crude oil (mixture of hydrocarbons), kerogen, bitumen	Refined into fuels like gasoline, diesel, kerosene; also used in petrochemical production
Natural gas	A flammable, colorless, and nearly odorless gas made of light hydrocarbons	Found in pockets above petroleum or on its own in sedimentary rock layers. Formed from kerogen exposed to high pressure and temperature	Mostly methane, but also contains ethane, propane, and butane	Methane: energy generation Ethane: used to produce ethylene and polyethylene (plastics) Propane: fuel for engines, cooking, heating Butane: used in lighters, portable stoves, and fuel additives

3.1 Coal as a Fossil Fuel

Coal is a fossil fuel formed from a combustible sedimentary rock that occurs in layers or veins known as coal beds or coal seams. It originates when dead plant material is gradually transformed into peat, which then undergoes further changes to become lignite, followed by subbituminous coal, then bituminous coal, and finally anthracite. This transformation is driven by a combination of biological and geological processes that take place over millions of years. The harder forms of coal—such as anthracite—are classified as metamorphic rock because they have been subjected to high temperatures and pressures during formation. Coal is primarily made up of carbon, but it also contains varying amounts of hydrogen, sulfur, oxygen, and nitrogen.

3.1.1 Formation and Types of Coal

At several points in Earth's geological history, dense forests thrived in wetland regions. Over time, these forests were buried beneath layers of soil due to natural events such as flooding. As additional sediment accumulated on top, the buried plant material was compressed further. With increasing depth, both temperature and pressure rose significantly. It is believed that muddy or acidic conditions helped preserve the deeply buried plant matter by preventing biological decay and oxidation. Over millions of years, this material underwent a slow transformation under high pressure and temperature into coal—a process known as carbonization, named for coal's high carbon content [24].

Coal deposits are found from most geological periods, with the Carboniferous period offering the most favorable conditions due to its widespread shallow seas. However, coal is rare from the Permian–Triassic extinction event. Interestingly, some coal even dates back to the Precambrian era—before the evolution of land plants—suggesting it formed from algal remains [25]. As buried organic material is subjected to increasing temperature and pressure over geological time, it passes through successive stages of metamorphic transformation [26]:

Peat: Considered the precursor to coal, peat has industrial importance in some countries (e.g., Ireland and Finland) as a fuel and is also used as a soil conditioner.

Lignite: Also known as *brown coal*, lignite is the lowest rank of coal and is primarily used for electricity generation.

Subbituminous coal: Falls between lignite and bituminous coal in terms of properties. It is primarily used as fuel for steam-electric power generation.

Bituminous coal: A dense, usually black (sometimes dark brown) sedimentary rock. It is widely used for electric power generation and industrial heating in manufacturing.

Steam coal: A grade between bituminous coal and anthracite. It is formerly used as fuel for steam locomotives and for domestic water heating.

Anthracite: The highest rank of coal, characterized by its hard, glossy black appearance. It is mainly used for residential and commercial space heating. Anthracite includes both metamorphosed bituminous coal and petrified oil.

Graphite: A difficult form of coal to ignite, not typically used as a fuel. Instead, it finds uses in industrial applications like lubricants, batteries, and pencils.

3.1.2 Coal Reserves

Of the three main fossil fuels, coal has the most widely distributed reserves. It is mined in over 100 countries, across every continent except Antarctica. Despite rising environmental concerns and the global push toward renewable energy, coal remains a key player in the world's energy landscape. In 2023, global coal production surged to an estimated 8940 million tonnes, highlighting the industry's resilience and continued demand amid expanding climate policies and sustainable energy efforts. Other studies suggest that coal deposits could last over 250 years, which is significantly longer than current estimates for oil and natural gas reserves [10]. It is important to note that these estimates include only proved reserves. As mining companies continue exploration—particularly in under-explored regions—additional reserves are discovered each year.

The largest coal reserves in the world are found in a few key countries. The top five coal-producing countries contribute over 83% of the world's total coal output, underscoring a significant concentration of this vital resource. This dominance is driven by a combination of geological abundance and energy strategies that continue to prioritize coal as a cornerstone of national power generation and industrial activity. Figure 4 illustrates the coal production and consumption trends from 1981 to 2019 for the six major coal-producing countries, excluding China [27]. In 2023, China led global coal production with 4710 million tonnes, accounting for 52.7% of the total. India followed with 1010 million tonnes (11.3%), while Indonesia produced 775 million tonnes (8.7%), maintaining its status as the top exporter of thermal coal, driven by rising demand in Asia. The United States ranked fourth with 527 million tonnes (5.9%), reflecting a continued shift toward natural gas and renewables. Australia placed fifth with 456 million tonnes (5.1%), and despite its small population, it leads in per capita coal production at 17.5 tonnes, highlighting its key role in the Asia-Pacific coal trade.

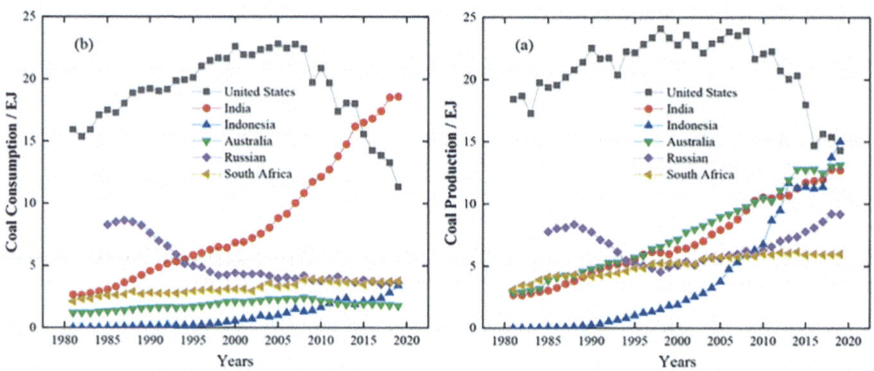

Fig. 4 Coal production and consumption of other major coal producers during the period of 1981–2019. Reprinted from Liu and Song [27]

3.2 Petroleum (Crude Oil)

Crude oil is a liquid fossil fuel found underground, formed from the remains of ancient marine organisms like plankton and algae. It is a hydrocarbon-rich mixture that must be refined to produce usable fuels and other products. Crude oil formed millions of years ago when tiny marine organisms died and settled on the seabed. Over time, they were buried under layers of sediment and rock. Heat and pressure transformed the organic material into oil. The oil migrated through porous rocks until trapped in reservoirs. This process is known as diagenesis, catagenesis, and metagenesis, depending on the depth and temperature. After refining, crude oil provides a wide range of products such as

Fuels: Gasoline (petrol), diesel, jet fuel, kerosene

Lubricants: Motor oil, greases

Asphalt: For road construction

Petrochemicals: Plastics, synthetic rubber, detergents, fertilizers, pharmaceuticals

Heating oil: For homes and buildings.

Oil is derived from both conventional and unconventional sources, though these terms are not consistently defined in the literature, and their meanings can vary significantly. For example, in his 1956 peak oil prediction, Hubbert [28] limited his analysis to crude oil that could be extracted using the technologies available at the time. In his later work, he incorporated anticipated advancements in exploration and production methods, but his peak oil assessments consistently excluded oil from oil sands and oil shale. A 2013 study serves as another example—it forecasted an early peak in oil production but excluded sources such as deepwater oil, tight oil, and oil located near the polar regions, categorizing all of these as unconventional sources [29]. The primary sources of energy today are carbon-based fossil fuels stored underground. The daily extraction from these underground sources is referred to as fossil fuel production.

3.2.1 Oil Reservoir

The total estimated amount of oil present in a reservoir, including both recoverable and non-recoverable oil, is termed oil-in-place. Due to technological limitations and specific characteristics of reservoirs, only a portion of this oil can be extracted and brought to the surface. This recoverable portion is classified as reserves. On the other hand, oil resources refer to all the oil that could technically be extracted, regardless of current market price or economic feasibility. Therefore, unlike oil resources, oil reserves fluctuate with changes in oil prices. Reserves may be reported at various levels—such as a single well, an oil reservoir, an oil field, a country, or even globally.

Venezuela and Canada had the highest reserves-to-production (R/P) ratios: 341.1 years and 105.1 years, respectively. The global R/P ratio stood at 50.6 years.

Table 2 World's countries by proven oil reserves, 2020

No.	Country	Reserves, billion barr	Reserves, billion tons	Share, %	The year of exhaustion
1	Venezuela	300,878.0	41.0	16.9	2380
2	Saudi Arabia	297,700.0	40.6	16.7	2368
3	Iran	211,600.0	28.9	11.9	2163
4	Canada	169,709.0	23.1	9.5	2144
5	Iraq	142,503.0	19.4	8.0	2106

Reprinted from Borodin et al. [30]

Based on data from Table 2, compiled using the 2020 Statistical Review of World Energy by British Petroleum, the total proven global oil reserves are estimated at 242.17 billion tons. Notably, OPEC member countries hold a dominant share, accounting for 70.5% of these reserves. The top countries with the largest proven oil reserves include Venezuela: 300.9 billion barrels (19.9% of global reserves), Saudi Arabia: 297.7 billion barrels (16.7%), and Iran: 211.6 billion barrels (11.9%). The table not only presents the volume of proven oil reserves but also includes the projected year of depletion for each country, based on current production rates. This provides valuable insight into how long a nation's oil reserves are expected to last if production continues at present levels. Additionally, the table details the annual oil production levels of various countries, offering a comprehensive overview of both their resource longevity and their role in global oil supply [30].

3.3 Natural Gas

Natural gas is primarily composed of methane (CH_4), with smaller amounts of ethane, propane, butane, and other hydrocarbons. It is a cleaner-burning fuel compared to coal and oil and is used extensively for heating, electricity generation, and industrial applications. It has been reported that natural gas was accidentally discovered in ancient China during the drilling of brine wells. The Chinese began using this gas to boil brine for salt production as early as 500 BC, and possibly even around 1000 BC [31]. Ingeniously, they developed a method to transport naturally seeping gas through simple bamboo pipelines to salt production sites, particularly in the Ziliujing District of Sichuan [32].

The first industrial extraction of natural gas, however, began much later—in 1825, in Fredonia, New York. In the nineteenth century, natural gas was typically obtained as a by-product of oil production. As oil was extracted and pressure decreased from the reservoir to the surface, lighter hydrocarbon gases would come out of solution. At the time, this gas was considered a waste product and posed a disposal problem in active oil fields. It was commonly burned off at the wellhead. Today, associated natural gas that is not immediately marketable is often reinjected into the reservoir through injection wells, preserving it for future use. When it becomes economically viable to deliver natural gas from the well site to consumers, pipelines are constructed—especially in regions with high demand for gas-powered electricity generation. Other common uses include exporting gas as liquefied natural gas or converting it into liquid fuels through gas-to-liquids technologies, which produce gasoline, diesel, or jet fuel.

One key advancement that has improved natural gas transport is liquefaction. This process involves removing contaminants like heavy metals, carbon dioxide, and hydrogen sulfide, then cooling and compressing the gas into LNG through advanced refrigeration techniques [33]. Adequate regasification terminals are essential to support this system. As a result, LNG usage is growing worldwide, and the market is expected to continue expanding in the near future [34]. Moreover, many studies predict that natural gas will play an increasingly significant role in future electricity generation and heating needs. The natural gas industry is also extracting growing volumes from challenging and unconventional sources, such as coal bed methane, tight gas, and shale gas.

Natural gas reserves are broadly distributed across the globe, but not evenly. There is ongoing disagreement among major organizations regarding which country holds the largest proved natural gas reserves. Figure 5 represents the top five countries producing natural gas.

Fig. 5 Proved natural gas reserves in top five countries

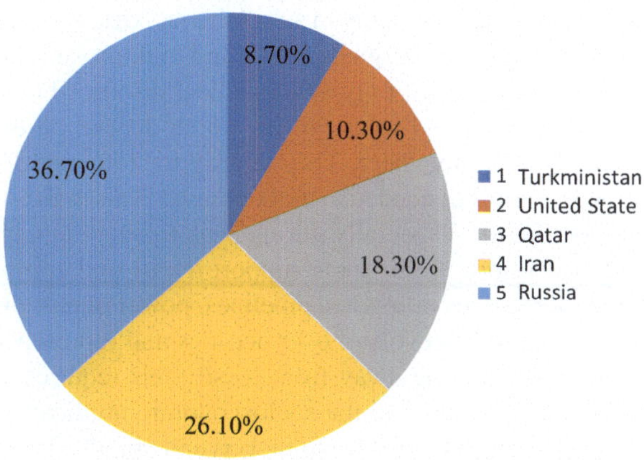

4 What Are the Advantages of Fossil Fuels?

Fossil fuels—namely coal, oil, and natural gas—have been central to global economic development and energy systems for over two centuries. These energy sources are formed from the remains of ancient plants and animals buried under the Earth's surface for millions of years. Their high energy density and widespread availability have made them the primary fuel sources for electricity generation, industrial processes, heating, and transportation across the globe. One of the key advantages of fossil fuels is their energy efficiency and reliability. Unlike intermittent renewable sources like solar or wind, fossil fuels can provide a continuous and stable energy supply, making them essential for base-load power generation. This has been especially important in supporting the growth of cities, industries, and modern infrastructure. As of 2023, fossil fuels still account for approximately 80% of the world's primary energy consumption, despite increasing investments in renewable energy sources [35].

The global transportation sector is another area heavily dependent on fossil fuels. Petroleum-based fuels such as gasoline, diesel, and jet fuel power nearly all internal combustion engine vehicles, aircraft, and ships. Until alternative fuel infrastructure becomes more developed and economically viable, fossil fuels remain indispensable for global mobility and logistics.

Fossil fuels have also played a crucial role in economic development, particularly in emerging economies. The Industrial Revolution was powered by coal, which helped Britain become a global superpower. Later, oil took over as the main energy source, especially in the twentieth century, driving the economic boom in countries. Fossil fuel exports have made some countries extremely wealthy. The fossil fuel industry also supports millions of jobs and contributes trillions of dollars to the global economy. Access to affordable and reliable energy is a cornerstone for industrialization, job creation, and poverty reduction. For example, countries such as India, China, and many in sub-Saharan Africa continue to rely on coal and natural gas to meet rising energy demands and to improve living standards [36]. In this way, fossil fuels serve not only as a technical resource but also as a developmental tool.

Unlike wind and solar energy, which depend on the weather, fossil fuels can be used anytime, anywhere. They're easy to store, and fossil fuel power plants can be turned on or off as needed to meet demand. This makes them essential for maintaining energy security, especially during peak times or emergencies.

Furthermore, fossil fuels support a vast and complex global energy infrastructure that includes refineries, pipelines, power plants, and distribution networks. This infrastructure has evolved over decades and represents trillions of dollars in investment. Transitioning away from fossil fuels requires careful consideration of these embedded systems and the socioeconomic implications of energy shifts. However, while the importance of fossil fuels is undeniable, their continued use presents significant environmental and climate-related challenges, including greenhouse gas emissions, air pollution, and habitat degradation. These concerns are driving the current global push for cleaner alternatives and carbon–neutral energy systems [37]. Fossil fuels have played an undeniable role in shaping modern civilization—powering industries, transportation, urban development, and lifting millions out of poverty. Their reliability, accessibility, and energy density have made them essential to global progress. However, their continued use comes with significant environmental costs, including climate change, air pollution, and ecological degradation.

As the world moves forward, the challenge lies not in completely abandoning fossil fuels overnight, but in strategically reducing our dependence on them while investing in cleaner, more sustainable alternatives. Through innovation, policy reform, and global cooperation, we can work toward an energy future that maintains reliability and economic growth—while minimizing harm to the planet for future generations.

5 Challenges and Future Perspectives

The future trajectory of fossil fuels is closely tied to global energy transition goals and international climate commitments. As efforts to meet the objectives of the Paris Agreement and similar accords intensify, decarbonizing the energy sector has become a pressing priority. Several key trends are expected to shape the role of fossil fuels in the coming decades. Advancements in carbon capture, utilization, and storage (CCS) technologies may allow for the continued use of fossil fuels with significantly reduced emissions. Additionally, the integration of cleaner alternatives such as hydrogen and biofuels into existing energy systems could help decrease dependence on traditional fossil fuels. Regulatory and policy shifts are anticipated, with stricter emissions regulations and incentives for clean technologies driving a shift away from carbon-intensive energy. Countries are also likely to diversify their energy portfolios by incorporating renewables, nuclear power, and transitional fuels to enhance sustainability and resilience. In the context of global energy equity, fossil fuels may remain essential in developing nations where access to energy is still limited, requiring policies that balance sustainability with equitable development. While fossil fuels are expected to remain part of the energy mix—especially in hard-to-electrify sectors like heavy industry and aviation—their share in global energy production is projected to decline as renewable technologies continue to advance and become more cost-competitive. Ultimately, the long-term outlook for fossil fuels will depend on strategic investment, cross-sectoral collaboration, and global policy alignment aimed at achieving a sustainable, just, and secure energy future.

6 Conclusion

Fossil fuels have underpinned global economic and technological advancement for over a century. Their abundance and energy density have made them the backbone of industrialization, contributing significantly to national economies and improving standards of living. However, this reliance has come at a growing environmental cost, including air and water pollution, ecosystem degradation, and escalating carbon emissions contributing to global warming. The scientific consensus on climate change and the urgent call for environmental responsibility have driven the global community to reconsider energy strategies. Many nations have enacted policies to diversify energy sources and promote renewable alternatives. However, transitioning away from fossil fuels presents complex challenges involving technological feasibility, economic impact, political will, and energy equity across developing and developed nations.

In summary, while fossil fuels will continue to play a role in meeting global energy demand in the short to medium term, it is imperative to accelerate innovation, policy reform, and investment in sustainable energy systems. The conclusion reaffirms that a deliberate, collaborative global effort is essential to balance energy security with long-term environmental and economic sustainability.

References

1. Zhang D, Rong C, Ahmad T, Xie H, Zhu H, Li X, Wu T (2023) Review and outlook of global energy use under the impact of COVID-19. Eng Rep 5(3):e12584. https://doi.org/10.1002/eng2.12584
2. Sato M (1990) Thermochemistry of fossil fuel formation. The Geological Society, Special Publication No. 02, pp 271–284. https://docslib.org/doc/977301/thermochemistry-of-the-formation-of-fossil-fuels
3. Novaczek I (2007). Canada's fossil fuel dependency. Available from: http://www.elements.nb.ca/theme/fuels/irene/novaczek.htm
4. EPA Fossil fuels (2007)
5. Mann P, Gahagan, Gordon M.B (2003) Tectonic setting of the world's giant oil and gas fields. In Halbouty MT (ed) Giant oil and gas fields of the decade 1990–1999, AAPG Memoir 78, pp 15–105. https://www.researchgate.net/publication/236235439_Tectonic_setting_of_the_world%27s_giant_oil_and_gas_fields
6. Patel NK, Shah SN (2015) Biodiesel from plant oils. In: Ahuja S (ed) Food, energy, and water: the chemistry connection. Elsevier, Amsterdam, p 278. https://doi.org/10.1016/B978-0-12-800211-7.00011-9
7. World Energy Council (2016) Survey of world energy resources. London. www.worldenergy.org and @WECouncil
8. International Energy Agency (2016) Oil market report. OECD, Paris. https://doi.org/10.1787/oilmar-2016-en
9. EIA (2017) Energy and the environment explained: green house gases' effect on the climate. U.S. Department of Energy, Washington, DC. https://www.eia.gov/energyexplained/index.cfm?page=environment_how_ghg_affect_climate
10. Ritchie J, Dowlatabadi H (2015) Fossil fuel divestment: review in arguments, implications and policy opportunities. Pac Inst Clim Solut 6:5174. https://www.researchgate.net/publication/277132436_Fossil_Fuel_Divestment_Reviewing_Arguments_Implications_Policy_Opportunities
11. Shukla P, Skea J, Slade R, Al Khourdajie A, van Diemen R, McCollum D (2022) Climate change: mitigation of climate change. In: Contribution of working group III to the sixth assessment report of the intergovernmental panel on climate change. Cambridge University Press, Cambridge, UK. https://doi.org/10.1017/9781009157926.002

12. Rosenboom JG, Langer R, Traverso G (2022) Bioplastics for a circular economy. Nat Rev Mater 7(2):117–137. https://doi.org/10.1038/s41578-021-00407-8
13. Min Y, Brinkerink M, Jenkins J, Mayfield E (2023) Effects of renewable energy provisions of the inflation reduction act on technology costs, materials demand, and employment. https://www.bluegreenalliance.org/wp-content/uploads/2023/06/Working-Paper_6-12-23.pdf
14. Shrestha R, Rajpurohit S, Saha D (2023) Redefining America's 'energy communities' can boost clean energy investment where it's needed most. World Resources Institute. https://www.wri.org/insights/redefining-americas-energy-communities
15. Penn I (2022) Expansion of clean energy loans is 'sleeping giant' of climate bill. The New York Times. https://www.nytimes.com/2022/08/22/business/energy-environment/biden-climate-bill-energy-loans.html
16. Martin C, Crowley K (2018) Exxon will use wind, solar to produce crude oil in Texas
17. Varghese A (2019) The Permian paradox: Texas Shale players go green to drill more. Reuters, https://www.reuters.com/article/us-usa-energy-renewables-idUSKBN1XI1HH
18. Mitsubishi Power (2021) Bakken Energy and Mitsubishi Power establish clean hydrogen partnership. Mitsubishi Heavy Industries Group. https://power.mhi.com/regions/amer/news/20210602.html
19. Nilsen E (2022) DOE will build nation's first large-scale facility to turn fossil fuel waste into rare materials for tech. CNN. https://www.cnn.com/2022/02/14/politics/energy-coal-waste-recycling-batteries-climate/index.html
20. Hsu CS, Robinson PR (2017). Springer handbook of petroleum technology, 2nd, illustrated ed. Springer, p 360. https://doi.org/10.1007/978-3-319-49347-3
21. Neumann C, Lewis W (1759) The chemical works of Caspar Neumann, pp 1683–1773
22. Ahuja S (2015) Food, energy, and water: the chemistry connection. Elsevier, pp 261–275. https://www.sciencedirect.com/book/9780128002117/food-energy-and-water
23. Miller G, Spoolman, Scott (2007) Environmental science: problems, connections and solutions. Brooks Cole, Belmont, CA, pp 1–551. ISBN 978-0-495-38337-6
24. Taylor TN, Taylor EL, Krings M (2009) Paleobotany: the biology and evolution of fossil plants. Elsevier/Academic, Burlington, MA; London; San Diego, CA, New York, NY. ISBN 978-0-12-373972-8
25. Tyler SA, Barghoorn ES, Barrett LP (1957) Anthracitic coal from Precambrian Upper Huronian Black Shale of the Iron River District, Northern Michigan. Bull Geol Soc Am 68(10):1293
26. Mancuso JJ, Seavoy RE (1981) Precambrian coal or anthraxolite; a source for graphite in high-grades schists and gneisses. Economic Geol 76(4):951–954. https://doi.org/10.2113/gsecongeo.76.4.951
27. Liu X, Song G (2024) Development status of coal industries in the world's major coal producing countries. J South Afr Inst Min Metal 124(7):383–386. https://scielo.org.za/pdf/jsaimm/v124n7/05.pdf
28. Hubbert MK (1956) Nuclear energy and the fossil fuels. Shell Development Company Publication No. 95. Drilling and Production Practice, American Petroleum Institute. https://www.spec2000.net/freepubs/B1956-Peak-Oil-Hubbert.pdf
29. White S, Cordell D (2008) Peak phosphorus: the sequel to peak oil. Global Phosphorus Research Initiative (GPRI), pp 1–2. https://phosphorusfutures.net/the-phosphorus-challenge/peak-phosphorus-the-sequel-to-peak-oil/
30. Borodin A, Panaedova G, Ilyina I, Harputlu M, Kiseleva N (2023) Overview of the Russian oil and petroleum products market in crisis conditions: economic aspects, technology and problems. Energies 16:1614. https://doi.org/10.3390/en16041614
31. Abbott M (2016) The economics of the gas supply industry. Routledge, Abingdon, p 185
32. Natural Gas.org (2017) History. http://naturalgas.org/overview/history
33. Vakylabad AB (2024) Absorption processes for CO_2 removal from CO_2-rich natural gas. In: Rahimpour MR, Makarem MA, Meshksar M (eds) Advances in natural gas: formation, processing, and applications. Volume 2: Natural gas sweetening. Elsevier, pp 207–257. https://doi.org/10.1016/B978-0-443-19217-3.00002-7

34. Jalilvand DR (2013) Iran's gas exports: can past failure become future Success? Oxford Institute for Energy Studies, Oxford. https://doi.org/10.26889/97819075557
35. International Energy Agency (IEA) (2023) World energy outlook 2023. https://www.iea.org/reports/world-energy-outlook-2023
36. British Petroleum (2023) Statistical review of world energy 2023. https://www.bp.com/en/global/corporate/energy-economics/statistical-review-of-world-energy.html
37. Alagoz E, Alghawı Y (2024) The future of fossil fuels: challenges and opportunities in a low-carbon. Int J Earth Sci Knowl Appl 5(3):381–388

The Future of Energy and Mobility

Abdul Jabbar Khan and Muhammad Sajjad

Abstract The transition toward a low-carbon, sustainable future is driving profound changes in the global energy and transportation sectors. This chapter explores the dynamic intersection of these two domains, focusing on the accelerating shift from fossil fuel dependency to integrated systems powered by renewable energy and electrified mobility. It begins by tracing the evolution of energy systems from traditional coal and oil to decentralized, smart grids supported by energy storage and digital infrastructure. The narrative then focuses on electric vehicles (EVs), highlighting their rapid global utilization, technological advancements in battery systems, and critical role in reshaping the mobility landscape. Furthermore, this chapter examines the growing integration of EVs into the energy grid through innovations such as vehicle-to-grid (V2G) systems and smart charging infrastructure, offering new models of energy resilience and flexibility. Through detailed analysis, global data, and current policy frameworks, this chapter outlines the challenges, opportunities, and emerging pathways that define the future of energy mobility and forward-looking a cleaner, smarter, and more equitable transportation-energy ecosystem.

Keywords Fossil fuels · Electric vehicles (EVs) · Transition · Mobility · Energy storage systems

1 Introduction

The development of mobility and transportation systems is deeply intertwined with the growth of modern society. The efficient, timely, and cost-effective movement of goods and individuals is essential in shaping economic activities and improving the quality of life. Furthermore, transportation is responsible for approximately one-third of total energy consumption in the United States and about one-quarter of total

A. J. Khan (✉) · M. Sajjad
College of Chemistry and Chemical Engineering, Huanggang Normal University, Huanggang, China
e-mail: ajkhan.au@gmail.com

J. A. Siddique et al. (eds.), *Energy Sources*, Green Energy and Technology,
https://doi.org/10.1007/978-981-95-1209-6_2

greenhouse gas emissions [1–3], rendering it a pivotal sector in the pursuit of energy security and sustainable development.

In the twenty-first century, the convergence of energy innovation and transportation transformation stands at the forefront of global progress. As the world faces mounting challenges, including climate change, urban congestion, diminishing fossil fuel reserves, and an increasing demand for sustainable solutions, the ways in which we produce, store, and consume energy, along with how we transport people and goods, are undergoing a profound transformation [4–6]. This chapter explores the interconnected future of energy and mobility, where cleaner energy systems power smarter, more efficient, and more connected modes of transportation.

For over a century, petroleum has been the dominant transportation fuel for passenger vehicles, trucks, trains, ships, and airplanes. In this connection, energy and mobility have been tightly linked through the widespread use of petroleum-based [7]. However, the growing urgency to decarbonize the economy, reduce emissions, and enhance energy resilience has accelerated a paradigm shift toward electrification, digitalization, and decentralization [8, 9]. Advances in renewable energy sources, battery technologies, hydrogen fuel cells, and smart grids are reshaping the mobility landscape, making electric vehicles (EVs), autonomous systems, and next-generation transport platforms increasingly viable and impactful [10–13].

With the advent of these technologies and the increasing availability of alternative fuels, the transportation sector globally faces a pivotal moment. Although the precise configuration of the future transport system remains indeterminate, it is anticipated to be intricately integrated with smart buildings, the electric grid, and digital ecosystems, as depicted in Fig. 1. These connections present unprecedented opportunities for optimizing systems at the intersection of mobility and energy domains.

This chapter aims to examine the technological, environmental, and societal forces shaping the future of energy and mobility, specifically EVs. By analyzing current trends, emerging innovations, and global initiatives, it uncovers how the integration of

Fig. 1 Schematic depiction of energy and its mobility

these sectors can contribute to a cleaner, more equitable, and intelligent transportation ecosystem. Through this perspective, this chapter provides a forward-looking analysis of future developments and the requirements necessary to navigate the transition toward a more sustainable future.

2 The Evolution of Energy Systems

The global energy landscape has undergone significant transformations over the past two centuries, transitioning from biomass and coal-based systems to petroleum-based infrastructure and, more recently, to low-carbon and renewable alternatives. Each energy transition has exerted profound impacts on industrial growth, social progress, and environmental sustainability. The current transition, arguably the most critical in human history, focuses on reducing carbon emissions and enhancing energy efficiency to meet climate goals and ensure long-term energy security [14, 15].

2.1 From Fossil Fuels to Renewable Energy

The industrial era was primarily powered by coal, then successively by oil and natural gas. These fossil fuels offered enormous energy density and the convenience of transport, enabling fast urbanization and the advancement of modern transportation and manufacturing systems [16]. But their extensive utilization has also been a major driver of greenhouse gas emissions, air pollution, and geopolitical conflicts [17].

In response to these trends, the twenty-first century has seen a significant surge in the adoption of renewable energy technologies. More specifically, solar photovoltaic (PV) and wind power, green hydrogen, and grid modernization have seen incredible expansion, spurred on by falling costs, technology advances, and supportive policy environments. The International Energy Agency (IEA) documents that renewables accounted for over 80% of global new electricity capacity additions in 2022, led predominantly by China, the United States, and the European Union [18].

2.2 The Role of Energy Storage and Grid Modernization

Due to the naturally intermittent nature of renewable energy resources such as solar and wind, their incorporation into current power networks necessitates the utilization of effective energy storage systems. Lithium-ion batteries have emerged as the dominant storage technology, mainly due to declining costs and rising energy density [19]. In addition, grid-scale battery storage plants, as well as emerging-generation technologies, including solid-state batteries, flow batteries, and hydrogen-based storage, are also gaining prominence in grid-scale applications [20, 21].

At the same time, smart grid technologies have emerged as essential tools for managing distributed energy resources (DERs). Smart grids leverage sensors, real-time data analytics, and automated control systems to optimize electricity generation, delivery, and consumption efficiency, making the system more reliable [22].

3 Mobility in Transition: The Rise of Electric Vehicles (EVs)

The international transport industry is undergoing a revolutionary change with EVs emerging as the cornerstone of decarbonized, sustainable mobility systems. With the environment becoming an increasing concern, evolving consumer preferences, and rapid technological innovation, EVs have transitioned from a niche market to a central focus in national and industrial decarbonization strategies [23]. This section explores the pivotal role of EVs in the future of mobility, examining their current status, driving technologies, market growth, and infrastructure development [24].

3.1 Global Growth and Market Penetration

The adoption of EVs has accelerated significantly in the past decade. In 2022 alone, more than 10 million EVs were sold globally, with China, Europe, and the United States accounting for the majority of sales. As a result, EVs represented around 14% of all new car sales worldwide, compared to just 4% in 2020. This growth is driven by several converging factors, including falling battery prices, tighter emissions regulations, fuel economy standards, and increased consumer awareness of climate change [25, 26]. Governments across the globe are playing a critical role in supporting EV adoption through subsidies, tax incentives, fuel economy mandates, and zero-emission vehicle (ZEV) targets.

3.2 Advances in Battery Technology

The performance and cost of EVs are closely tied to battery technology. Lithium-ion batteries remain the dominant technology due to their high energy density, long cycle life, and scalability [27, 28]. However, the industry is rapidly evolving toward next-generation technologies such as solid-state batteries, lithium iron phosphate (LFP), and sodium-ion batteries, which offer improved safety, sustainability, and affordability [29].

Ongoing research is also focused on enhancing fast-charging capabilities, increasing the recyclability of battery components, and reducing reliance on critical materials such as cobalt and nickel. These advancements are key to achieving parity with internal combustion engine (ICE) vehicles and scaling mass-market adoption [30].

3.3 *Environmental and Economic Impacts*

EVs produce zero exhaust emissions, rendering them an effective instrument in reducing urban air pollution and mitigating climate change. Their overall life cycle emissions are considerably lower than those of ICE vehicles, particularly when powered by clean electricity. Nonetheless, challenges persist concerning the sustainability of battery supply chains, raw material extraction, and end-of-life battery management [31].

From an economic perspective, the EV sector has created new opportunities in areas such as manufacturing, battery production, software development, and energy services. The transition to EVs represents not only a technological shift but also a reconfiguration of the automotive value chain and labor market [32].

4 Integration of Energy and Mobility

The integration of energy and mobility systems constitutes a pivotal enabler of sustainable urban development and decarbonization. Historically regarded as distinct sectors, energy and transportation are increasingly interconnected through the rapid adoption of EVs, smart grids, and renewable energy technologies. This section examines the bidirectional relationship between EVs and energy infrastructure, emphasizing innovations such as vehicle-to-grid (V_2G) systems, smart charging technologies, and the role of digital platforms in fostering intelligent, flexible, and resilient mobility-energy ecosystems [32, 33].

4.1 *Electric Vehicles as Distributed Energy Resources (DERs)*

EVs have transitioned from being mere consumers of electricity to becoming mobile energy storage units that can interact dynamically with the electrical grid. When aggregated across neighborhoods or fleets, these vehicles can offer demand response, frequency regulation, and backup power services, thereby functioning as effective DERs [34].

V2G systems enable the bidirectional exchange of energy between EV batteries and the power grid. During periods of peak demand, energy stored in EVs can be discharged back to the grid, whereas during periods of low demand or surplus renewable energy generation, vehicles can be recharged, thereby facilitating the balance of supply and demand. This integration not only enhances grid stability but also offers potential revenue opportunities for EV owners participating in energy markets [35].

4.2 Smart Charging and Grid Management

Smart charging infrastructure is crucial for aligning EV charging with grid capacity and the availability of renewable energy. In contrast to conventional chargers, smart chargers possess the capability to adjust power levels, slow charging during periods of grid bottleneck, and optimize charging schedules based on price signals, carbon intensity, or grid constraints [36].

Advanced smart charging systems utilize artificial intelligence (AI) and machine learning algorithms to analyze historical consumption patterns and forecast demand, thereby improving energy planning and distribution. These systems are particularly beneficial when integrated with renewable energy sources such as solar PV and wind, which are variable and require adaptable load management strategies [37].

4.3 Infrastructure Synergies and Urban Energy Planning

Urban centers and energy providers are increasingly collaborating to design infrastructure that concurrently enhances mobility and energy resilience. The strategic colocation of EV charging stations with solar panels, battery energy storage systems (BESS), and public transit hubs yields synergistic advantages in land utilization, energy efficiency, and carbon emission reduction. Contemporary urban energy planning incorporates scenarios that simulate EV adoption, grid load impacts, and the integration of renewable energy sources. Platforms facilitating coordination among transportation authorities, utilities, and private stakeholders are becoming essential in developing Mobility-Energy Integration (MEI) frameworks that are both responsive and adaptive [38, 39].

4.4 Policy and Standards for Interoperability for the Effective Scaling of Energy

Mobility integration is essential to establish standardized communication protocols, safety regulations, and policy frameworks. The development of standards such as ISO 15118 (pertaining to V2G communication) and Open Charge Point Protocol (OCPP) facilitates seamless interaction between diverse charging equipment and

Table 1 The comparison summary of energy and its mobility from traditional system to future

Aspect	Traditional system	Future-oriented system
Energy source	Fossil fuels (coal, oil, gas)	Renewables (solar, wind, hydrogen)
Grid structure	Centralized power generation	Decentralized smart grids
Transportation mode	Internal combustion engine vehicles	Electric and autonomous vehicles
Fuel type	Gasoline, diesel	Electricity, green hydrogen
Environmental impact	High emissions, resource depletion	Low emissions, circular economy
Technology integration	Limited digital integration	AI, IoT, smart control systems
User role	Consumer	Prosumer (producer + consumer)
Infrastructure	Fuel stations, centralized power plants	Charging networks, V2G infrastructure, microgrids
Policy focus	Energy security, fuel subsidies	Decarbonization, equity, digital innovation

vehicles [40]. National and regional governments play a pivotal role in incentivizing grid-friendly EV behavior, funding pilot programs, and supporting research into battery interoperability and recycling. Coordinated regulation is instrumental in overcoming fragmentation in EV-grid integration and ensuring system-wide efficiency. The comparison of traditional vs. future energy and mobility systems is outlined in Table 1.

5 Conclusion and Outlook

The global energy and mobility sectors are undergoing a significant transformation, driven by the convergence of clean energy technologies and EV innovation. This shift is characterized by a transition from fossil fuels to renewable energy sources and the rise of EVs as a central component of modern mobility strategies. The integration of EVs into the broader energy ecosystem through smart charging and V2G technology is reshaping the relationship between transportation and energy infrastructure. However, this transformation faces several challenges, including global disparities in EV adoption, environmental concerns related to battery materials, infrastructure limitations, and the need for harmonized standards. Addressing these issues requires collaboration across government, industry, and academia. Future trends point toward a deepening interplay between mobility and energy through digitization, decentralized energy generation, and AI integration. Emerging technologies such as solid-state batteries, green hydrogen, and autonomous electric transport systems are poised to further revolutionize the sector. As this transformation progresses, it is crucial to

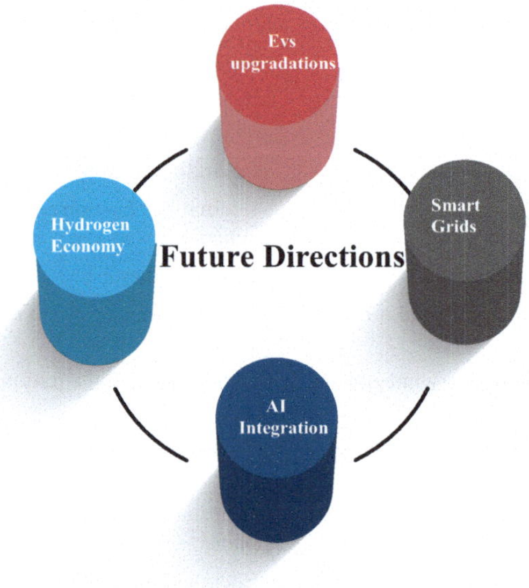

Fig. 2 The prospective pathways shaping the future performance of energy

align technological innovation with broader societal goals, considering economic implications, social and ethical considerations, and the role of policy in shaping this transition. The global implications of these changes, particularly for developing nations, must also be carefully examined. While this rapid transformation offers immense potential for sustainable development, it is essential to explore potential risks and unintended consequences to ensure a balanced and equitable transition toward a cleaner, more efficient energy and mobility future. Underlying the energy and its mobility for future outcomes is summarized in Fig. 2.

References

1. Tol R (2020) Sources of greenhouse gas emissions. Department of Economics, University of Sussex Business School
2. Muratori M et al (2020) Future integrated mobility-energy systems: a modeling perspective. Renew Sustain Energy Rev 119:109541
3. Mai TT et al (2018) Electrification futures study: scenarios of electric technology adoption and power consumption for the United States. National Renewable Energy Lab. (NREL), Golden, CO (United States)
4. Daniels D, Yeh S (2022) Complexities in the energy-transport co-transformation. Prog Energy 4(4)
5. Kremzow-Tennie S, Mecit H (2024) On the convergence of electric mobility and energy systems—potentials and challenges. In: 2024 22nd international conference on research and education in mechatronics (REM). IEEE

6. Gupta JG et al (2017) Introduction to sustainable energy, transportation technologies, and policy. Sustainable energy and transportation: technologies and policy. Springer, pp 3–7
7. Eia U (2012) Annual energy outlook 2012. US Energy Information Administration, Washington, DC
8. Creutzig F et al (2015) Transport: a roadblock to climate change mitigation? Science 350(6263):911–912
9. Kashem MA, Shamsuddoha M, Nasir T (2024) Sustainable transportation solutions for intelligent mobility: a focus on renewable energy and technological advancements for electric vehicles (EVs) and flying cars. Future Transport 4(3):874–890
10. Muratori M, Mai T (2020) The shape of electrified transportation. Environ Res Lett 16(1):011003
11. de Mendonca HV et al (2021) Corrigendum to "Microalgae in a global world: new solutions for old problems?" [Renewable Energy volume 165, part 1, (2021) 842–862]. Renew Energy 167:962–965
12. Crabtree G (2019) The coming electric vehicle transformation. Science 366(6464):422–424
13. Song J et al (2022) Shaping future low-carbon energy and transportation systems: digital technologies and applications. IEnergy 1(3):285–305
14. IEA P (2022) World energy outlook 2022. International Energy Agency (IEA), Paris, France
15. Haegel NM, Kurtz SR (2021) Global progress toward renewable electricity: tracking the role of solar. IEEE J Photovoltaics 11(6):1335–1342
16. Nuhu SK, Emmanuel EO, Manan ZA (2023) Global challenges to sustainable renewable energy expansion: the funding and incentives structures. Chem Eng Trans 106:553–558
17. Zakeri B, Syri S (2015) Electrical energy storage systems: a comparative life cycle cost analysis. Renew Sustain Energy Rev 2015(42):569–596
18. Dechezleprêtre A et al (2024) A comprehensive overview of the renewable energy industrial ecosystem
19. Deguenon L, Yamegueu D, Gomna A (2023) Overcoming the challenges of integrating variable renewable energy to the grid: a comprehensive review of electrochemical battery storage systems. J Power Sources 580:233343
20. Khan AJ et al (2025) Advancements and challenges in Si-based solid-state batteries: from anode design to manufacturing processes. Next Mater 7:100371
21. Behabtu HA et al (2020) A review of energy storage technologies' application potentials in renewable energy sources grid integration. Sustainability 12(24):10511
22. Odero HZ, Wekesa CW, Irungu GK (2022) Comprehensive review of energy storage technologies: types, applications, optimal sizing and siting in power systems. IEEE PES/IAS PowerAfrica 2022:1–5
23. Soni L, Kaur A (2023) Why electric vehicles are the future of transportation. In: 2023 IEEE international conference on ICT in business industry & government (ICTBIG). IEEE
24. Sperling D (2023) Future drive: electric vehicles and sustainable transportation. Island Press
25. Pamidimukkala A et al (2024) Barriers and motivators to the adoption of electric vehicles: a global review. Green Energy Intell Transport 3(2):100153
26. Bryła P, Chatterjee S, Ciabiada-Bryła B (2022) Consumer adoption of electric vehicles: a systematic literature review. Energies 16(1):205
27. Jerry A (2023) Madrid electric and hybrid vehicle battery technologies: advancements and challenges. Int J Adv Res Sci Commun Technol 3:803–807
28. Celadon A et al (2024) Batteries for electric vehicles: technical advancements, environmental challenges, and market perspectives. SusMat 4(5):e234
29. Zhang Y et al (2024) Chlorine-rich $Na_{6-x}PS_{5-x}Cl_{1+x}$: a promising sodium solid electrolyte for all-solid-state sodium batteries. Materials 17(9):1980
30. Aphale S et al (2020) Li-ion batteries for electric vehicles: requirements, state of art, challenges and future perspectives. In: 2020 IEEE international conference on power and energy (PECon). IEEE
31. Rajaeifar MA et al (2022) Challenges and recent developments in supply and value chains of electric vehicle batteries: a sustainability perspective. Elsevier, p 106144

32. Koniak M, Jaskowski P, Tomczuk K (2024) Review of economic, technical and environmental aspects of electric vehicles. Sustainability 16(22):9849
33. Barman P et al (2023) Renewable energy integration with electric vehicle technology: a review of the existing smart charging approaches. Renew Sustain Energy Rev 183:113518
34. Štogl O et al (2024) Electric vehicles as facilitators of grid stability and flexibility: a multidisciplinary overview. Wiley Interdiscip Rev: Energy Environ 13(5):e536
35. Giordano F, Diaz-Londono C, Gruosso G (2023) Comprehensive aggregator methodology for EVs in V2G operations and electricity markets. IEEE Open J Vehic Technol 4:809–819
36. Qu T (2024) Application of artificial intelligence optimization in hybrid vehicles. In: 2024 4th international signal processing, communications and engineering management conference (ISPCEM). IEEE
37. Ahsan MS et al (2023) Integration of electric vehicles (EVs) with electrical grid and impact on smart charging. Int J Multidiscip Sci Arts 2(4):225–234
38. Garau M, Torsæter BN (2024) A methodology for optimal placement of energy hubs with electric vehicle charging stations and renewable generation. Energy 304:132068
39. Zhao AP et al (2024) Electric vehicle charging planning: a complex systems perspective. IEEE Trans Smart Grid
40. Kumar G, Mikkili S (2024) Advancements in EV international standards: charging, safety and grid integration with challenges and impacts. Int J Green Energy 21(12):2672–2698

Composition, Classification, and Properties of Petroleum

Rajeev Kumar, Riya Chugh, and Indrani

Abstract Crude oil, referred to as petroleum, is a naturally occurring fossil fuel that was created over millions of years by the decomposition of ancient sea animals like zooplankton and algae that were buried under sediment and exposed to heat and pressure. It is a complicated blend of various organic and hydrocarbon molecules that is a significant source of energy globally. It mainly includes aromatic and aliphatic hydrocarbons and a very small amount of S, N, O, and trace amounts of Ni and Fe. Petroleum is extracted by drilling wells and then transported to refineries to go through chemical and distillation processes. Gasoline, diesel, petrol, jet fuel, and grease are the main products of petroleum. Petroleum affects trade and geopolitics, and it is a major engine of the world economy. Petroleum is categorized and described according to a number of factors, such as its physical characteristics, chemical composition, API gravity, sulfur content, and place of origin. These elements affect how it is handled and applied in different ways. Flash, fire, pour and cloud points, carbon and hydrogen ratio, density and API gravity, viscosity, chemical composition, and boiling point of the petroleum determine the properties of petroleum. This chapter provides the complete information about the composition, classification, and properties of petroleum, which helps to enhance the characteristics of petroleum.

Keywords Crude oil · Hydrocarbon · Petroleum composition · Petroleum classification · Petroleum properties

R. Kumar (✉) · R. Chugh · Indrani
Department of Applied Sciences, Manav Rachna International Institute of Research and Studies, Faridabad, Haryana, India
e-mail: rajeevkumar.set@mriu.edu.in

R. Chugh
e-mail: riyachugh.set@mriu.edu.in

Indrani
e-mail: indrani.set@mriu.edu.in

J. A. Siddique et al. (eds.), *Energy Sources*, Green Energy and Technology,
https://doi.org/10.1007/978-981-95-1209-6_3

1 Introduction

Petroleum, also known as "black gold," is necessary for the operation of this contemporary society. Crude oil (petroleum) is a naturally occurring liquid found beneath the Earth's surface. Hydrocarbons and other organic molecules make up the majority of its composition. Petroleum is primarily composed of alkanes, cycloalkanes, and aromatic hydrocarbons [42]. One of the most vital natural resources, petroleum, is used as a raw material for many different industries and as a major source of energy. The remains of extinct marine life, buried beneath sedimentary layers and exposed to heat and pressure over millions of years, give rise to petroleum. Its numerous uses in a variety of industries account for its significance. Chemicals, including plastics, synthetic rubber, fertilizers, and medications, are made using petroleum as a basic source. The main energy sources for automobiles, trucks, aircraft, and ships are petroleum derivatives such as gasoline, diesel, and jet fuel [25, 55].

Sedimentary rock formations, where organic matter has been buried and changed over time, are the primary locations for petroleum. The main sources of sedimentary petroleum basins are the Arabian Basin (Middle East), the Permian Basin (USA), and the Western Siberian Basin (Russia). The primary onshore petroleum deposits include the Texas oil fields (USA), the Ghawar field (Saudi Arabia), and the Burgan field (Kuwait). On the other hand, the primary offshore petroleum reserves are in the North Sea (UK and Norway), the Gulf of Mexico (USA and Mexico), and the Campos Basin (Brazil). The Gulf of Guinea (West Africa) and pre-salt basins (Brazil) are deepwater and ultra-deepwater petroleum resource areas. Pitch Lake in Trinidad and Tobago and the La Brea tar pits in the United States are examples of natural seeps where small amounts of petroleum organically seep to the surface of the Earth [13, 17, 49, 51].

The global economy is significantly influenced by petroleum, which also has an impact on trade ties and geopolitics. Petroleum is essential to construction since it provides materials for infrastructure building and asphalt for roadways [22]. It demonstrates the vital significance of each stage. Petroleum exports are a major source of revenue for several nations.

Numerous issues pertaining to the environment, economy, society, and technology confront the petroleum sector [16]. One of the main causes of greenhouse gas emissions that contribute to global warming is the burning of petroleum products. The main factor causing harm to both marine and terrestrial ecosystems has been petroleum spills that occur during drilling, transportation, or refining. Particulate matter, sulfur dioxide (SO_2), and nitrogen oxides (NO_x) are among the pollutants released during the refining and burning of petroleum [34]. Due to the deterioration of many large oil fields, expensive and ineffective enhanced oil recovery (EOR) techniques are needed. Conflicts and interruptions in supply are frequently caused by the concentration of oil reserves in politically unstable areas. Through advancements in sustainable energy and technology, efforts are being made to lessen reliance on petroleum [27].

Advances in sustainable fuel production, changes in global consumption trends, and the use of artificial intelligence in exploration and extraction are all examples of recent innovations in petroleum research. Significant progress has been achieved by researchers in turning lignin-based agricultural waste into sustainable aviation fuel [36]. A continuous technique that directly turns lignin into jet fuel has been devised, providing a sustainable substitute for conventional fuels derived from petroleum. AI is being used more and more in the petroleum sector to increase productivity and profitability. AI helps locate the best drilling locations and oversee operations from a distance in areas like the Permian Basin, which results in more effective extraction procedures and lower operating expenses [4, 21]. With an emphasis on sustainability, technology integration, and the finding of new reserves to meet the world's energy demands, these advancements demonstrate the dynamic character of petroleum research.

Global energy regulations, environmental concerns, and technological advancements are all influencing the future demand for petroleum. Petroleum's conventional uses, especially in energy generation and transportation, are predicted to decline in the future. Nonetheless, its importance in specialized applications and petrochemicals guarantees that it will continue to be a valuable resource for many years to come [10, 54]. Technological developments, legislative frameworks, and international collaboration on sustainability objectives will all have a significant impact on how quickly the shift proceeds. The demand for petroleum-based fuels like gasoline and diesel is predicted to gradually decrease as a result of the global drive toward electric vehicles (EVs) and developments in battery technology. Fossil fuels are gradually being replaced by renewable energy sources, including solar, wind, and hydrogen, in the generation of power [56]. Petroleum will continue to be vital for the manufacturing of petrochemicals, which are necessary for plastics, synthetic textiles, fertilizers, and medications, even though its conventional applications are dwindling. Petroleum consumption may continue with less of an impact on the environment because of technologies like carbon capture and storage and cleaner refining techniques.

2 Composition of Petroleum

Petroleum has a complex structure of compounds. Due to the presence of various hydrocarbons and other chemicals, it has versatile properties and characteristics. Alkanes, cycloalkanes, and aromatic hydrocarbons are the prime components of petroleum. However, there are some more components present in petroleum [39]. Figure 1 shows the main composition of petroleum.

Petroleum is composed of various types of hydrocarbons, non-hydrocarbons, and other components. The percentages of various components in petroleum vary on the basis of type of petroleum. Generally, the percentage of carbon is 83–87%, hydrogen 10–14%, sulfur 0.05–6%, nitrogen 0.1–2%, and oxygen 0.1–1.5%. The main component of petroleum is saturated alkanes or paraffins having the formula C_nH_{2n+2} [53]. These hydrocarbons are generally linear or branched-chain

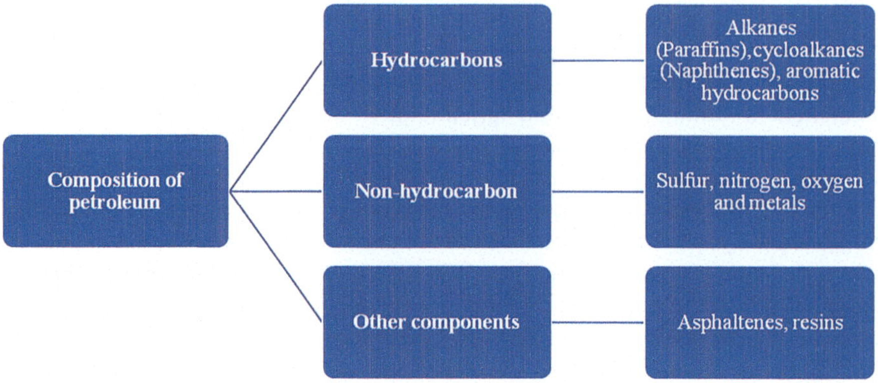

Fig. 1 Composition of the petroleum

hydrocarbons. Other types of hydrocarbons are cyclic hydrocarbons having the formula C_nH_{2n}. Petroleum also contains aromatic hydrocarbons such as benzene or benzene-associated compounds. Petroleum has some percentage of non-hydrocarbon compounds such as S, N, and O compounds. H_2S and thiol are the main types of sulfur compounds present in petroleum. The percentage of sulfur in sweet crude is 0.1 and 5% in sour crude. Pyridines and quinolines are nitrogen compounds present in petroleum in trace amounts. Carboxylic acids, phenols, and ketones are oxygen-containing compounds found in petroleum [33]. Trace amounts of various types of metals are found in petroleum such as iron, vanadium, and nickel. Some parts of petroleum contain large molecules of hydrocarbon with S, N, and O known as asphaltenes. High molecular weight resin, which acts as a stabilizer for asphaltenes, is also present in petroleum.

3 Classification of Petroleum

Petroleum is a complicated mixture of hydrocarbons that is usually categorized according to its density, sulfur concentration, formulation, and preparation techniques (Fig. 2).

3.1 Classification of Petroleum by Its Origin

Petroleum is categorized by origin, which often refers to the kind of raw petroleum resource and where it was extracted [18]. These are the primary groups according to place of origin.

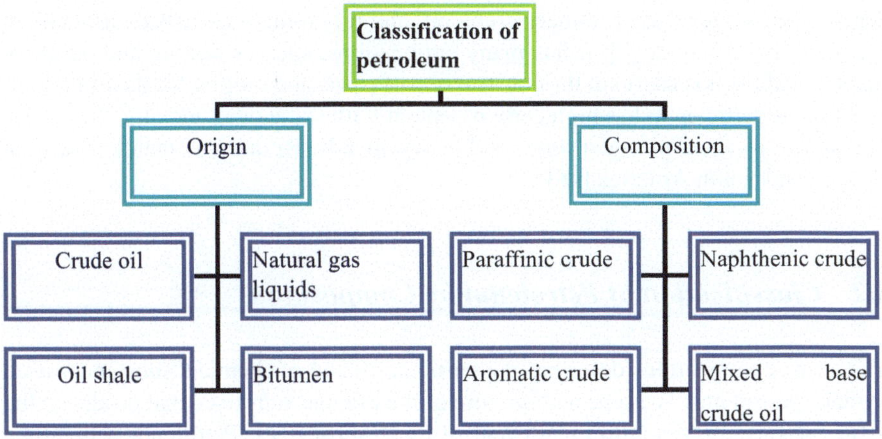

Fig. 2 Classification of petroleum

3.1.1 Crude Oil

This is petroleum that has been recovered straight from subterranean reserves without any processing. In addition to hydrocarbons, it also consists of some metallic impurities like nitrogen, oxygen, and sulfur. Crude oil can be of different types, including light crude oil, heavy, and bitumen crude [44].

3.1.2 Natural Gas Liquids

Hydrocarbons recovered as liquids from natural gas are known as natural gas liquids (NGLs). Like crude oil and natural gas, they belong to the same molecular family and are byproducts of natural gas. NGLs include propane, butane, isobutane, pentane, butadiene, and condensate [11].

3.1.3 Oil Shale

Oil and gas can be extracted from oil shale, a sedimentary rock that includes biologically active or organic materials. One kind of sedimentary rock that includes kerogen, an oil precursor, is called shale oil [41]. It is found in Russia, India, United States, Estonia, and China.

3.1.4 Bitumen

Bitumen, which is typically dark brown or black, is a strongly viscous, heavy, and tenacious type of petroleum. It is found mostly in natural deposits such as oil-sand deposits or as a residue of refinery crude oil, and it is sometimes called "asphalt" in specific circumstances [8]. It has many applications, such as sealing and adhesive traits, and thus, it is useful in the construction of roads and roofing for the binding of aggregated particles. It has high carbon content with some other metallic impurities like sulfur, oxygen, nitrogen, iron, and heavy metals. Bitumen is mainly found in Canada and South America [48].

3.2 Classification of Petroleum by Composition

The term "composition" describes the particular blend of chemical substances used to make petroleum. The type and mixture of the organic substance that produced the hydrocarbons contribute to the content of these substances. Petroleum, which is a traditional crude oil, is often brown-green to black with a specific gravity that varies from 0.75 to 1.0 at a temperature of 15 °C. The majority of crude oils have a specific gravity that ranges from 0.80 to 0.95. Petroleum generally boils around 20 and 350 °C, depending upon the chemical constituents. Beyond this, distillation attempts result in the active breakdown of the molecules. It contains 1.0–40% of gasoline, as well as some amount of kerosene. Thus, the constitution of petroleum differs from one oil to another depending upon the type and molecular weight of the molecules present in it [53]. Petroleum is regarded as a naturally occurring hydrocarbon because it comprises compounds that are entirely made of carbon and hydrogen and devoid of any heteroatom like nitrogen, oxygen, or sulfur, and other metallic components like nickel, copper, and iron. Some categories of petroleum hydrocarbons are given below [39].

3.2.1 Paraffinic Crude

It is mainly alkanes, saturated molecules with straight chains called n-paraffins or branching (isoparaffins). This paraffinic crude is highly preferred due to its high-yield production of lightweight and intermediate distillates, including kerosene, diesel, lubricants, and gasoline [52]. It contains low aromatic content with respect to naphthenic and aromatic crudes and is also less viscous. The American Petroleum Institute (API) gravity of the paraffinic crude is usually greater than 30°, which makes it a very light crude oil. Paraffinic crude oils are mainly obtained from West Africa, the Middle East, and North American regions [19].

3.2.2 Naphthenic Crude

The naphthenic crude oil is basically formed of a significant proportion of cyclic alkanes or cycloparaffins, which are also called naphthenes. Cyclic saturated hydrocarbons, generally called naphthene, are made of one or multiple rings with side chains (alkyl chains). Unlike aromatic crude, which has a high concentration of (cyclic alkenes/alkynes) unsaturated cyclic hydrocarbons, and paraffinic one is composed of a high concentration of straight chain hydrocarbons, the naphthenic crude is in between both, containing hydrocarbon rings [6]. It has low wax content as compared to paraffinic crude oil and has a higher solvency performance than that of paraffinic crude oil. Some examples of naphthenic crude are cyclohexane, methyl-cyclohexane, and cyclopentane [43]. The API gravity of naphthenic crude oil varies between 22° and 28°, which is a bit less than paraffinic crude oil. This oil is generally obtained from the United States, India, and China.

3.2.3 Aromatic Crude

The aromatic crude oils are generally unsaturated hydrocarbons or high in aromatic hydrocarbons with high sulfur content, which contain one or multiple benzene rings like toluene, dimethyl benzene, and derivatives, xylene or cyclic structures [33]. The aromatic crude has low paraffinic content (low concentration of saturated hydrocarbons such as alkanes). It may have a moderate concentration of naphthene oil (i.e., cycloalkenes, like cyclopentane, cyclobutene, and cyclohexane). Mainly, aromatic crude oils indicate higher density and have lower API gravity. Aromatic crude oils contain aromatic compounds, which are generally complicated molecules; thus, on refining, they leave a high amount of carbon residue and contribute to darker color oil. Aromatic crude oil contains different types of hydrocarbons, out of which it contains 30–50% aromatic components (major component), a moderate amount of naphthene 20–40% and a lower amount of paraffin, which constitutes just 10–20% (secondary or minor component). Aromatic crude oil is obtained mainly from China, New Zealand, Iraq, and Western Australia. This crude oil is used in the petrochemical industry and generally utilized as a raw material or raw feedstock for the synthesis of compounds like toluene, xylene, and benzene, which are required for the synthesis of artificial fibers, plastics, and raisins as well [23].

3.2.4 Mixed Base Crude Oil

One of the significant forms of petroleum, which is known as mixed base crude oil, is comprised of a variety of hydrocarbons from different chemical groups, including different properties of aromatic crude, naphthene, and paraffinic crude oil. The combination of hydrocarbons in the composition, rather than the dominance of any one kind, makes it suitable for a range of refining applications. The capability of a mixture of crude oil to balance the benefits of both paraffinic and naphthenic crude types while meeting a variety of demands in production and refining makes it economically viable [5].

4 Properties of Petroleum

Crude oil, another name for petroleum, is a naturally occurring liquid made up of various organic molecules and hydrocarbons. Crude oil or petroleum may appear as viscous, semisolid, or exceptionally fluid and volatile. It is frequently black or blackish-green in color. It could be greenish or reddish sometimes and occasionally transparent, pale yellow, or yellow. Although alkanes make up the majority of petroleum or crude oil, they also contain aromatic hydrocarbon compounds such as cyclohexene and cyclopentanes, along with more complicated and complex hydrocarbons, including asphaltenes. Due to the presence of different molecules and compounds, it constitutes different properties that may change extensively and, thus, are characterized by physical and chemical properties (Fig. 3) [37].

4.1 Physical Properties of Petroleum

On the basis of physical properties, petroleum is classified into six main parts as follows.

4.1.1 Appearance

Different types of petroleum and their products vary in color. Some petroleum products or oils appear to be black, red, green, or brown, depending on the composition, which is a mixture of hydrocarbons and impurities present in it. Not all petroleum oils or their products exhibit color, as the crude form or lighter ones are generally

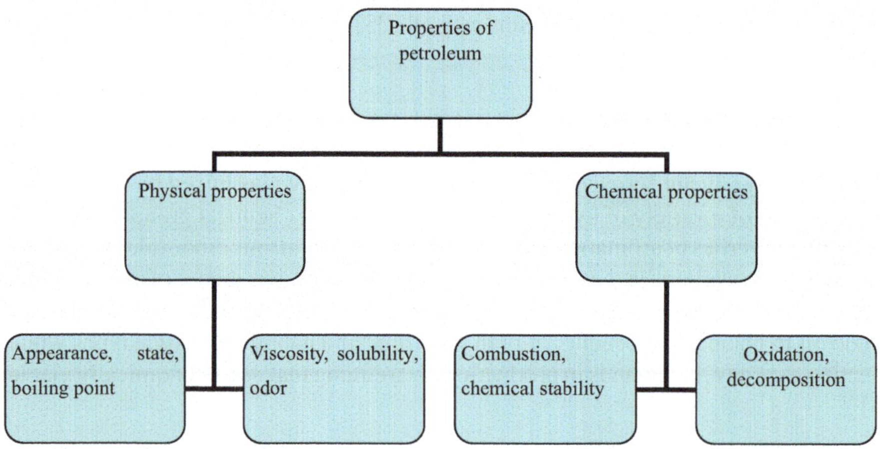

Fig. 3 Classification of properties of petroleum

transparent or slightly translucent. Generally, oils or petroleum products with high molecular weight are highly viscous or thick, having a sticky consistency. In contrast, lighter ones are generally softer and have a liquid consistency. Due to their viscous or oily nature, they appear to have shiny surfaces in appearance [15].

4.1.2 State

By considering its physical and chemical composition, the surrounding environment, and the state of existence of its constituents, petroleum can exist in a variety of states, including solid, liquid, gaseous, and semisolid. If any of the petroleum-based products consist of heavy components, then they may exist in a solid or semisolid state, but under certain conditions. Solid-based petroleum products, such as bitumen and paraffinic crude oil, are generally found in sands or left as wax residue after refining. As liquid is one of the most common forms of petroleum or crude oil, which is a mixture of hydrocarbons [31]. However, liquid petroleum sometimes contains solid suspended particles and dissolved gases like hydrogen, methane, and acetylene. Liquid crude oil or liquid petroleum can be classified as light crude oil and heavy crude oil. Light crude oil or petroleum generally has an API gravity of 30°–32°, which means this oil has a lower density than water. The lower density of this oil leads to easy flow and has low viscosity. It is rich in alkanes and exhibits a lower content of sulfur, nitrogen, and other heavy metal compounds. These oils are generally yellow to light brown in color. This oil has a great advantage in that it is not much expensive than that of other petroleum products like gasoline, kerosene, diesel, and jet fuel. This oil is easily available and exhibits good market value, as it can be easily processed or refined [38].

In contrast to light petroleum or crude oil, heavy crude oil has a lower API gravity value of 20°–23°, which means it is denser than water and light crude oil. The higher density and lower API gravity make this oil highly viscous, thick, and sticky. It is rich in heavy hydrocarbons like aromatic compounds having multiple rings. Large amounts of impurities are found in this heavy crude oil, including heavy metals, sulfur, and nitrogen. Generally, the color of this heavy crude oil varies from dark brown to black. The processing and refining of these heavy crude oils are not easy; they require very complicated and complex procedures, like hydroprocessing, which finally leads to their lower market value. Heavy crude oil is used for the production of heavy oils, heavier fuels, industrial lubricants, etc. [35].

4.1.3 Boiling Point

One important physical characteristic of crude oil or petroleum that illustrates its composition and the variety of hydrocarbons it contains is its boiling point. Petroleum is a complex combination that has boiling points that vary, rather than a single or fixed boiling point, in contrast to some raw or pure substances that have a fixed boiling point. The boiling point of crude oil or petroleum typically ranges from

40 °C to 600 °C, depending upon the different types of hydrocarbons and molecules present in it. Generally, lighter oils boil at a lower temperature, ranging from 40 °C to 200 °C, such as kerosene and gasoline, whereas heavier ones boil at a high temperature, ranging from 200 °C to 600 °C, for example, lubricating oil, castor oil, diesel, and bitumen. Hydrocarbons or petroleum with different ranges of boiling points serve some significant commercial and industrial purposes. The branched or cyclic hydrocarbon-based petroleum oils tend to exhibit a lower boiling point than that of straight-chain hydrocarbon-based petroleum oils. The boiling point of crude oil or petroleum can also be influenced by the impurities present in it, like oxygen, sulfur, nitrogen, or other metallic compounds [20].

4.1.4 Viscosity

One of the important physical characteristics of petroleum that characterizes its inability to flow is consistency or viscosity. It has a big impact on how petroleum is handled, transported, refined, and used in numerous types of applications. Light crude oil has a lower viscosity and can flow like water, whereas heavy crude oil has high viscosity (thicker as compared to light crude oil), so it requires high-temperature conditions or dilution to flow [45]. Viscosity is dependent on temperature and pressure as well; as the temperature increases, the viscosity or consistency of any fluid decreases. That is, crude oil or petroleum becomes less viscous at high temperatures (upon heating). At high pressure, the viscosity of the oil will be high. Petroleum or heavy crude oils contain high molecular weight hydrocarbons or resins increase the viscosity, but in contrast, petroleum or crude oil containing lower or light molecular weight hydrocarbons decreases the viscosity [1]. The petroleum or light-weight crude oils are generally easier and require less expense for transportation than heavy-weight crude oils.

4.1.5 Solubility

One significant physical characteristic of petroleum is solubility, which refers to its capability for dissolution in a variety of solvents. The ability of crude oil/petroleum or its constituents to dissolve in a particular solvent medium without separating into distinct phases is known as solubility [28]. As solubility is attributed to the principle "like dissolves like, and it is well known that petroleum is hydrophobic in nature and, thus, does not dissolve in water as water is polar in nature, whereas petroleum is nonpolar in nature [40]. There is very little interaction between the nonpolar petroleum hydrocarbons and polar water molecules. Being nonpolar in nature, petroleum hydrocarbons readily dissolve in nonpolar or slightly polar solvents such as ether, benzene, chloroform, toluene, hexane, and xylene [14]. Lighter crude oil hydrocarbons can readily dissolve gases like carbon dioxide, methane, and ethane under low temperatures and high-pressure conditions. Generally, heavy crude oils like bitumen, kerosene, diesel, and gasoline show less solubility in solvents and precipitate out in some solvents. In contrast, lighter-weight hydrocarbons are slightly more soluble in organic solvents like benzene, ethanol, hexane, toluene, ethyl acetate, and methanol.

4.1.6 Odor

One of the vital properties of petroleum or crude oil, out of all the properties described above, is the odor, whether pleasant or unpleasant, which gives information about the composition and impurities present in the oil. Depending on the composition, impurities (like sulfur, oxygen, and nitrogen), and the type of crude oil or hydrocarbon present in it, the odor of the petroleum can vary. Typically, petroleum oil can have a very strong, pungent odor or sweet smell/odor. Light crude oils with light hydrocarbons (such as aromatics and alkanes) contribute to gasoline or sweet odor, while the other crude oils smell like rotten eggs due to the presence of high sulfur compounds or sulfide content in them. A stale or musty smell contributes to the presence of nitrogen-containing components like amines in petroleum oil. Generally, light crude oils tend to have a sweet or less intense odor due to the presence of light-weight hydrocarbons in them. On the other hand, heavy crude oils tend to have a very intense and unpleasant smell due to the presence of heavily weighted hydrocarbons, impurities, and other metallic components. Temperature also affects the odor of any crude oil or petroleum. As the temperature rises, volatile petroleum substances evaporate more easily, intensifying the odor of crude oil or petroleum products. Petroleum odors can also help to identify whether there is any leakage in the pipelines, tanks, or other instruments. Along with this, the prolonged inhalation of petroleum smell can lead to various health and environment-related issues like headaches, respiratory issues, skin infections, nausea, cancer, vomiting, and memory loss, especially if some toxic components (like sulfides) are present in it and can also lead to pollution (air pollution), which directly impacts surrounding societies and communities located near the petroleum production plants. In addition to being a distinguishing feature, the odor serves as a crucial signal for monitoring the environment, safety, and quality evaluation [26].

4.2 *Chemical Properties of Petroleum*

On the basis of chemical properties, petroleum is classified into five main parts as follows.

4.2.1 Combustion

An essential chemical characteristic of petroleum is its combustion, which is the basis for its use as a fuel in a variety of applications, such as electricity production, heating, and transportation. Combustion is the capacity of petroleum to catch fire in the presence of oxygen, emitting energy in the form of both light and heat [29]. Combustion of hydrocarbons includes a general chemical reaction described below:

$$C_aH_b + O_2 \rightarrow CO_2 + H_2O + \text{energy (in form of light and heat)}$$

Petroleum hydrocarbons generally react with oxygen in the presence of air to produce carbon dioxide and water, along with the release of energy in the form of heat and light. On burning, petroleum or crude oil releases a significant amount of energy in the form of heat and light as it has high energy density, which makes it a practical, effective, and efficient source of energy for engines and power plants. The flash point of any petroleum product varies depending on the type of hydrocarbons present in the petroleum. For example, gasoline has a lower flash point than that of diesel. Generally, combustion took place as ideal/complete or nonideal/incomplete combustion. The former includes gases like carbon dioxide and water vapors, and the latter includes carbon monoxide, particulate matter, and a few unburnt or noncombustible hydrocarbons [50]. The combustibility of petroleum includes some disadvantages: it produces greenhouse gases like carbon monoxide, carbon dioxide, nitrogen oxides, and chlorofluorocarbons, which finally lead to global warming and various environmental issues. Crude oil or petroleum containing impurities like sulfur-containing hydrocarbons and sulfides releases toxic sulfur dioxide gas on combustion, which contributes to acid rain.

4.2.2 Chemical Stability

Chemical stability is one of the most vital properties of petroleum, which is defined as the capability to resist any chemical changes that take place while transporting or storing under certain conditions or normally [32]. The chemical stability of petroleum is the property that determines the shelf life of petroleum, its safety, and performance. The chemical stability of petroleum also gets reduced due to exposure to air and moisture when stored in poor conditions.

4.2.3 Oxidation

The hydrocarbons present in petroleum, when exposed to oxygen present in the air, lead to an oxidation reaction, which is also known as rancidity, affected by conditions such as heat, light, and the presence of moisture. On oxidation, it produces peroxides, which decrease the quality of petroleum and also cause gum issues and acidity. On exposure to UV radiation, the hydrocarbons present in the petroleum break down and produce unwanted toxic by-products [46].

4.2.4 Decomposition

The decomposition of hydrocarbon petroleum refers to the breakdown of larger molecules or compounds into smaller ones, and the decomposition of petroleum hydrocarbons is influenced by various factors like temperature, pressure, and the presence of other compounds. Hydrocarbon breakdown takes place when it is exposed to high temperature [9]. For example, the decomposition of octane into smaller molecules like butane and butene.

$$C_8H_{18} \xrightarrow{\text{Heat}} C_4H_{10} + C_4H_8\text{(in presence of heat)}$$

Oxidative decomposition of methane: It occurs in the presence of oxygen and heat conditions.

$$CH_4 + 2O_2 \rightarrow CO_2 + 2H_2O \quad \text{(in presence of oxygen)}$$

4.2.5 Acidity

Acidity is considered an essential property of petroleum, which gives a complete idea about how many acidic substances are present in the petroleum or crude oil/refined by-products. The purity, quality, processing, and transportation of petroleum crude oil or its by-products can be influenced by the presence of acidic substances in it. The organic acid, mainly naphthenic acid, is most commonly found in petroleum or crude oil. The naphthenic acids consist of carboxylic acid groups attached to hydrocarbon chains, which are mainly responsible for the acidity of petroleum. Phenol, which is a weak acid, is sometimes also found in crude oil [47]. Typically, the presence of these acidic substances in the crude oil can cause corrosion on the metallic surface of pipelines and tanks but can be controlled by neutralization technique and blending, which is very important to maintain the safety and quality of petroleum crude oil [24].

5 Challenges and Future Scope of Petroleum

Numerous issues spanning the social, technological, environmental, and economic spheres confront the petroleum sector. The different difficulties encountered in the petroleum business are depicted in Fig. 4.

Fossil fuel combustion contributes significantly to greenhouse gas emissions, the main contributor to global warming. The primary source of soil, marine, and other environmental contamination that seriously damages ecosystems is unintentional oil spills that occur during extraction, transportation, or storage [12]. Petroleum resources are under a lot of strain due to the growing demand for petroleum in

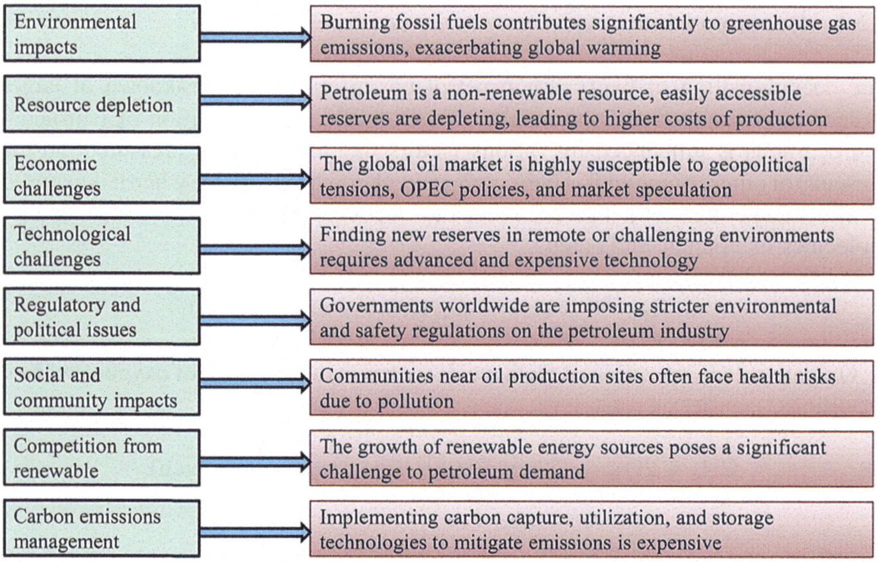

Fig. 4 Challenges faced in petroleum industry

various industries. It is a nonrenewable resource, and readily available supplies are running out daily, which raises the price of production and exploration. Reliance on unconventional sources, such as shale oil or deepwater drilling, which are more expensive to extract, grows as conventional reserves diminish. To maintain efficiency and safety, many oil facilities and pipelines need to be upgraded due to their aging [2].

In order to lessen dependency on petroleum, the transition to renewable energy necessitates novel technologies. Market uncertainty and supply chain disruption can result from political unrest in oil-rich areas. Growing opposition to new pipelines and oil projects is a result of growing environmental consciousness. Cleaner energy is being invested in by businesses and governments, which may lessen the dominance of petroleum in the energy mix. For the petroleum industry to make the switch to low-carbon operations, substantial investment is needed [30]. In order to combine economic growth with environmental and social obligations, addressing these issues calls for a combination of technological innovation, sustainable habits, and aggressive policymaking [7].

Petroleum's position in the future is changing as the globe moves toward renewable and sustainable energy sources. Although petroleum will continue to be used for decades to meet the world's material and energy demands, its industries and applications will probably change dramatically [10, 54]. Investments in renewable energy sources, including solar, wind, and hydrogen, are being propelled by governments and businesses' commitment to net-zero emissions targets. The demand for petroleum in the transportation sector is anticipated to decrease as EVs increase in market share and energy efficiency advances. Petrochemicals, which are essential for producing plastics, medications, and other goods, are attracting the attention of many oil firms. Businesses are branching out into clean energy initiatives, such as renewable energy production, biofuels, and carbon capture, use, and storage [56].

Oil extraction is becoming more efficient thanks to innovations, which also lower costs and waste. Governments are penalizing operations that rely heavily on fossil fuels, enforcing stronger emission rules, and providing incentives for low-carbon solutions. Because of market imbalances, geopolitical conflicts, and changes in the world's energy consumption, the business is vulnerable to fluctuations in oil prices. Petroleum is used to cover short-term demands while switching to cleaner alternatives in areas like South Asia and Africa, where energy demand is still growing. In many applications, blue and green hydrogen produced from renewable resources or natural gas might take the place of petroleum [3]. Emerging technologies make better use of petroleum waste and recycle plastics. Although it is unlikely to go away very soon, the petroleum industry will play a smaller role as the globe transitions to more sustainable energy sources. It is anticipated that businesses that embrace innovation and sustainability will prosper, while others might encounter existential difficulties.

6 Conclusion

The worldwide petroleum industry handles the exploration, extraction, refinement, transportation, and marketing of petroleum products. It is vital to the world economy since it provides raw materials and energy to many sectors. Through its operations across the supply chain, such as exploration, extraction, refinement, transportation, and combustion of petroleum products, the petroleum industry has a substantial environmental impact. It significantly affects many areas of the world economy and shapes geopolitical processes. The petroleum sector contributes to economic expansion while also presenting difficulties, particularly regarding price volatility, environmental sustainability, and the worldwide switch to renewable energy. Despite being at a turning point, the petroleum sector is nevertheless essential to the world economy. Its evolution depends on embracing technological improvements while addressing sustainability and environmental issues. It will remain relevant in the future with carbon constraints if cleaner energy methods are integrated and efficiency is prioritized. A balanced transition to sustainably fulfill global energy demands can be made possible by strategic innovation and cross-sector collaboration. In order to guarantee safe, ecologically conscious, and economically equitable operations, the petroleum sector is governed by a complicated system of national and international laws and regulations. Regulations frequently aim to reduce greenhouse gas emissions, air pollution, and the effects on water bodies (such as oil spills). Regulations governing emissions from refineries, drilling rigs, and transportation are enforced by organizations such as the U.S. Environmental Protection Agency. Hazardous waste is produced by petroleum operations (e.g., drilling mud and contaminated water). To avoid contaminating the soil and water, regulations make sure that things are disposed of properly. Rules issued by the U.S. Occupational Safety and Health Administration

(OSHA) guarantee worker safety in potentially dangerous settings, such as refineries, pipelines, and drilling sites. Global frameworks for petroleum sector practices are provided by international organizations such as the International Energy Agency (IEA) and the International Maritime Organization (IMO). Although regulations differ greatly from one nation to the next, the general objective is to strike a balance between the necessity of energy production and public health and safety, economic justice, and environmental preservation.

Acknowledgements This work was supported by the management of SET, MRIIRS, Faridabad, Haryana, India.

References

1. Abdulla MMHI, Pokharel S (2022) Analytical models for predicting oil recovery from immiscible CO_2 injection: a literature review. J Petrol Sci Eng 219:111131. https://doi.org/10.1016/J.PETROL.2022.111131
2. Abudu H, Sai R (2020) Examining prospects and challenges of Ghana's petroleum industry: a systematic review. Energy Rep 6:841–858. https://doi.org/10.1016/J.EGYR.2020.04.009
3. Al-Khelaiwi MS et al (2024) Evaluation of green and blue hydrogen production potential in Saudi Arabia. Energy Convers Manage X 24:100742. https://doi.org/10.1016/J.ECMX.2024.100742
4. Alipour Bonab S et al (2024) Machine learning-powered performance monitoring of proton exchange membrane water electrolyzers for enhancing green hydrogen production as a sustainable fuel for aviation industry. Energy Rep 12:2270–2282. https://doi.org/10.1016/J.EGYR.2024.08.028
5. Anthony M, Chinenyeze J, Ekene UR (2015) Physical and chemical properties of crude oils and their geologic significances. Int J Sci Res (IJSR). https://doi.org/10.21275/ART20174603
6. Barros EV et al (2022) Characterization of naphthenic acids in crude oil samples—a literature review. Fuel 319:123775. https://doi.org/10.1016/J.FUEL.2022.123775
7. Bathrinath S et al (2021) An initiative towards sustainability in the petroleum industry: a review. Mater Today Proc 46:7798–7802. https://doi.org/10.1016/J.MATPR.2021.02.330
8. Belyaev PS et al (2021) Petroleum bitumen and polymer-bitumen binders: current state and Russian specifics. Review. AIP Conf Proc 2412(1):060001. https://doi.org/10.1063/5.0075420/1000099
9. Berger CV, Appell HR (1956) Decomposition of hydrocarbons. Ind Eng Chem 48(9):1566–1575. https://doi.org/10.1021/IE51401A003/ASSET/IE51401A003.FP.PNG_V03
10. Bhattacherjee P et al (2024) ESG, clean energy, and petroleum futures markets: asymmetric return connectedness and hedging effectiveness. Int Rev Econ Financ 94:103375. https://doi.org/10.1016/J.IREF.2024.103375
11. Brignole E, Pereda S (2013) Phase equilibrium engineering principles. Supercrit Fluid Sci Technol 3:127–151. https://doi.org/10.1016/B978-0-444-56364-4.00006-6
12. Chen K et al (2024) Review of carbonated water injection as a promising technology to enhance oil recovery in the petroleum industry: challenges and prospects. Pet Sci. https://doi.org/10.1016/J.PETSCI.2024.07.009
13. da Silva EB, Severiano Ribeiro HJP, de Souza ES (2025) Assessing the geochemical correlation between petroleum source rocks in the Brazilian Equatorial margin basins and global oceanic anoxic events. Mar Pet Geol 172:107205. https://doi.org/10.1016/J.MARPETGEO.2024.107205

14. Davis MC et al (2019) Solubility of hydrocarbon oils in alcohols (≤C 6) and synthesis of difusel carbonate for degreasing. RSC Adv 9(40):22891–22899. https://doi.org/10.1039/C9RA04220B
15. Delvigne GAL (2002) Physical appearance of oil in oil-contaminated sediment. Spill Sci Technol Bull 8(1):55–63. https://doi.org/10.1016/S1353-2561(02)00121-4
16. Elnourani M et al (2024) Khartoum war's echoes in oil and energy sectors: economic and environmental implications for Sudan and South Sudan. Heliyon 10(15):e34739. https://doi.org/10.1016/J.HELIYON.2024.E34739
17. Famelli N et al (2024) Understanding the petrophysical properties, seismic responses and impacts of the basalt-sediment transition in prospective sedimentary basins. J S Am Earth Sci 148:105113. https://doi.org/10.1016/J.JSAMES.2024.105113
18. Guzmán-Osorio FJ et al (2021) Classification of petroleum origin and integrity by FTIR. Egypt J Pet 30(2):63–67. https://doi.org/10.1016/J.EJPE.2021.03.003
19. Hakimi MH et al (2022) The origins of paraffinic oils collected from oilfields in the western Siberian Basin, Russia: implications from geochemical and physical characteristics. J Pet Explor Prod Technol 12(1):35–49. https://doi.org/10.1007/S13202-021-01325-9/FIGURES/12
20. Hill JB, Ferris SW (1925) Relation between boiling point and some other properties of petroleum products. Ind Eng Chem 17(12):1250–1252. https://doi.org/10.1021/IE50192A018/ASSET/IE50192A018.FP.PNG_V03
21. Khosravi M et al (2024) AI-assisted optimal energy conversion for cost-effective and sustainable power production from biomass-fueled SOFC equipped with hydrogen production/injection. Process Saf Environ Prot 192:1151–1171. https://doi.org/10.1016/J.PSEP.2024.08.045
22. Lentschig H, Patonia A, Quitzow R (2025) Multilateral governance in a global hydrogen economy: an overview of main actors and institutions, key challenges and future pathways. Int J Hydrogen Energy 97:76–87. https://doi.org/10.1016/J.IJHYDENE.2024.11.393
23. Li Y et al (2023) 'Geochemical characteristics and significance of aromatic hydrocarbons in crude oil from the East Fukang Sag, Junggar Basin, NW China. Front Earth Sci 10:1031367. https://doi.org/10.3389/FEART.2022.1031367/BIBTEX
24. Liu S et al (2023) Analysis of the contribution of petroleum acid components to the viscosity of heavy oils with high TAN. ACS Omega 8(31):28866–28876. https://doi.org/10.1021/ACSOMEGA.3C04098/ASSET/IMAGES/MEDIUM/AO3C04098_0014.GIF
25. Mafat IH et al (2024) Exploring machine learning applications in chemical production through valorization of biomass, plastics, and petroleum resources: a comprehensive review. J Anal Appl Pyrol 180:106512. https://doi.org/10.1016/J.JAAP.2024.106512
26. Motalebi Damuchali A, Guo H (2020) Evaluation of odour properties, their relationships, and impact of an oil refinery plant on the surrounding environment using field measurements. Atmos Environ. https://doi.org/10.1016/J.ATMOSENV.2020.117480
27. Olivares-Rubio HF, Ponce-Vélez G (2025) Polycyclic aromatic hydrocarbons during the 21st Century in Southern Gulf of Mexico, a prominent petroleum area: a review. Mar Pollut Bull 210:117343. https://doi.org/10.1016/J.MARPOLBUL.2024.117343
28. Page CA et al (2000) Solubility of petroleum hydrocarbons in oil/water systems. Mar Chem 70(1–3):79–87. https://doi.org/10.1016/S0304-4203(00)00016-5
29. Petty SE (1983) Combustion of crude oil on water. Fire Saf J 5(2):123–134. https://doi.org/10.1016/0379-7112(83)90005-X
30. Popkova EG, Sergi BS (2024) Energy infrastructure: investment, sustainability and AI. Resour Policy 91:104807. https://doi.org/10.1016/J.RESOURPOL.2024.104807
31. Raisbeck MF (2013) Petroleum hydrocarbons. In: Small animal toxicology, 3rd ed, pp 755–761. https://doi.org/10.1016/B978-1-4557-0717-1.00071-5
32. Raya SA et al (2020) A critical review of development and demulsification mechanisms of crude oil emulsion in the petroleum industry. J Pet Explor Prod Technol 10(4):1711–1728. https://doi.org/10.1007/S13202-020-00830-7/TABLES/6
33. Romanczyk M, Pfutzner CJ, Tuttle SG (2024) Compositional analysis and fate of aromatic hydrocarbons in weathered crude oil in a non-aqueous environment by using a high-resolution orbitrap mass spectrometer. Fuel 373:132379. https://doi.org/10.1016/J.FUEL.2024.132379

34. Sahoo AK et al (2024) Network-based investigation of petroleum hydrocarbons-induced ecotoxicological effects and their risk assessment. Environ Int 194:109163. https://doi.org/10.1016/J.ENVINT.2024.109163
35. Santos RG et al (2014) An overview of heavy oil properties and its recovery and transportation methods. Braz J Chem Eng 31(3):571–590. https://doi.org/10.1590/0104-6632.20140313S00001853
36. Segovia-Hernández JG (2025) Advancing E-fuels production through process intensification: overcoming challenges and seizing opportunities for a sustainable energy future—a critical review. Chem Eng Process Process Intensif 208:110107. https://doi.org/10.1016/J.CEP.2024.110107
37. Selley RC, Sonnenberg SA (2023) The physical and chemical properties of petroleum. In: Elements of petroleum geology, pp 15–42. https://doi.org/10.1016/B978-0-12-822316-1.00002-1
38. Sheng JJ (2020) Air injection. In: Enhanced oil recovery in shale and tight reservoirs, pp 401–455. https://doi.org/10.1016/B978-0-12-815905-7.00013-X
39. Shishkova I et al (2022) Challenges in petroleum characterization—a review. Energies 15(20):7765. https://doi.org/10.3390/EN15207765/S1
40. Shiu WY et al (1990) The water solubility of crude oils and petroleum products. Oil Chem Pollut 7(1):57–84. https://doi.org/10.1016/S0269-8579(05)80034-6
41. So Li Chian S et al (2016) A review of shale oil production methodologies and its impact to global energy and economy. Res J Appl Sci Eng Technol 13(7), 555–568. https://doi.org/10.19026/RJASET.13.3015
42. Song Y et al (2024) Connotation and research strategy of the whole petroleum system. Pet Explor Dev 51(6):1386–1401. https://doi.org/10.1016/S1876-3804(25)60548-9
43. Sotoudeh Y et al (2023) Review on naphthenic acids: an important environmental pollutants caused by oil extraction and industries. Pollution 9(1):254–270. https://doi.org/10.22059/poll.2022.344876.1532
44. Speight JG (2019) Heavy oil, extra heavy oil, and tar sand bitumen. In: Heavy oil recovery and upgrading, pp 3–47. https://doi.org/10.1016/B978-0-12-813025-4.00001-5
45. Tahmasbi Y, Moghaddam MS, Kheshti MF (2024) Prediction of the viscosity of saturated liquid and unsaturated gaseous chlorofluorocarbons as well as hydrocarbon mixtures using the Peng–Robinson equation of state. Pet Chem. https://doi.org/10.1134/S0965544124070132
46. Thomson NR (2024) In situ chemical oxidation of petroleum hydrocarbons, pp 503–547. https://doi.org/10.1007/978-3-031-34447-3_15
47. Tomczyk NA et al (2001) On the nature and origin of acidic species in petroleum. 1. Detailed acid type distribution in a California crude oil. Energy Fuels 15(6):1498–1504. https://doi.org/10.1021/EF010106V
48. Tripathi RP, Bahl JS, Purohit RC (1974) Studies on bitumen production from North Rumaila crude oil. J Appl Chem Biotech 24(9):505–514. https://doi.org/10.1002/JCTB.2720240906
49. Wang X et al (2024) Detailed oil-source correlation within the sequence and sedimentary framework in the Fushan Depression, Beibuwan Basin, South China Sea. Pet Sci. https://doi.org/10.1016/J.PETSCI.2024.08.005
50. Wang Z et al (2022) Investigation of oxygen-enriched atmosphere combustion of oily sludge: performance, mechanism, emission of the S/N-containing compound, and residue characteristics. J Clean Prod 378:134233. https://doi.org/10.1016/J.JCLEPRO.2022.134233
51. Xie X et al (2024) Sedimentary characteristics and geological significance of a mixed-process delta for petroleum exploration in the Zhuhai formation of the Baiyun Depression, Pearl River Mouth Basin, China. Pet Res 9(4):599–609. https://doi.org/10.1016/J.PTLRS.2024.06.004
52. Yu Y et al (2023) Critical assessment of carbon pastes for carbon electrode-based perovskite solar cells. Carbon 205:270–293. https://doi.org/10.1016/J.CARBON.2023.01.046
53. Zhang W et al (2024) Classification method of heavy oil based on chemical composition and bulk properties. Energies 17(15):3733. https://doi.org/10.3390/EN17153733/S1
54. Zhang Y et al (2024) Petrochemical industry for the future. Engineering. https://doi.org/10.1016/J.ENG.2024.06.017

55. Zhang Z et al (2024) Cryogenian–Cambrian tectono-sedimentary evolution, paleoclimate and environment effects, and formation of petroleum resources in the Tarim Block. Earth Sci Rev 248:104632. https://doi.org/10.1016/J.EARSCIREV.2023.104632
56. Zhu J et al (2024) Recent progress in microencapsulation technology and its applications in petroleum industry. J Mol Liq 407:125162. https://doi.org/10.1016/J.MOLLIQ.2024.125162

The Role of Fossil Fuels in a Sustainable Energy System

Babita Saroha, Anand Kumar, Naresh Kumar, Deepali Singhal, Rakesh Dhoundiyal, H. V. Pant, Sandeep Negi, Harish Chandra, and Shyam Vir Singh

Abstract Fossil fuels have historically driven global economic growth and technological advancement. However, their environmental externalities and contribution to climate change present serious challenges to sustainability. This chapter explores the complex and evolving role of fossil fuels in the transition to sustainable energy systems. It examines the historical context, environmental impacts, current global energy reliance, and emerging technologies aimed at reducing the carbon intensity of fossil fuels. By integrating fossil fuels with clean energy technologies and circular carbon approaches, it is possible to bridge the gap between current energy demands and future sustainability goals. However, achieving this balance will require coordinated global efforts, substantial investments, and a robust commitment to environmental stewardship.

Keywords Fossil fuels · Sustainable energy transition · Clean energy technologies · Carbon capture and storage · Energy sustainability

1 Introduction

Despite the growing urgency to decarbonize the global energy system, fossil fuels—coal, oil, and natural gas—remain the backbone of global energy supply. As of 2022, fossil fuels accounted for approximately 80% of the world's primary energy consumption, underscoring their entrenched role in powering industries, transport,

B. Saroha
Department of Zoology,, Roorkee Degree College, Latherdeva Shekh, PO Iqbapur, Roorkee, Uttarakhand, India

A. Kumar (✉) · D. Singhal · R. Dhoundiyal · H. V. Pant · S. Negi · H. Chandra · S. V. Singh
Department of Chemistry, SGRR (PG) College, Dehradun, Uttarakhand, India
e-mail: anandkciitd17@gmail.com

N. Kumar
Department of Applied Sciences and Humanities, School of Engineering and Technology, H.N.B. Garhwal University, Srinagar Garhwal, Uttarakhand, India

J. A. Siddique et al. (eds.), *Energy Sources*, Green Energy and Technology,
https://doi.org/10.1007/978-981-95-1209-6_4

and households worldwide [41]. However, their continued dominance poses significant challenges to the realization of a sustainable energy future. The combustion of fossil fuels is the leading source of anthropogenic greenhouse gas (GHG) emissions, particularly carbon dioxide (CO_2), which is the primary driver of climate change [61]. Additionally, fossil fuel extraction and usage contribute to severe environmental degradation, including habitat destruction, water and soil contamination, and ambient air pollution—affecting human health and ecosystem integrity [23].

Fossil fuels also impose substantial external costs on societies. These include healthcare burdens due to pollution-related diseases, infrastructure damage from climate-induced disasters, and ecological losses from oil spills, mining, and land degradation [61]. In response to these threats, international climate agreements, such as the Paris Agreement, have set an ambitious target to limit the global temperature rise to well below 2 °C above pre-industrial levels. This goal is crucial in mitigating the worst effects of climate change and is supported by numerous studies, including those by Alli et al. [4–6]. Furthermore, the finite and geographically uneven nature of hydrocarbon reserves creates geopolitical vulnerabilities and heightens concerns about long-term energy security [43].

Nonetheless, the role of fossil fuels in a sustainable energy system is not entirely negative—particularly in the context of a managed transition. In the near to medium term, fossil fuels will continue to play a role in supporting energy reliability, especially in regions where renewable energy infrastructure is not yet mature or economically viable. Natural gas, in particular, has been touted as a "bridge fuel" due to its lower-carbon intensity compared to coal and oil, and its ability to provide firm capacity to balance intermittent renewable energy sources like solar and wind [45]. Combined-cycle gas turbines can ramp up and down quickly, making them valuable assets for grid stability in hybrid energy systems [52].

Furthermore, the integration of Carbon Capture, Utilization, and Storage (CCUS) technologies offers a pathway to mitigate the environmental impact of fossil fuel use. These technologies can capture CO_2 emissions from power plants and industrial facilities and either store them underground or repurpose them for use in value-added products such as synthetic fuels, building materials, and chemicals [37, 32]. In this way, fossil fuels may support the transition toward net-zero energy systems if paired with carbon abatement strategies.

Advanced co-firing technologies, where coal is blended with biomass, and the development of blue hydrogen (produced from natural gas with carbon capture) are additional avenues through which fossil fuels may contribute to lower-emission pathways [32]. However, the long-term sustainability of such approaches depends on robust regulatory frameworks, international cooperation, investment in innovation, and the concurrent scaling of renewable energy technologies and energy storage systems [32].

In conclusion, while fossil fuels are largely incompatible with long-term sustainability due to their environmental and climate impacts, they may play a transitional role in facilitating the global energy shift—provided their use is tightly coupled with decarbonization technologies and policies. The challenge lies not only in reducing fossil fuel dependency but also in transforming their role within a broader, multi-faceted strategy for a just and secure energy transition.

2 Historical Context and Current Dependence

The use of fossil fuels dates back thousands of years; coal was used for cooking and heating in China as early as 1000 BC [4]. However, it was during the Industrial Revolution in the eighteenth and nineteenth centuries that fossil fuels gained widespread importance. The invention of the steam engine, powered by coal, revolutionized transportation and manufacturing, leading to a significant surge in fossil fuel consumption.

Today, fossil fuels remain the dominant source of energy worldwide. According to the International Energy Agency (IEA), fossil fuels accounted for approximately 84% of global primary energy consumption, with oil contributing 33%, coal 27%, and natural gas 24% [41]. This trend is unlikely to change in the near future, as the transition to renewable energy sources is both gradual and complex.

Although renewable technologies such as solar, wind, hydroelectric, and geothermal energy have gained momentum in recent years, they still represent only a small share of global energy use. Moreover, Ali points out that the cost of renewable energy production remains relatively high compared to fossil fuels, posing barriers for many countries [4]. In addition, nations with abundant fossil fuel reserves often rely on them for economic growth, further complicating the energy transition. Nevertheless, initiatives to promote renewable energy adoption are increasing. Government subsidies, tax incentives, and expanded research and development efforts are being implemented globally. It is hoped that these efforts will accelerate the shift to a cleaner, more sustainable energy system in the coming decades.

3 Environmental and Health Impacts

Fossil fuel combustion releases CO_2, methane, and other greenhouse gases, contributing to global warming [39]. The sector also emits sulfur dioxide, nitrogen oxides, and particulate matter, contributing to air pollution and public health crises. According to WHO (2021), ambient air pollution causes over 4 million premature deaths annually. Moreover, oil spills, coal mining, and gas flaring result in ecosystem degradation, biodiversity loss, and water contamination.

Khan et al. investigated the intricate relationship between fossil fuel intensity and sustainable development goals across 24 countries from 2000 to 2020, including 12 high-income and 12 low- and lower-middle-income nations. The findings indicate that higher fossil fuel intensity in high-income countries is linked to reduced mortality rates and better health outcomes, primarily due to advanced technologies and robust healthcare systems. In contrast, low- and lower-middle-income countries experience increased death rates and sustainability challenges, largely attributed to weak pollution control mechanisms and underdeveloped healthcare infrastructure [81].

4 Fossil Fuels and Energy Security

Fossil fuels continue to play a crucial role in global energy security by providing reliable and dispatchable energy, particularly through natural gas, which helps stabilize power grids with intermittent renewables like solar and wind [44]. However, their use raises significant environmental concerns, including CO_2 emissions and air pollution. To address these issues, CCUS technologies are being developed to capture CO_2 emissions from fossil fuel-based industries, preventing their release into the atmosphere. Captured CO_2 can be utilized in the production of synthetic fuels, chemicals, and hydrogen, advancing the transition to cleaner energy sources [10]. Hydrogen, particularly when produced via low-carbon methods like electrolysis (green hydrogen) or from fossil fuel reforming with CCUS (blue hydrogen), is increasingly seen as a key energy carrier for decarbonizing sectors such as transport and heavy industry, while enhancing long-term energy security. These technologies allow for continued reliance on fossil fuels while reducing their environmental impact, creating a more sustainable and resilient energy system.

5 How Fossil Fuels Can Play a Role in a Sustainable System

5.1 Carbon Capture, Utilization, and Storage (CCUS)

CCUS technologies are pivotal in mitigating the climate impacts of fossil fuels. CCUS can capture up to 90% of CO_2 emissions from power plants and industrial processes [40]. Captured carbon can be stored underground or utilized in products like synthetic fuels and construction materials [32]. While promising, CCUS is capital-intensive and requires supportive policy and infrastructure. Table 1 provides a comprehensive overview of CCUS technologies, detailing methods across chemical, biological, and industrial sectors. It highlights the higher efficiency of pre-combustion capture compared to post-combustion, albeit with greater capital costs, and discusses the use of chemical and physical solvents, particularly amine-based ones, for CO_2 scrubbing. Rapid advancements in CCUS technologies—including oxyfuel combustion, cryogenic separation, and membrane systems—are improving energy efficiency, though integration still demands significant energy input. The chemical conversion of CO_2 into value-added products such as urea, methanol, and polycarbonates relies on advanced catalytic systems (electrochemical, photocatalytic, thermocatalytic), while CO_2-based copolymers offer a sustainable alternative to petrochemical plastics. Biological utilization through microalgae showcases high CO_2 absorption efficiency, converting carbon into useful biomass and biochemical compounds. Finally, CO_2 use in beverage carbonation not only serves functional roles in soft drinks but also presents a consumer-driven model for sustainable CO_2 reuse, especially when sourced from industrial emissions or direct air capture systems.

5.1.1 Pre-combustion Carbon Capture

It is generally more efficient than post-combustion due to higher CO_2 concentrations and pressures in the synthesis gas, making separation easier [60]. However, it involves higher capital costs and significant infrastructure changes compared to conventional power plants [46, 84]. Hydrogen produced can be used as a clean fuel. CO_2 and H_2 are typically separated using chemical or physical solvents. Amine-based chemical solvents are effective at low partial pressures, while physical solvents perform better under high pressures and low temperatures, although they come with issues like low selectivity, high density, toxicity, and flammability [11, 57].

5.1.2 Post-combustion Carbon Capture

Post-combustion carbon capture refers to the removal of CO_2 from flue gases after fossil fuels have been burned. As outlined by [14], this method typically involves the use of chemical solvents—such as amines—that selectively absorb CO_2 from the exhaust stream. The captured CO_2 is then separated from the solvent for storage or utilization, and the solvent is regenerated for reuse. This approach is widely applicable to existing power plants and industrial facilities due to its retrofit capability.

5.2 Alternative Methods for CO_2 Capture and Utilization

5.2.1 CO_2 to-Jet Fuel Conversion

Fe and Co-Based Catalysts for CO_2 to-Jet Fuel Conversion.

The direct hydrogenation of CO_2 to jet fuel-range hydrocarbons has garnered significant attention, particularly through the use of traditional Fischer–Tropsch synthesis (FTS) catalysts such as Fe and cobalt Co, owing to their high catalytic activity, selectivity, and thermal stability. While Ru-based systems have demonstrated some potential for this transformation, their tendency toward methane production limits their applicability in producing liquid hydrocarbons [59, 67]. Among Fe- and Co-based catalysts, substantial formation of jet fuel-range hydrocarbons (C_8–C_{16}) has been consistently observed under optimized conditions. For instance, Fe–K/Al_2O_3 catalysts demonstrated one of the earliest and most effective results, achieving

Table 1 Comprehensive overview of pre-combustion and post- combustion carbon capture technologies

Aspect	Details	References
Post-combustion carbon capture	More efficient than post-combustion due to high-pressure,CO_2-rich synthesis gas	[60]
	Higher capital expenditure due to complex gasification systems and infrastructure needs	[46, 84]
	Enables the use of hydrogen as a clean fuel	[84]
	Syngas scrubbing using chemical or physical solvents	[46]
	Amine-based solvents effective at low partial pressures with high CO_2 absorption capacity	[57]
	Work best at high CO_2 partial pressures and low temperatures; follow Henry's Law	[57]
	Include low CO_2 H_2 specificity, high solvent density, corrosivity, toxicity, flammability, and limited heat stability	[11]
Advancements in capture technologies	Rapid technological advances and population growth have increased fossil fuel consumption, which negatively affects the environment due to limited supply and CO_2 emissions	[2, 24]
	Carbon capture and storage (CCS) involves separating CO_2 from transportation and industrial sectors and storing it in controlled environments, offering a solution for global warming	[65]
	CCS affects both energy generation costs and the speed of adoption; it is crucial to develop practical strategies for carbon sequestration in power plants	[31]
	Integrating CCS requires additional energy (10–40% more), with energy needs arising from CO_2 capture and compression. It can reduce emissions by 80–90%	[27, 75, 79]
	Demonstration projects for oxyfuel combustion technology are ongoing, but their capacities have generally been below 100 MWth	[27]
	Biomass usage in oxyfuel combustion, through supercritical CO_2 cycles, has shown a yearly CO_2 reduction of approximately 3.7 mega tons	[77]
	Alkanolamine solvents, particularly monoethanolamine (MEA), are widely used for their high absorption efficiency but are costly and environmentally damaging during desorption	[23]
	Rectisol, DMC, CAP, and MEA methods have varying energy requirements, with Rectisolrequiring the least (1.88 GJ/t CO_2). MEA requires the most energy (5.20 GJ/t CO_2)	[56]
	The cost of CO_2 capture follows the sequence: Rectisol < DMC < CAP < MEA, with ongoing research focused on reducing costs	[13, 56]

(continued)

Table 1 (continued)

Aspect	Details	References
	Novel solvents and sorbents, including amino-functionalized ionic liquids and metal–organic frameworks (MOFs), have improved CO_2 capture efficiency	[13, 34]
	Membrane-based capture technologies, such as assisted transport and mixed-matrix membranes, are gaining interest for their reduced energy use and easy operation	[22, 64, 74]
	Cryogenic separation improves energy efficiency by cooling flue gases to condense CO_2, with hybrid systems combining this with other techniques to improve efficacy and lower energy	[66, 68]
Chemical CO_2 Conversion		
Catalytic conversion	Urea (180 Mt), inorganic carbonates (60 Mt), polyurethane (15 Mt), polycarbonates (5 Mt), salicylic acid (0.17 Mt)	[16, 15]
	High energy demand due to CO_2 stability ($\Delta G^f = -394$ kJ/mol); requires effective catalysts	[3]
	Chemicals and synthetic fuels: urea, formic acid, methanol, carbonates, salicylic acid	[73, 82]
	Electrochemical: e.g., Cu → hydrocarbons; Au/Ag → CO thermocatalytic: high-temp catalysis; RWGS reaction Photocatalytic: TiO_2, MOFs under sunlight Biochemical and hybrid: microbial/electro-photo synergies	[20, 26, 54, 83]
	Single-atom catalysts Bimetallic alloys Machine-learning-optimized catalysts Advanced materials: doped graphene, MOFs, core–shell structures	[30, 55, 83], [9]
	Use of rare/expensive materials, uncertain kinetics, competition with hydrogen evolution	[19, 54]
Fertilizer use in agriculture	Synthetic nitrogen fertilizer accounts for 70% of global fertilizer use. Urea is a neutral fertilizer with high nitrogen content (46%)	[63]
Urea production method	Urea is typically produced through the reaction of CO_2 with ammonia in the presence of a catalyst	[21, 83]
CO_2 *based polymers*		
	CO_2-based polymers, mainly aliphatic polycarbonates	[53, 70]
	CO_2 content in copolymers: 31–50%	[53, 70]

(continued)

Table 1 (continued)

Aspect	Details	References
	Reduces reliance on petrochemical plastics; more eco-friendly alternative	[70]
	First developed by [38] via epoxide copolymerization	[38, 83]
	Sequential copolymerization of CO_2 with epoxides	[83]
	Homogeneous, heterogeneous, supported catalysts; typically organometallic compounds	[69, 83]
	Cheap, efficient catalysts are essential for scale-up and industrial application	[69, 83]
Biological utilization		
Microalgal value-added product	Utilization refers to using biomass as raw material for various biomass-based products, including microalgae. Microalgae convert CO_2 into valuable compounds	[18, 47]

(continued)

Table 1 (continued)

Aspect	Details	References
	Microalgal cells absorb CO_2 and convert it into macromolecules, metabolites, and biochemical compounds like lipids, proteins, polysaccharides, pigments, etc.	[18]
	Microalgae capture CO_2 through photosynthesis, incorporating carbon into their biomass	[35]
	Microalgae grow rapidly, doubling their biomass in under 24 h. 1 ton of microalgae can absorb 1.8 tons of CO_2. CO_2 conversion rates are 36 g/m^2/day	[35]
	To produce 1 ton of microalgae, approximately 50,000 m^2 of land or a 3000 m^3 reactor is needed, capturing 1.8 tons of CO_2 in the process	[35]
	Photosynthesis in microalgae promotes $CaCO_3$ formation, which can serve as a stable long-term CO_2 storage	[35]
Beverage carbonation		

(continued)

Table 1 (continued)

Aspect	Details	References
Role of CO_2 *in soft drinks*	CO_2 is essential in soft drink production; acts as an acidifying agent, flavor stabilizer, and antibacterial during carbonation	[33, 83]
	Beverage carbonation supports CO_2 reuse and promotes industrial carbon capture. Encourages consumer engagement with sustainability	[33]
	CO_2 used in beverages can be sourced from industrial emissions or direct air capture systems, reducing environmental impact	[33, 36]

yields of C_5^+ hydrocarbons of up to 22%, with gas chromatography confirming C_9 as the predominant product [80]. A critical challenge in early catalyst development was suppressing methane formation, which was successfully mitigated through potassium promotion. Potassium (K), typically added in the form of K_2CO_3, CH_3COOK, or $KHCO_3$, acts predominantly as an electronic promoter by modulating the electronic environment of the Fe active sites. It reduces hydrogen adsorption while enhancing CO dissociation, thereby shifting product selectivity toward olefins and long-chain hydrocarbons. Furthermore, these carbonated K promoters facilitate the formation of active Fe_3C_2–K_2CO_3 interfaces, essential for boosting olefin production in CO_2 hydrogenation. The $Fe_2O_3@K_2CO_3$ catalyst, for example, exhibits exceptional performance with CO_2 conversion rates reaching 44% and selectivity for C_2–C_{10} olefins as high as 63% [51]. In contrast, non-carbonated potassium sources are less effective in stabilizing these active phases. Optimal performance is associated with precise spatial distribution between K and Fe species, enabling enhanced carbon–carbon coupling, suppressed methane formation, and improved catalyst durability, marking a promising avenue for sustainable aviation fuel production via CO_2 valorization.

Direct CO_2 utilization via modified Fischer–Tropsch synthesis (FTS) shows promise for carbon–neutral jet fuel production but faces challenges of low CO_2 conversion and yield. A techno-economic assessment by [17] demonstrated that incorporating 90% tail gas recycling in CO_2 FTS can boost CO_2 conversion to 85.8%, increase jet fuel yield to 35.6%, and reduce the minimum selling price to US$2.6/kg, highlighting its potential for commercial viability.

5.2.2 Solar-Driven Thermochemical Conversion of H_2O and CO_2 into Sustainable Fuels

Thermochemical redox materials for CO_2 and H_2O splitting are broadly classified into three types—volatile, non-volatile stoichiometric, and non-stoichiometric materials—each with distinct characteristics and challenges. Volatile redox pairs (e.g., ZnO/Zn, CdO/Cd) undergo phase changes during cycling, which can complicate product separation and cause recombination issues [49]. Non-volatile stoichiometric materials (e.g., Fe_3O_4/FeO, CeO_2/Ce_2O_3) offer high oxygen exchange capacity but typically require very high temperatures, leading to phase instability and sintering problems [1, 50]. Non-stoichiometric materials, such as doped ferrites and ceria-based compounds, maintain structural integrity through partial reduction and operate at relatively moderate temperatures, offering enhanced redox stability, cycling efficiency, and scalability [25, 62]. These features make them promising candidates for solar thermochemical fuel production systems. Fig. 1 represents a solar-driven thermochemical fuel production system where concentrated solar power (CSP) heats a receiver-reactor to split H_2O and CO_2 into H_2 and CO via a redox cycle. These syngas products are stored and later converted into liquid hydrocarbons through Fischer–Tropsch synthesis, producing sustainable fuels [78].

5.2.3 Electrochemical CO_2 Conversion to Ethanol

The electrochemical reduction of CO_2 to ethanol is a promising strategy for carbon neutrality and renewable energy storage due to ethanol's high energy density and versatility in fuels and chemicals. However, achieving high selectivity and efficiency for ethanol remains challenging, primarily because its formation competes with closely related pathways that produce products such as ethylene, and because most current catalysts exhibit low energy efficiency, limited stability, and insufficient control over intermediate adsorption and C–C coupling steps necessary to direct the reaction toward ethanol [83]. To address this, recent strategies focus on engineering catalyst morphology, employing copper oxides, forming Cu-metal alloys, designing tandem and composite systems, and exploring non-copper materials [12]. These approaches aim to enhance intermediate stabilization, facilitate C–C coupling,

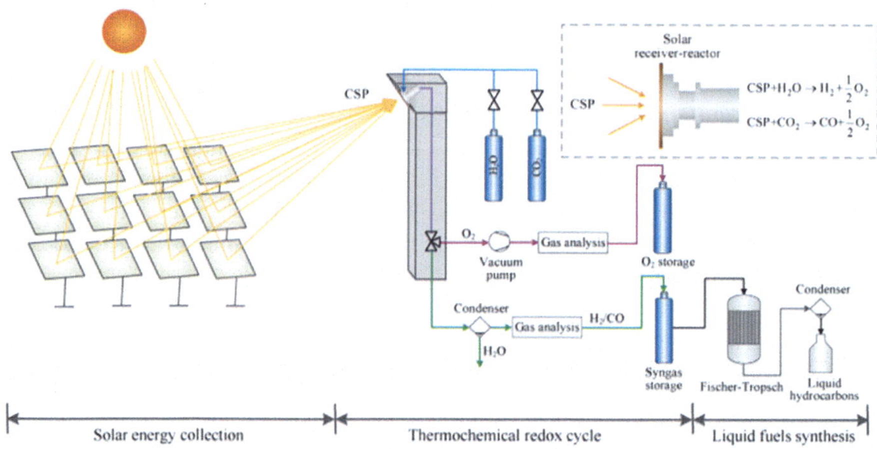

Fig. 1 Solar-driven thermochemical fuel production chain from H_2O and CO_2 to sustainable fuels (Reprinted from ref. [78])

and improve catalyst durability. Continued advances in catalyst design, mechanistic understanding, and scalable system integration are essential to bridge the gap between laboratory research and industrial-scale CO_2-to-ethanol conversion [48]. Figure 2 shows strategies for enhancing selective ethanol production over ethylene in the electrochemical CO_2 reduction process. By tailoring catalyst design—including surface morphology, copper oxides, Cu-metal alloys, tandem catalysts, composites, and non-copper materials—research aims to direct CO_2 conversion toward ethanol, which has diverse applications in fuel, medicine, and the chemical industry [76].

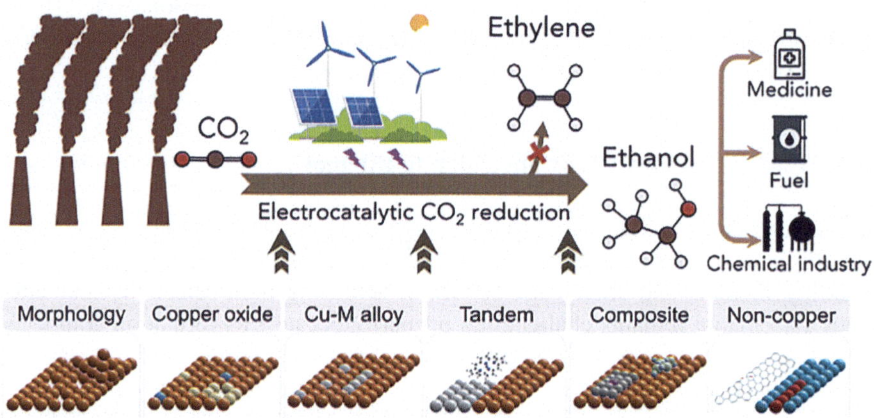

Fig. 2 Schematic of strategies for selective ethanol over ethylene production in electrochemical CO_2 reduction and its application (Reprinted from ref. [76])

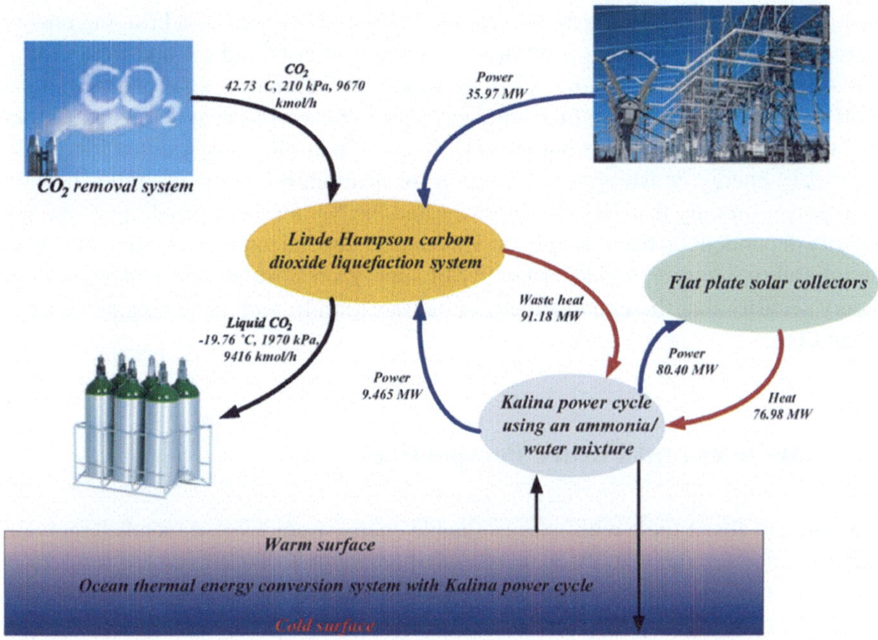

Fig. 3 Block diagram of energy transfer and streams in the integrated structure of CO_2 liquefaction (Reprinted from ref. [29])

5.3 *Blue Hydrogen*

Blue hydrogen, produced from natural gas with CCS, offers a lower-carbon fuel alternative. It is considered a transitional solution toward green hydrogen derived from electrolysis using renewable electricity [42]. Similarly, co-firing biomass with coal can reduce net CO_2 emissions in existing plants. These hybrid solutions offer pathways to reduce emissions while leveraging existing fossil infrastructure [58]. In Fig. 3, the CO_2 removal system represents the carbon capture stage crucial for blue hydrogen. Capturing CO_2 prevents it from being released into the atmosphere, making hydrogen production "low carbon."

5.4 *The Role of Natural Gas in Achieving a Sustainable Future*

Germany's energy policy has been significantly shaped by its commitment to the Paris Agreement of 2016, aiming to limit global temperature rise to below 2 °C [72]. While the country is expected to miss its 2020 climate targets, it is working toward expanding its renewable energy supply and enhancing energy efficiency in the long term [28]. Natural gas, though a fossil fuel, is viewed as a bridging energy source due to its lower CO_2 emissions compared to coal and oil and the existing infrastructure supporting its use [71]. However, in Germany's goal of achieving a 100% renewable energy system, natural gas will not have a place, as it still contributes to climate-damaging CO_2 emissions [71]. In 2017, natural gas accounted for 22.7% of primary energy consumption, primarily for heating, but Germany relies heavily on imports—mainly from Russia, Norway, and the Netherlands—making its energy security dependent on these suppliers. This import dependency raises concerns about the reliability, security, and affordability of supply, encapsulating broader issues of energy security that the country faces in its transition to a more sustainable energy future [71].

6 Challenges and Future Perspectives

The future of fossil fuels within a sustainable energy framework depends largely on the successful deployment of advanced decarbonization technologies and supportive policy mechanisms. Innovations such as CCS and CCU are poised to significantly reduce the carbon footprint of fossil fuel usage, allowing these resources to play a cleaner role in the global energy mix. Furthermore, the integration of fossil fuels with renewable energy sources through hybrid systems can offer flexible and reliable energy solutions that help stabilize grids increasingly reliant on intermittent renewable power. Natural gas, in particular, holds promise for the production of low-carbon hydrogen, positioning it as a transitional fuel in the emerging hydrogen economy. Policy-driven transformations, including stricter emission regulations, carbon pricing, and investment incentives, could accelerate the adoption of sustainable fossil fuel practices, while advances in digitalization and artificial intelligence will optimize operational efficiency and environmental performance.

However, numerous challenges must be addressed to realize these prospects. High costs associated with carbon management technologies remain a significant barrier to widespread implementation, especially in regions with limited financial resources. Policy and regulatory uncertainty across different countries create investment risks and slow down the transition toward cleaner fossil fuel use. Additionally, societal pressure for a complete phase-out of fossil fuels could undermine efforts to develop cleaner fossil fuel technologies, making it harder to secure public and political support. Resource depletion, particularly the declining quality of conventional oil and gas reserves, poses further environmental and economic risks. Finally, the entrenched nature of existing fossil fuel infrastructure—built over decades with enormous capital investments—presents inertia that challenges rapid transformation. Addressing these economic, social, and technological hurdles will be critical for fossil fuels to play a constructive role in a sustainable energy future.

7 Conclusion

Fossil fuels, while central to modern civilization's development, present profound environmental and societal challenges. In the journey toward a sustainable energy system, outright elimination of fossil fuels is neither immediately feasible nor necessarily optimal. Instead, their role must evolve through innovation, integration with clean energy technologies, and stringent carbon management strategies. Emerging technologies such as CCU, hydrogen production from natural gas with carbon capture, and hybrid renewable-fossil systems offer pathways to reduce emissions while meeting global energy demands. Nevertheless, realizing this vision requires overcoming significant economic, political, and technological hurdles. A balanced, pragmatic approach—where fossil fuels serve as a transitional component within a broader decarbonization agenda—will be vital. The future energy landscape must be shaped by both bold innovation and cautious stewardship, ensuring that fossil fuels, if used, contribute positively to sustainable development goals rather than hindering them.

References

1. Abanades S, Charvin P, Flamant G, Neveu P (2006) Screening of water splitting thermochemical cycles potentially attractive for hydrogen production by concentrated solar energy. Energy 31(15):2805–2822. https://doi.org/10.1016/j.energy.2005.11.002
2. Abdelkareem MA, Elsaid K, Wilberforce T, Kamil M, Sayed ET, Olabi A (2021) Environmental aspects of fuel cells: a review. Sci Total Environ 752:141803. https://doi.org/10.1016/j.scitotenv.2020.141803
3. Al-Mamoori A, Krishnamurthy A, Rownaghi AA, Rezaei F (2017) Carbon capture and utilization update. Energ Technol 5(6):834–849. https://doi.org/10.1002/ente.201600747
4. Ali U (2019) The history of the oil and gas industry from 347 AD to today. Offshore Technol. Retrieved 15 June 2025, from https://www.offshore-technology.com/comment/history-oil-gas/
5. Alli YA, Magida NE, Matebese F, Romero N, Ogunlaja AS, Philippot K (2023a) Biomimetic photocatalysts for the transformation of CO_2: design, properties and mechanistic insights. Mater Today Energy 37:101310. https://doi.org/10.1016/j.mtener.2023.101310
6. Alli YA, Oladoye PO, Ejeromedoghene O, Bankole OM, Alimi OA, Omotola EO, Olanrewaju CA, Philippot K, Adeleye AS, Ogunlaja AS (2023d) Nanomaterials as catalysts for CO_2 transformation into value-added products: a review. Sci Total Environ 868:161547. https://doi.org/10.1016/j.scitotenv.2023.161547
7. Alli YA, Oladoye PO, Matebese F, Senehi NL, Alimi OA, Ogunlaja AS, Abdi FF, Anuar H, Adeleye AS, Philippot K (2023c) Step-scheme photocatalysts: promising hybrid nanomaterials for optimum conversion of CO_2. Nano Today 53:102006. https://doi.org/10.1016/j.nantod.2023.102006
8. Alli YA, Oladoye PO, Onawole AT, Anuar H, Adewuyi S, Ogunbiyi OD, Philippot K (2023b) Photocatalysts for CO_2 reduction and computational insights. Fuel 344:128101. https://doi.org/10.1016/j.fuel.2023.128101
9. An L, De La Torre P, Smith PT, Narouz MR, Chang CJ (2023) Synergistic porosity and charge effects in a supramolecular porphyrin cage promote efficient photocatalytic CO_2 reduction. Angew Chem Int Ed 62(12):e202209396. https://doi.org/10.1002/anie.202209396

10. Artz J, Müller TE, Thenert K, Kleinekorte J, Meys R, Sternberg A, Bardow A, Leitner W (2018) Sustainable conversion of carbon dioxide: an integrated review of catalysis and life cycle assessment. Chem Rev 118(2):434–504. https://doi.org/10.1021/acs.chemrev.7b00435
11. Ashkanani HE, Wang R, Shi W, Siefert NS, Thompson RL, Smith K, Steckel JA, Gamwo IK, Hopkinson D, Resnik K, Morsi BI (2020) Levelized cost of CO_2 captured using five physical solvents in pre-combustion applications. Int J Greenhouse Gas Control 101:103135. https://doi.org/10.1016/j.ijggc.2020.103135
12. Cao B, Li FZ, Gu J (2022) Designing Cu-based tandem catalysts for CO_2 electroreduction based on mass transport of CO intermediate. ACS Catal 12(16):9735–9752. https://doi.org/10.1021/acscatal.2c02579
13. Castro-Pardo S, Bhattacharyya S, Yadav RM, de Carvalho Teixeira AP, Campos Mata MA, Prasankumar T, Kabbani MA, Kibria MG, Xu T, Roy S, Ajayan PM (2022) A comprehensive overview of carbon dioxide capture: from materials, methods to industrial status. Mater Today 60:227–270. https://doi.org/10.1016/j.mattod.2022.08.018
14. Chao C, Deng Y, Dewil R, Baeyens J, Fan X (2021) Post-combustion carbon capture. Renew Sustain Energy Rev 138:110490. https://doi.org/10.1016/j.rser.2020.110490
15. Chauvy R, Meunier N, Thomas D, De Weireld G (2019) Selecting emerging CO_2 utilization products for short- to mid-term deployment. Appl Energy 236:662–680. https://doi.org/10.1016/j.apenergy.2018.11.096
16. Chauvy R, De Weireld G (2020) CO_2 utilization technologies in Europe: a short review. Energ Technol 8(12):2000627. https://doi.org/10.1002/ente.202000627
17. Chen Y, Wang L, Zhang T, Liu X (2023) Techno-economic analysis of CO_2-based Fischer-Tropsch synthesis for sustainable jet fuel production. J Clean Prod 408:137563. https://doi.org/10.1016/j.jclepro.2023.137563
18. Daneshvar E, Wicker RJ, Show PL, Bhatnagar A (2022) Biologically-mediated carbon capture and utilization by microalgae towards sustainable CO_2 biofixation and biomass valorization: a review. Chem Eng J 427:130884. https://doi.org/10.1016/j.cej.2021.130884
19. Das S, Pérez-Ramírez J, Gong J, Dewangan N, Hidajat K, Gates BC, Kawi S (2020) Core–shell structured catalysts for thermocatalytic, photocatalytic, and electrocatalytic conversion of CO_2. Chem Soc Rev 49(9):2937–3004. https://doi.org/10.1039/c9cs00713j
20. Dasireddy VDBC, Vengust D, Likozar B, Kovač J, Mrzel A (2021) Production of syngas by CO_2 reduction through reverse water–gas shift (RWGS) over catalytically-active molybdenum-based carbide, nitride and composite nanowires. Renew Energy 176:251–261. https://doi.org/10.1016/j.renene.2021.05.051
21. Deacy AC, Phanopoulos A, Lindeboom W, Buchard A, Williams CK (2022) Insights into the mechanism of carbon dioxide and propylene oxide ring-opening copolymerization using a Co(III)/K(I) heterodinuclear catalyst. J Am Chem Soc 144(41):17929–17938. https://doi.org/10.1021/jacs.2c06921
22. Demir H, Aksu GO, Gulbalkan HC, Keskin S (2022) MOF membranes for CO_2 capture: past, present and future. Carbon Capture Sci Technol 2:100026. https://doi.org/10.1016/j.ccst.2021.100026
23. Dubey A, Arora A (2022) Advancements in carbon capture technologies: a review. J Clean Prod 373:133932. https://doi.org/10.1016/j.jclepro.2022.133932
24. Elsaid K, Kamil M, Sayed ET, Abdelkareem MA, Wilberforce T, Olabi A (2020) Environmental impact of desalination technologies: a review. Sci Total Environ 748:141528. https://doi.org/10.1016/j.scitotenv.2020.141528
25. Ermanoski I, Siegel NP, Stechel EB (2013) A new reactor concept for efficient solar-thermochemical fuel production. J SolEnergy Eng 135(3):031002. https://doi.org/10.1115/1.4023356
26. Etafo NO, Bamidele MO, Bamisaye A, Alli YA (2024) Revolutionizing photocatalysis: unveiling efficient alternatives to titanium (IV) oxide and zinc oxide for comprehensive environmental remediation. J Water Process Eng 62:105369. https://doi.org/10.1016/j.jwpe.2024.105369

27. Garcia JA, Villen-Guzman M, Rodriguez-Maroto JM, Paz-Garcia JM (2022) Technical analysis of CO_2 capture pathways and technologies. J Environ Chem Eng 10(6):108470. https://doi.org/10.1016/j.jece.2022.108470
28. German Environment Agency (2020) Germany 2020. International energy agency. https://www.iea.org/reports/germany-2020
29. Ghorbani B, Jouybari AK (2021) Development of a new CO_2 liquefaction system using the Linde-Hampson process, Kalina power generation unit, and flat plate solar collectors with Al_2O_3/H_2O nanofluid. Energy Sci Eng 10(11):4252–4272. https://doi.org/10.1002/ese3.1034
30. Guo Y, Gao Y, Guo B, Luo Y, Zhao G, Sun J, Li W, Wang R, Zhao C (2024) One-pot hydrothermal synthesis of transition metal sulfides-decorated CuS microflower-like structures for electrochemical CO_2 reduction to CO. Carbon Neutrality 3(1):21. https://doi.org/10.1007/s43979-024-00097-5
31. Gür TM (2022) Carbon dioxide emissions, capture, storage and utilization: review of materials, processes and technologies. Prog Energy Combust Sci 89:100965. https://doi.org/10.1016/j.pecs.2021.100965
32. Hanson E, Nwakile C, Hammed VO (2025) Carbon capture, utilization, and storage (CCUS) technologies: evaluating the effectiveness of advanced CCUS solutions for reducing CO_2 emissions. Results Surf Interfaces 18:100381. https://doi.org/10.1016/j.rsurfi.2024.100381
33. Hosseini SM, Aslani A, Kasaeian A (2023) Life cycle cost and environmental assessment of CO_2 utilization in the beverage industry: a natural gas-fired power plant equipped with post-combustion CO_2 capture. Energy Rep 9:414–436. https://doi.org/10.1016/j.egyr.2022.11.200
34. Huang Z, Karami D, Mahinpey N (2021) Study on the efficiency of multiple amino groups in ionic liquids on their sorbent performance for low-temperature CO_2 capture. Chem Eng Res Des 167:198–206. https://doi.org/10.1016/j.cherd.2021.01.016
35. Huang C-H, Tan C-S (2014) A review: CO_2 utilization. Aerosol Air Qual Res 14(2):480–499. https://doi.org/10.4209/aaqr.2013.10.0326
36. Hunt AJ, Sin EHK, Marriott R, Clark JH (2010) Generation, capture, and utilization of industrial carbon dioxide. Chemsuschem 3(3):306–322. https://doi.org/10.1002/cssc.200900169
37. Husain FM, Khan A, Khan RA, Siddique JA, Oves M, Khan AAP, Ansari MO, Dzudzevic Cancar H (2021) Bio-based aerogels and their environment applications: an overview. In: Khan AAP, Ansari MO, Khan A, Asiri AM (eds) Advances in aerogel composites for environmental remediation. Elsevier, pp 347–356. https://doi.org/10.1016/B978-0-12-820732-1.00018-7
38. Inoue S, Koinuma H, Tsuruta T (1969) Copolymerization of carbon dioxide and epoxide. J Polym Sci, Part C: Polym Lett 7(4):287–292. https://doi.org/10.1002/pol.1969.110070408
39. Intergovernmental Panel on Climate Change (IPCC) (2021) AR6 climate change 2021: the physical science basis. https://www.ipcc.ch/2021/08/09/ar6-wg1-20210809-pr/
40. International Energy Agency (IEA) (2020) Global energy review 2025: CO_2 emissions. https://www.iea.org/reports/global-energy-review-2025/co2-emissions
41. International Energy Agency (IEA) (2022) World energy outlook 2022. https://doi.org/10.1787/4b2e47f9-en
42. International Renewable Energy Agency (IRENA) (2022) Green hydrogen derivatives trade. https://www.irena.org/-/media/Files/IRENA/Agency/Publication/2024/Oct/IRENA_Green_hydrogen_derivatives_trade_2024.pdf
43. Jaiswal KK, Chowdhury CR, Yadav D, Verma R, Dutta S, Jaiswal KS, Sangmesh B, Karuppasamy KSK (2022) Renewable and sustainable clean energy development and impact on social, economic, and environmental health. Energy Nexus 7:100118. https://doi.org/10.1016/j.nexus.2022.100118
44. Jenkins C (2020) The state of the art in monitoring and verification: an update five years on. Int J Greenhouse Gas Control 100:103118. https://doi.org/10.1016/j.ijggc.2020.103118
45. Jordaan SM (2021) Natural gas as a bridge fuel? In: Wells to wire. Springer, pp 149–172. https://doi.org/10.1007/978-3-030-71971-5_7
46. Kheirinik M, Ahmed S, Rahmanian N (2021) Comparative techno-economic analysis of carbon capture processes: pre-combustion, post-combustion, and oxy-fuel combustion operations. Sustainability 13(24):13567. https://doi.org/10.3390/su132413567

47. Khoo KS, Ahmad I, Chew KW, Iwamoto K, Bhatnagar A, Show PL (2023) Enhanced microalgal lipid production for biofuel using different strategies including genetic modification of microalgae: a review. Prog Energy Combust Sci 96:101071. https://doi.org/10.1016/j.pecs.2023.101071
48. Kim YE, Lee W, Youn MH, Jeong SK, Kim HJ, Park JC, Park KT (2019) Leaching-resistant $SnO_2/\gamma\text{-}Al_2O_3$ nanocatalyst for stable electrochemical CO_2 reduction into formate. J Ind Eng Chem 78:73–78. https://doi.org/10.1016/j.jiec.2019.05.042
49. Kim J, Miller JE, Maravelias CT, Stechel EB (2013) Comparative analysis of environmental impact of S2P (Sunshine to Petrol) system for transportation fuel production. Appl Energy 111:1089–1098. https://doi.org/10.1016/j.apenergy.2013.06.035
50. Lapp J, Davidson JH, Lipinski W (2012) Efficiency of two-step solar thermochemical non-stoichiometric redox cycles with heat recovery. Energy 37(1):591–600. https://doi.org/10.1016/j.energy.2011.10.027
51. Lee MH, Park S, Kim DH (2023) Enhanced olefin selectivity in CO_2 hydrogenation via potassium-promoted iron oxide nanostructures. Appl Catal B 323:122180. https://doi.org/10.1016/j.apcatb.2022.122180
52. Li N, Aksoy M, Nutakki TUK, Singh PK, El-Shorbagy MA, Dahari M, Alkhalaf S, Alotaibi KO, Ali HE (2024) Development of a modified gas turbine-based sustainable power generation and water treatment system: economical/environmental considerations and data-driven optimization. J Clean Prod 450:141904. https://doi.org/10.1016/j.jclepro.2024.141904
53. Liu Y, Lu X-B (2023) Current challenges and perspectives in CO_2-based polymers. Macromolecules 56(6):1759–1777. https://doi.org/10.1021/acs.macromol.2c02483
54. Liu H, Song T, Wang Y, Zhang X, Cui L, Liu T, Yuan Z (2024) Efficient electrocatalytic reduction of CO_2 to CO via mechanochemical synthesized copper-based composite metallic oxide catalyst. J Porous Mater 31:437–448. https://doi.org/10.1007/s10934-023-01524-1
55. Ma X, Li Z, Achenie LEK, Xin H (2015) Machine-learning-augmented chemisorption model for CO_2 electroreduction catalyst screening. J Phys Chem Lett 6(17):3528–3533. https://doi.org/10.1021/acs.jpclett.5b01660
56. Ma Y, Liao Y, Su Y, Wang B, Yang Y, Wang D, Ji D, Li H, Zhou H (2021) Comparative investigation of different CO_2 capture technologies for coal to ethylene glycol process. Processes 9(2):207. https://doi.org/10.3390/pr9020207
57. Osman AI, Hefny M, Abdel Maksoud MIA, Elgarahy AM, Rooney DW (2021) Recent advances in carbon capture storage and utilisation technologies: a review. Clean Technol Environ Policy 19:797–849. https://doi.org/10.1007/s10311-020-01133-3
58. Quang DV, Milani D, Abu Zahra M (2023) A review of potential routes to zero and negative emission technologies via the integration of renewable energies with CO_2 capture processes. Int J Greenhouse Gas Control 124:103862. https://doi.org/10.1016/j.ijggc.2023.103862
59. Rao A, Tan L, Meunier F (2022) Catalytic pathways in CO_2 hydrogenation for fuel production. Green Chem 24(2):456–470. https://doi.org/10.1039/D1GC03820C
60. Sarafraz MM, Christo FC (2021) Sustainable three-stage chemical looping ammonia production (3CLAP) process. Energy Convers Manage 229:113735. https://doi.org/10.1016/j.enconman.2020.113735
61. Saravanan A, Kumar PS (2022) Biochar derived carbonaceous material for various environmental applications: systematic review. Environ Res 214:113857. https://doi.org/10.1016/j.envres.2022.113857
62. Scheffe J, Steinfeld A (2014) Oxygen exchange materials for solar thermochemical splitting of H_2O and CO_2: A review. Materials Today 17(7): 341–348. https://doi.org/10.1016/j.mattod.2014.04.025
63. Shafreen M, Vishwakarma K, Shrivastava N, Kumar N (2021) Physiology and distribution of nitrogen in soils. In: Cruz C, Vishwakarma K, Choudhary DK, Varma A (eds) Soil nitrogen ecology. Springer, pp 3–31. https://doi.org/10.1007/978-3-030-71206-8_1
64. Shah S, Shah M, Shah A, Shah M (2020) Evolution in the membrane-based materials and comprehensive review on carbon capture and storage in industries. Emergent Mater 3(1):33–44. https://doi.org/10.1007/s42247-020-00069-2

65. Shen M, Kong F, Tong L, Luo Y, Yin S, Liu C, Zhang P, Wang L, Chu PK, Ding Y (2022a) Carbon capture and storage (CCS): development path based on carbon neutrality and economic policy. Carbon Neutrality 1(1):37. https://doi.org/10.1007/s43979-022-00039-z
66. Shen M, Tong L, Yin S, Liu C, Wang L, Feng W, Ding Y (2022b) Cryogenic technology progress for CO_2 capture under carbon neutrality goals: a review. Sep Purif Technol 299:121734. https://doi.org/10.1016/j.seppur.2022.121734
67. Smith JP, Muller T (2021) Methane selectivity in CO_2 hydrogenation using Ru-based catalysts. Catal Today 375:113–122. https://doi.org/10.1016/j.cattod.2020.05.018
68. Song C, Run L, Liu Q, Deng S, Li H, Sun J (2024) Membrane-cryogenic hybrid CO_2 capture—a review. Carbon Capture Sci Technol 13:100249. https://doi.org/10.1016/j.ccst.2024.100249
69. Song X, Wang J, Yang L, Pan H, Zheng B (2020) The transformation strategies between homogeneous and heterogeneous catalysts for the coupling reactions of CO_2 and epoxides/olefins. Inorg Chem Commun 121:108197. https://doi.org/10.1016/j.inoche.2020.108197
70. Tang S, Nozaki K (2022) Advances in the synthesis of copolymers from carbon dioxide, dienes, and olefins. Acc Chem Res 55(13):1524–1532. https://doi.org/10.1021/acs.accounts.2c00162
71. Ullah S, Boulanouar Z, Meo MS, Ali K (2025) Navigating sustainable mobility: wavelet insights into carbon emissions in Germany's transportation sector. Transp Policy 170:120–136. https://doi.org/10.1016/j.tranpol.2025.05.017
72. United Nations (2016) The Paris agreement. UNFCCC. https://unfccc.int/process-and-meetings/the-paris-agreement
73. Vakylabad AB (2024) Carboxylic acid production from CO_2. In: Rahimpour MR, Makarem MA, Meshksar M (eds) Advances and technology development in greenhouse gases: emission, capture and conversion. Elsevier, pp 349–378. https://doi.org/10.1016/B978-0-443-19235-7.00015-4
74. Vermaak L, Neomagus HWJP, Bessarabov DG (2021) Recent advances in membrane-based electrochemical hydrogen separation: a review. Membranes 11(2):127. https://doi.org/10.3390/membranes11020127
75. Wang J, Pu Q, Ning P, Lu S (2021) Activated carbon-based composites for capturing CO_2: a review. Greenhouse Gases: Sci Technol 11(3):377–393. https://doi.org/10.1002/ghg.2051
76. Wang Z, Li Y, Ma Z, Wang D, Ren X (2024) Strategies for overcoming challenges in selective electrochemical CO_2 conversion to ethanol. iScience 27(8):110437. https://doi.org/10.1016/j.isci.2024.110437
77. Wei X, Manovic V, Hanak DP (2020) Techno-economic assessment of coal- or biomass-fired oxy-combustion power plants with supercritical carbon dioxide cycle. Energy Convers Manage 221:113143. https://doi.org/10.1016/j.enconman.2020.113143
78. Wei L, Pan Z, Shi X, Esan OC, Li G, Qi H, Wu Q, An L (2023) Solar-driven thermochemical conversion of H_2O and CO_2 into sustainable fuels. iScience 26(11):108127. https://doi.org/10.1016/j.isci.2023.108127
79. Wilberforce T, Olabi AG, Sayed ET, Elsaid K, Abdelkareem MA (2021) Progress in carbon capture technologies. Sci Total Environ 761:143203. https://doi.org/10.1016/j.scitotenv.2020.143203
80. Xiong L, Liu S, Men Y, Li L, Niu X, Guo K, Xu J, An W, Wang J, Cong Y (2022) Highly selective hydrogenation of CO_2 to C_5^+ hydrocarbons over Fe catalysts copromoted by K with Pd. J Environ Chem Eng 10(5):108407. https://doi.org/10.1016/j.jece.2022.108407
81. Yang X (2025) Is fossil fuel intensity adversely affecting health improvement and sustainability? Resour Policy 101:105388. https://doi.org/10.1016/j.resourpol.2024.105388
82. Yusuf N, Almomani F, Qiblawey H (2023) Catalytic CO_2 conversion to C_1 value-added products: review on latest catalytic and process developments. Fuel 345:128178. https://doi.org/10.1016/j.fuel.2023.128178
83. Zhang Z, Pan S-Y, Li H, Cai J, Olabi AG, Anthony EJ, Manovic V (2020) Recent advances in carbon dioxide utilization. Renew Sustain Energy Rev 125:109799. https://doi.org/10.1016/j.rser.2020.109799
84. Zhao F, Cui C, Dong S, Xu X, Liu H (2023) An overview on the corrosion mechanisms and inhibition techniques for amine-based post-combustion carbon capture process. Sep Purif Technol 304:122091. https://doi.org/10.1016/j.seppur.2022.122091

Coal Energy: Carbon Capture, Utilization, and Storage

Shahid Pervez Ansari, Mohammad Ayub Husain Khan, Deepika Varshney, Mohammad Omaish Ansari, and Jamal Akhter Siddique

Abstract Coal is a carbonaceous material, an important energy source primarily composed of carbon and hydrocarbons. There are many different varieties of coal found around the world; each type of coal is unique in nature and is a result of complex reactions. It releases high energy on combustion in the presence of oxygen. Coal is so important to the world that it is the primary energy source as well as the second most highly utilized energy source globally. When the coal is burned for energy demand, various chemicals are released, which affect the environment and living beings. Coal burning is a major contributor to the greenhouse effect and global warming. There is an urgent need to control emissions and their adverse effects. Decarbonization of the atmosphere offers ways to reduce or eliminate CO_2 from the atmosphere to contribute positively to controlling the harmful impacts of climate change. This chapter presents a detailed discussion of coal emission, carbon capture, and utilization.

Keywords Coal · Carbon emission · Decarbonization · CO_2 utilization

---

S. P. Ansari (✉)
Department of Applied Chemistry, Faculty of Engineering and Technology, Aligarh Muslim University, Aligarh, India
e-mail: shahidzahir@gmail.com

M. A. H. Khan
Global Institute of Engineering and Technology, Chilkur, Moinabad, Hyderabad, Telangana, India

D. Varshney
Department of Engineering, University of Technology and Applied Sciences, Al Musannah, Oman

M. O. Ansari
Center of Nanotechnology, King Abdulaziz University, Jeddah, Saudi Arabia

J. A. Siddique
Department of Chemistry, Gandhi Faiz-E-Aam College (M.J.P. Rohilkhand University), Shahjahanpur, India

J. A. Siddique et al. (eds.), *Energy Sources*, Green Energy and Technology,
https://doi.org/10.1007/978-981-95-1209-6_5

1 Introduction: Coal Energy and Carbon Emissions

Coal is an important source of energy and is primarily composed of carbon and hydrocarbon materials. It releases high energy on combustion in the presence of oxygen. Due to easy access and high energy content, coal is chiefly used to fulfill energy demand in many countries in the world [1]. Many developing nations still use coal as a primary energy source to meet their local energy demand. Coal is so important to the world that it is the primary energy as well as the second most highly utilized energy source in the world, and contributes approximately 40% of the global primary energy demand [2, 3]. Coal as an energy source is a revolutionary material on the basis of its economic considerations, but from the environmental point of view, it causes ambient air pollution due to its combustion and has become a threat to human beings. In December 2015 (Paris, France), all the member nations of the Conference of Parties (COP) agreed to raise investment and intensify the fight to control global warming within certain limits for a future that is sustainable as well as low carbon. The Paris Agreement was intended to reduce greenhouse gases (GHGs) by reducing their production. However, the goals agreed upon in the Paris Agreement (2015) are unlikely to be achieved very soon, as most of the developing nations are more concerned and forced to fulfill the immediate need for energy using coal rather than with the reduction of GHGs for the future [4]. When coal is burned for energy demand, a number of components are released, which may be classified as (i) gaseous components (mixture of many gaseous products formed during combustion of coal) and (ii) solid particles (particulate matter (PM) of definite size, specific morphology, and other characteristics) into the atmosphere. The living communities (animals, plants, and human beings) around coal-fueled thermal power plants are under continuous exposure to the emissions from these power plants, due to which their health and well-being are under constant threat [5]. The burning of coal is not the only factor that is responsible for significant health issues; the extraction, transport, and handling processes of coal release a tremendous amount of coal dust in the atmosphere and pose an alarming threat to the workers, the surrounding neighborhoods, and cause xenobiotic effects [6, 7]. Apart from the above concern, another important challenge is the ever-increasing concentration of critical elements in the atmosphere that may pose a bigger health threat to human beings and other living beings [8]. Coal is also a source of carbon nanoparticles and rare earth elements that can be utilized in advanced technologies (nano- and microtechnology). Therefore, coal mining methods need to be improved and evolved to avoid any further harm to the environment [9, 10].

1.1 Coal Composition, Emission, and Health Risks

Coal is a complex carbonaceous material that came into existence due to intricate reactions between many biological, geochemical, and mineralogical conditions and properties [10]. There are many different varieties of coal found around the world. Each type of coal is unique and results from complex reactions and complex conditions for millions of years for its formation [11]. Over the millions of years, different biopolymers from algae, microbes, plants (vascular and nonvascular), remains of prehistoric microorganisms, resins, and other inorganic and organic materials were slowly piled up in layered formation, commonly known as peat deposits, retain their most of the carbon while it underwent complex metamorphism (coalification), eventually to give the resultant material called as coal [10–14]. The peat deposits are also the source of the increased amount of toxic (S, As, Hg, etc.) components in the coal, and therefore, the impact on human health is eventual [10, 15–17]. The coal-fueled thermal power plant processes induce desirable and undesirable chemical changes in elements. As a result, significant amounts of gaseous oxides of carbon (CO and CO_2), oxides of sulfur (SO_2 and SO_3), and oxides of nitrogen (NO and NO_2) are released into the atmosphere [11, 18, 19]. The resultant mixture of these pollutants can potentially trigger and severely impact human health due to their chemically reactive nature [19–21]. The intricate mixture of various small particles emitted as the combustion product of coal contributes a large amount of PM. These PMs may be classified according to particle size (aerodynamic equivalent diameter), such as PM2.5 and PM10. PM2.5 has the ability to infiltrate and affect different regions of the respiratory system, which results in pulmonary diseases [22]. The smaller the size of PMs (PM2.5), the higher the threat, as these can penetrate deep into the lungs and may even (possibly) pass to the blood/nutrients circulation, influencing other organs in the body [23], while the larger PMs (PM10) will be filtered by the nose or may get deposited in the proximal airways [24]. The airborne PM emitted during coal combustion may also travel across a larger distance [25].

The composition and components of the coal depend on the genesis site of the naturally occurring coal deposits, which may vary significantly over a range of a few kilometers [26, 27]. China and the United States of America (USA) are the main coal miners in the world. However, Australia, Colombia, Indonesia, Russia, and South Africa also mine large amounts of coal [28]. In many countries, coal is burned to produce electricity, as in South Africa (93%), Poland (92%), China (79%), Australia (77%), India (69%), Germany (46%), and the United States (23.5%); and Brazil is the largest consumer of coal in South America [29–32]. The impacts of coal and its combustion on human health vary due to the variation in the compositional constituents of the coal [33]. According to the peat depositional environment reports, each coal deposit in a global location produces coal of a distinct composition [10, 33]. For example, the Indian and the Chinese coal-fuelled power plants emit large amounts of mercury (Hg) into the atmosphere [34]; furthermore, the negligence in SO_2 and NO_x control in India makes India one of the main contributing nations of global coal combustion-related human health impacts [28]. In South Africa, the coal-fuelled power plants contribute to high concentrations of lead (Pb) in the atmosphere of surrounding areas [35], while in Brazil, burning of coal in thermal power plants or for industrial purposes releases large amounts of arsenic (As) and zinc (Zn) in the atmosphere [15, 36]. The waste materials derived from these mining sources may contribute to environmental contamination if not correctly disposed of [31, 32, 37].

The emission from the power plants impacts all the living beings living near or a little far from the power plant. Therefore, the increasing human mortality, illness, and other adverse impacts on human health due to air pollution from all sources are evident [38].

2 Decarbonization

The urgency and significance of the challenges of climate change and its mitigation can be understood in terms of goals set during the Conference of Parties (COP29) meeting in 2024. COP29 concluded with an emphasis on addressing and controlling the disastrous impact of climate change on human health and their population. COP has reiterated several times that it will control the emission of CO_2 in the atmosphere and reduce its presence in the atmosphere [39, 40]. In a similar quest, decarbonization of the atmosphere offers excellent potential for reducing the amount of CO_2 in the atmosphere. The International Panel on Climate Change, namely: "The Intergovernmental Panel on Climate Change (IPCC)," has also recognized the process of decarbonization of the atmosphere as one of the most significant and promising strategies to limit global warming well below 1.5 °C above the preindustrial levels [39]. The decarbonization process (DP) of the atmosphere offers ways to reduce or eliminate CO_2 from the atmosphere with the aim of contributing positively to controlling the harmful impacts of climate change. It is established that CO_2 traps heat in the atmosphere and blocks the emission of heat away from the atmosphere, causing the greenhouse effect, which ultimately results in global warming. Therefore, CO_2 is looked upon as a GHG and a major contributor to global warming [41–44]. The DP is preferred for several important reasons; primarily, it is necessary to avoid and control the most hazardous impacts of climate change like high frequency of extreme weather, rising sea levels, disruption to the ecosystem, and severe health hazards to the human population [45–48]. The secondary significance of DP of atmosphere includes better air quality, energy security, economic growth, and improved public health. With time, the DP is becoming a significant and urgent option for limiting the temperature rise below 1.5 °C [49]. In this chapter, we will discuss the various options and strategies to mitigate the challenges of climate change and foster a sustainable future using DP through carbon capture, utilization, and storage (CCUS) technologies.

Carbon capture and storage (CCS) technologies involve segregating the CO_2 from the atmosphere for a long time by separating the CO_2 from relevant emission sources and transporting it to the storage sites. These technologies can reduce the amount of CO_2 entering the atmosphere by reducing its emissions from the source and also by removing the CO_2 already present in excess in the atmosphere on a large scale. These technologies offer a good method to alleviate the greenhouse effect caused by the presence of GHGs [50, 51]. However, due to the economic constraints of carbon capture technologies and the risk of leakage of geological storage, it is needed to look beyond the existing technologies and strategies. As a result, researchers shift the focus of their research to evolve new and improved technologies for CCUS to ensure long-term stability [52].

The concept of CCUS is getting accepted globally, and CO_2 capture and utilization technologies are being explored intensively. The CCUS technologies offer great potential in the reduction of CO_2 concentration from the source point and utilization of the captured CO_2 for the production of important molecules, therefore offering economic benefits as well. The CO_2 captured from one production site is transferred into another production process, where it is converted into different forms and utilized in different ways. The processes are based on the simple principle of recycling and reuse of CO_2 to reduce its concentration in the atmosphere [50, 53]. Our further discussion will be concerned with the various CO_2 capture technologies, utilization technologies, and several other recently reported frontier technologies of CO_2 resource utilization to provide some ideas toward mitigating the challenges in climate change and its trend.

2.1 Carbon Dioxide (CO_2) Capture Technologies

Most of the CO_2 capture technologies may be broadly classified into three groups: (i) precombustion CO_2 capture technologies, (ii) post-combustion CO_2 capture technologies, and (iii) oxy-fuel combustion technologies [50] (Fig. 1, Table 1). Post-combustion CO_2 capture is a mature technology and is widely used in the industry, with good CO_2 selectivity and CO_2 capture efficiency. In contrast, precombustion CO_2 capture and oxy-fuel combustion capture technologies are relatively less developed due to their limitation of specific requirements of materials and high-temperature conditions [9].

2.1.1 Post-combustion CO_2 Capture

Post-combustion CO_2 capture technologies capture the CO_2 from the exhaust gases of industrial operations or power plants after the combustion of fuel. Typically, these technologies use CO_2-absorbing solvents (such as amine-based solutions), and when the exhaust gases interact with these solvents, CO_2 is absorbed in these solvents from flue gas. The CO_2 is recovered from these solvents to produce a concentrated CO_2 stream. This recovered CO_2 is then compressed and transported for storage or utilization [54–57]. The post-combustion CO_2 capture technologies are the most mature and widely used CCUS technologies, and several large-scale projects and commercial applications are operational in different parts of the world [58]. Despite their significant potential to reduce the CO_2 emissions into the atmosphere from industrial processes and power plants, the post-combustion CO_2 capture technologies also come across several challenges, like energy and financial constraints to the CO_2 capture, its regeneration, disposal of solvents, and by-product wastes. At present, research and development are focused on enhancing the efficiency and reducing the

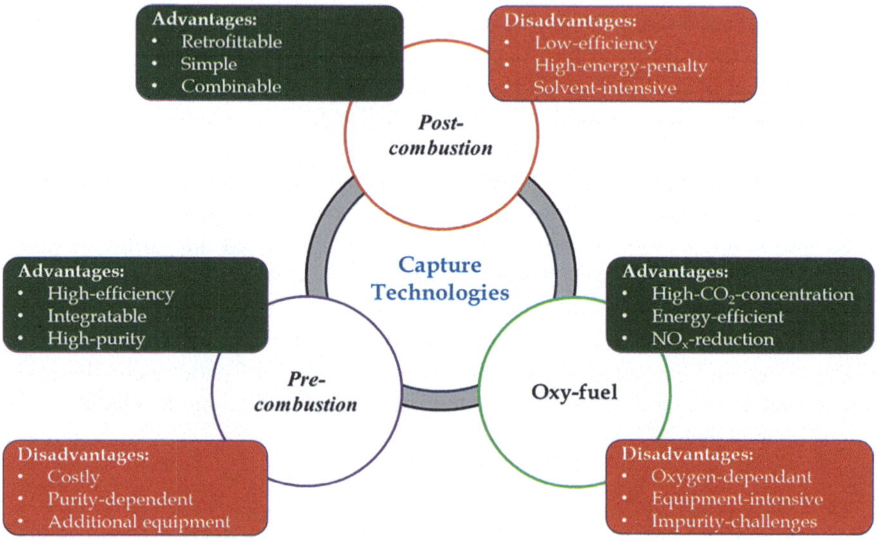

Fig. 1 CO_2 capture technologies. Courtesy: ACS (2024) (ACSEng. Au2024, 4, 22–48)

cost and the environmental impacts of post-combustion CO_2 capture, additionally exploring alternative technologies like membrane-based CO_2 capture and direct air capture (DAC) [59, 60].

2.1.2 Precombustion CO_2 Capture

The precombustion CO_2 capture technologies capture the CO_2 from a gas stream before the fuel is burned. During the process, the feedstock is transformed into synthesis gas (syngas) through the process of gasification, and the separation of the CO_2 syngas is performed using either chemical or physical absorption processes [61, 62]. The precombustion capture technologies have the potential to reduce the CO_2 emissions from many industries like iron-steel, chemicals, refining, where large amounts of hydrogen are used, and a significant amount of CO_2 is evolved in the form of CO_2 emissions [63–65]. Thus, captured CO_2 is transported to the storage site or for utilization, and the hydrogen (H_2) gas is used as a fuel [66–70]. One of the prime advantages of precombustion CO_2 capture technologies is that they are relatively more efficient than post-combustion CO_2 capture technologies. However, precombustion CO_2 capture technologies face challenges like high capital costs, the requirement of high-purity feedstocks, and technical complexities during operation [71]. Research and development for the precombustion CO_2 capture technologies is primarily concerned with improving efficiency, cost reduction, and solutions to technical challenges. Alternative technologies for precombustion CO_2 capture, like

Table 1 Comparative analysis of some CO_2 capture technologies [50]

CO_2 capture technology	Advantages	Limitations	Future outlooks
Chemical absorption	Relatively mature technology, superior CO_2 capture, excellent CO_2 selectivity	High corrosion rate of equipment, high energy consumption, high operating costs, and discharge causes environmental pollution	Development of new absorbent with good absorption performance and low corrosion; designing of equipment and processes with low energy consumption
Solid adsorption	Relatively simple, low energy consumption, noncorrosive, suitable for CO_2 capture at low concentration, multiple adsorption technologies	Low CO_2 capture capacity, low capture of industrial flue gas	To improve the performance of CO_2 capture new adsorbent with higher capacity should be developed
Membrane separation	Simple operation process, low cost, low energy consumption, high separation purity, friendly to environment	Easy damage to membranes, short service life, moisture affects the permeability of the membranes large area of membrane is required for industrial applications	To develop advanced membrane materials with enhanced CO_2 separation efficiency, more investment is to be diverted to industrial pilot projects
Cryogenic separation	Suitable for CO_2 capture at high concentration or high partial pressure, high separation purity	High operating cost, complex process, solidification of CO_2 at low temperatures leads to pipeline blockage, not suitable for the low CO_2 capture	To design a more effective and simple low-temperature process, to solve the blocking of the pipeline due to CO_2 solidification
Hydrate method	Simple process flow, low energy consumption, noncorrosive and nonpolluting	Few practical applications and immature technology; high requirements for equipment selection; CO_2 separation efficiency is affected by the hydrate formation method in a specific pressure vessel	Need to improve CO_2 separation efficiency, develop new equipment, and processes
Microbiological method	No pollution, low cost, high return value can directly convert CO_2 into organic matter	There are a few related studies, and it is still in the theoretical stage	More experiments are needed to develop improved microbiological methods for enhanced CO_2 capture

adsorption-based CO_2 capture and membrane-based CO_2 separation, are also being evaluated [50, 65, 72, 73].

2.1.3 Oxyfuel Combustion

Oxyfuel combustion technologies involve the combustion of fuel in the presence of a mixture of recycled flue gas and oxygen instead of air. The resultant flue gas stream is enriched with CO_2 and water vapor. The CO_2 is then captured and separated from the flue gas stream and transported for storage or further utilization. The oxyfuel combustion technologies offer several advantages, like higher flame temperatures and higher concentration of CO_2 in the flue gas stream compared to traditional combustion methods, and eventually offer enhanced energy efficiency and better economics of CO_2 capture [39, 74, 75]. Furthermore, the utilization of flue gas (recycled) also lessens the requirement of additional air separation setup and its cost and reduces NO_x emissions as well [76]. However, oxyfuel combustion technologies also have some challenges, such as the need for O_2 production and the technical complexities during the combustion process. Additionally, the impurities in the recycled flue gas affect the combustion efficiency and CO_2 capture performance [39, 76].

3 Utilization of CO_2

Several projects worldwide have been installed and are running to capture and store CO_2 (CCS). However, the high cost of CO_2 storage, the potential risk of leakage of stored CO_2, and the risk of small levels of earthquakes caused by CO_2 under pressure are some of the uncertainties that are associated with CO_2 storage. As a result, there is a shift in the interest and preferences for CCUS technologies that can reuse CO_2 [77]. Furthermore, it is necessary to mention that CO_2 utilization alone cannot be helpful in achieving the target of net-zero CO_2 emissions within the set time frame because, during the process of CO_2 utilization, some electricity will also be consumed, which further contributes to CO_2 emissions [78]. However, the economic benefits and long-term stability of CCUS technologies offer a promising future, and the utilization of CO_2 contributes to lowering the CO_2 concentration to address the problem of high CO_2. The resource utilization of CO_2 mainly includes biological, chemical, physical, and mineralization utilization [79, 80].

3.1 Physical Utilization of CO_2

Physical methods of CO_2 utilization are widely accepted and used in industries such as food, refrigeration, and petroleum. There should be no doubt that during the physical methods of CO_2 utilization, the release of CO_2 in the atmosphere is delayed only and, eventually, will be released into the atmosphere in due course of time [81]. The physical methods of CO_2 utilization do not belong to a DP or separate CO_2 from the environment. There are many areas (beer and carbonated beverages, fresh/frozen food, CO_2 flooding, CO_2 gas shielded welding, tobacco expansion technology, supercritical CO_2 in extraction, etc.) where these technologies are being used. However, the economic benefits of the physical utilization methods will pave the path for CCS-related projects in the future.

3.2 *Chemical Utilization of CO_2*

CO_2 is an essential raw material for the synthesis of some industrial chemicals. Industrial chemicals such as alcohols, amines, dimethyl ether, esters, salicylic acid, organic acids, and urea can be synthesized from CO_2 through appropriate chemical reactions [82]. The chemicals obtained by the chemical conversion of CO_2 can be broadly classified into three categories: (i) bulk chemicals, (ii) organic fuels, and (iii) polymer materials. During the chemical utilization of the CO_2, the products synthesized from the CO_2 will eventually be oxidized to CO_2, and it will be released into the atmosphere [83].

3.3 *Biological Utilization of CO_2*

The concept of the natural photosynthesis process, where plants use CO_2 and H_2O to produce glucose molecules, can be utilized to design novel systems for artificial photosynthesis to convert the CO_2 in the atmosphere to glucose. The concept is referred to as the biological utilization of CO_2. It can be used to design artificial photosynthesis systems, simulating the pathways of photosynthesis in plants or microorganisms in nature for efficient conversion of CO_2 into different products [84]. Presently, research around the world is focused on the bio-utilization of the CO_2 for the conversion of plant aerosol fertilizers into biofuels and biomineralization of the CO_2.

4 Frontier of the CO_2 Utilization

To further improve the utilization of CO_2, new avenues are being explored in research around the world [50]. A few of these efforts are discussed here in this section. Ma et al. [85] reported a novel concept of natural gas purification and efficient CO_2 conversion by simultaneously converting CO_2 and H_2S via a photoelectrochemical method. They utilized graphene and zinc oxide-based cathode catalysts, which effectively reduce the CO_2, while graphene acts as an anode catalyst to oxidize the EDTA-Fe^{2+} complex effectively. Thus, the oxidation product (EDTA-Fe^{3+}) was further reacted with H_2S and produced protons and elemental sulfur. These protons were then utilized for the electrochemical reduction of CO_2 in the cathodic section of the cell. Liu et al. [86] modified a "squaramide-derived framework (PMO-SAF)" to develop a novel catalytic system containing periodic mesoporous organosilicon to convert the captured CO_2 to a system of cyclic carbonates in a mild environment without any metal or solvent. In addition, the structural benefits of (PMO-SAF), like Lewis base nature, double hydrogen bond donor co-modification, and high specific surface area, are some of the favorable conditions that assist the CO_2 adsorption and its conversion. This novel catalyst and procedural design have the potential to capture and utilization of CO_2 simultaneously and, thus, provide further motivation to develop more novel catalytic materials. Cai et al. [87] reported, for the first time, the complete artificial synthesis of starch from CO_2. The artificial synthesis of the starch could be performed by reacting CO_2 and H_2 in suitable conditions. The structure of the synthetic starch was found to be the same as that of the naturally occurring starch, and the rate of synthesis of the artificial starch was ~ 8.5 times higher than that of corn starch formed naturally. Cai et al. proposed that the artificial synthesis of starch opened up a new method to utilize CO_2.

5 Potential for CO_2 Storage in India

The best possible solution for the captured CO_2 could be long-term storage. The collected CO_2 could be permanently stored in deep (underground) geological reservoirs, such as gas reservoirs, basaltic rock formations, and deep saline and depleted oil reserves [88]. Utilization can take place through different available methods, such as urea generation and conversion into other molecules (such as methane, ethanol, or methanol) [89]. In India, a preliminary evaluation of the storage capacity is presently underway. These are mainly focused on the viability of CO_2-enhanced oil recovery (CO_2-EOR), which is viable by conducting geomechanical, seismic, and reservoir investigations [90]. Some recent reports also suggest the possibilities of storing CO_2 in salty aquifers and basalts. The deep saline aquifers of India offer great potential for storing CO_2. Since storage in such aquifers is not the usual practice, sufficient data is missing. Therefore, data collection of such aquifers of India should be enhanced using modern characterization techniques to characterize the diverse onshore and offshore salty formations.

6 Future Scope and Conclusion

The emission of CO_2 from various sources should be controlled on a priority basis to save human beings from its harmful impacts. The decarbonization in various industrial sectors offers a promising future. There is a need to reconsider the scientific policies around the world. The investment should be enhanced to create awareness among the various sections of human society to minimize CO_2 emissions. Research and development should also be encouraged to upgrade existing technologies, develop novel technologies to tackle increasing concentrations of CO_2, and develop efficient equipment to harness renewable energy.

References

1. Zou et al (2016) Energy revolution: from a fossil energy era to a new energy era. Nat Gas Ind B (1):1
2. Smil V (2016) Energy transitions: global and national perspectives. In: BP statistical review of world energy, 2 edn. Praeger
3. World Energy Council (2016) World energy resources. https://www.worldenergyorg/assets/images/imported/2016/10/World-Energy-Resources-Full-report-2016.10.03.pdf
4. Gasparotto J, Martinello KDB (2021) Coal as an energy source and its impacts on human health. Energy Geosci 2:113
5. Hendryx M, Zullig KJ, Luo J (2020) Impacts of coal use on health. Annu Rev Publ Health 41:397
6. Espitia-Perez et al (2018) Cytogenetic instability in populations with residential proximity to open-pit coal mine in Northern Colombia in relation to PM. Ecotoxicol Environ Saf 148:453
7. Rovira et al (2019) Contamination by coal dust in the neighborhood of the Tarragona Harbor (Catalonia, Spain): a preliminary study. Open Atmos Sci J 12:14
8. Daia S, Finkelman RB (2018) Coal as a promising source of critical elements: progress and future prospects. Int J Coal Geol 186:155
9. Silva et al (2020) Rare earth elements and carbon nanotubes in coal mine around spontaneous combustions. J Clean Prod 253:120068
10. Dai et al (2020) Organic associations of non-mineral elements in coal: a review. Int J Coal Geol 218(1)
11. Orem WH, Finkelman RB (2003) Coal formation and geochemistry. Treat Geochem 7:191–222
12. Hower JC, Rimmer SM, Mastalerz M, Wagner NJ (2019) Notes on the mechanisms of coal metamorphism in the Pennsylvania Anthracite Fields. Int J Coal Geol 202
13. Hatcher PG, Clifford DJ (1997) The organic geochemistry of coal: from plant materials to coal. Org Geochem 27(5–6)
14. Hower JC, Gayer RA (2002) Mechanisms of coal metamorphism: case studies from Paleozoic coalfields. Int J Coal Geol 50:215
15. Silva et al (2010) Mineralogy and leaching characteristics of coal ash from a major Brazilian power plant. In: Coal combustion and gasification products
16. Oliveira et al (2018) Study of coal cleaning rejects by FIB and sample preparation for HR-TEM: mineral surface chemistry and nanoparticle-aggregation control for health studies. J Clean Prod 188:662
17. Dai et al (2012) Geochemistry of trace elements in Chinese coals: a review of abundances, genetic types, impacts on human health, and industrial utilization. Int J Coal Geol 94:3
18. Flores RM (2014) Coal composition and reservoir characterization, vol 2014. Elsevier Science. https://doi.org/10.1016/B978-0-12-396972-9.00005-7

19. Munawer ME (2018) Human health and environmental impacts of coal combustion and post-combustion wastes. J Sustain Min 17:87
20. Finkelman RB, Belkin HE, Zheng B (2002) Health impacts of coal domestic coal use in China. Proc Natl Acad Sci USA 7:3427
21. Finkelman et al (2002) Health impacts of coal and coal use: possible solutions. Int J Coal Geol 50:425
22. Anderson JO, Thundiyil JG, Stolbach A (2012) Clearing the air: a review of the effects of particulate matter air pollution on human health. J Med Toxicol 8(2):166
23. Gomes EC, Florida-James G (2014) Lung inflammation, oxidative stress and air pollution. IntechOpen. https://cdn.intechopen.com/pdfs/33894/InTech-Air_pollution_and_cardio vascular_diseases.pdf
24. Glencross et al (2020) Air pollution and its effects on the immune system. Free Radic Biol Med. https://doi.org/10.1016/j.freeradbiomed.2020.01.179
25. Verstraeten et al (2015) Rapid increases in tropospheric ozone production and export from China. Nat Geosci 8:690
26. Pires M, Querol X (2004) Characterization of candiota (South Brazil) coal and combustion by-product. Int J Coal Geol 60:57
27. O'Keefe et al (2013) On the fundamental difference between coal rank and coal type. Int J Coal Geol 58
28. Oberschelp et al (2019) Global emission hotspots of coal power generation. Nat Sustain 2:113
29. Tarafdar A, Sinha A (2018) Polycyclic aromatic hydrocarbons (PAHs) pollution generated from coal-fired thermal power plants: formation mechanism, characterization, and profiling (chapter 5): pollutants from energy sources
30. U.S. Energy Information Administration (EIA) (2020) Electricity power, Feb 2020. https://www.eia.gov/electricity/monthly/current_month/epm.pdf
31. Oliveira et al (2012) Mineralogy and leaching characteristics of beneficiated coal products from Santa Catarina, Brazil. Int J Coal Geol 94:314
32. Cutruneo et al (2014) A mineralogical and geochemical study of three Brazilian coal cleaning rejects: demonstration of electron beam applications. Int J Coal Geol 130:33
33. Rauner et al (2020) Coal-exit health and environmental damage reductions outweigh economic impacts. Nat Clim Change 10:308
34. Liu et al (2018) A highly resolved mercury emission inventory of Chinese coal-fired power plants. Environ Sci Technol 52(4):2400
35. Okedeyi et al (2014) Assessing the enrichment of heavy metals in surface soil and plant (*Digitaria eriantha*) around coal-fired power plants in South Africa. Environ Sci Pollut Res Int 21(6):4686
36. Flues et al (2013) Toxic elements mobility in coal and ashes of Figueira coal power plant, Brazil. Fuel 103:430
37. Silva et al (2020) The impact of air pollution on the rate of degradation of the fortress of Florianopolis Island, Brazil. Chemosphere 251:126838
38. Saikia et al (2017) Ambient nanoparticles/nanominerals and hazardous elements from coal-combustion activity: implications on energy challenges and healthhazards. Geosci Front 9:863
39. IPCC (2018) Global warming of 1.5 °C. Intergovernmental Panel on Climate Change, United Nations, Paris Agreement, United Nations Framework Convention on Climate Change
40. nagireddi-et-al-2023-carbon-dioxide-capture-utilization-and-sequestration-current-status-challenges-and-future.pdf
41. Hakstol R (2023) A holistic environmental investigation of comple mentary energy in Alberta. Phys Sci Rev 8:3003
42. Ren et al (2021) Decarbonizing China's iron and steel industry from the supply and demand sides for carbon neutrality. Appl Energy 298:117209
43. Cormos et al (2020) Techno-economic and environmental evaluations of decarbonized fossil-intensive industrial processes by reactive absorption & adsorption CO_2 capture systems. Energies 13:1268

44. Gładysz P, Bartela Ł, Sowizdzał A, Qvist S et al (2021) Retrofit decarbonization of coal power plants: a case study for Poland. Energies 14:120
45. United Nations (2015) Paris Agreement. United Nations Framework Convention on Climate Change
46. Baumgärtner et al (2021) Life-cycle assessment of sector-coupled national energy systems: environmental impacts of electricity, heat, and transportation in Germany till 2050. Front Energy Res 9:621502
47. Kashintseva et al (2018) Consumer attitudes towards industrial CO_2 capture and storage products and technologies. Energies 11:2787
48. Zhou et al (2019) A comparison of low carbon investment needs between China and Europe in stringent climate policy scenarios. Environ Res Lett 14(5):054017
49. Smith et al (2019) Human health: impacts, adaptation, and co-benefits. In: IPCC special report on the ocean and cryosphere in a changing climate
50. Fu et al (2022) Research progress on CO_2 capture and utilization technology. J CO_2 Util 66:102260
51. Heuberger et al (2016) Quantifying the value of CCS for the future electricity system. Energy Environ Sci 9(8):2497
52. Rodrigues CFA, Dinis MAP, Sousa MJL (2015) Review of European energy policies regarding the recent "carbon capture, utilization and storage" technologies scenario and the role of coal seams. Environ Earth Sci 74(3):2553
53. Xie H (2021) Application and research progress of CO_2 capture technologies. Chem Fert Des 59(06):1
54. Asif et al (2018) Post-combustion CO_2 capture with chemical absorption and hybrid system: current status and challenges. Greenh Gases Sci Technol 8(6):998
55. Léonard et al (2015) Influence of process operating conditions on solvent thermal and oxidative degradation in post-combustion CO_2 capture. Comput Chem Eng 83:121
56. Tobiesen FA, Haugen G, Hartono A (2018) Asystematic procedure for process energy evaluation for post combustion CO_2 capture: case study of two novel strong bicarbonate-forming solvents. Appl Energy 211:161
57. Kong et al (2019) Synthesis of porous organic polymers with tunable amine loadings for CO_2 capture: balanced physisorption and chemisorption. Nanomaterials 9:1020
58. Wu et al (2016) Progress and prospect of CCS in China: using learning curve to assess the cost viability of a 2 $\times$ 600 MW retrofitted oxyfuel power plant as a case study. Renew Sustain Energy Rev 60:1274
59. IEA (2021) Net zero by 2050: a roadmap for the global energy sector. International Energy Agency
60. IEA (2020) Energy technology perspectives 2020. International Energy Agency
61. Liu et al (2013) Zeolite apgiia for adsorption based carbon dioxide capture. Sep Sci Technol 48(3):388
62. Guo L, Tan Y (2012) Application of electric swing adsorption in capture of CO_2 from flue gas. Gas Heat 32(05):13
63. Wang et al (2012) Experimental and modeling investigation on post-combustion carbon dioxide capture using zeolite 13X-APG by hybrid VTSA process. Chem Eng J 197:151
64. Shen et al (2011) Capture of CO_2 from flue gas by vacuum pressure swing adsorption using activated carbon beads. Adsorption 17(1):179
65. Liu et al (2011) Multi-bed vacuum pressure swing adsorption for carbon dioxide capture from flue gas. Sep Purif Technol 81(3):307
66. Grande CA, Ribeiro RP, Rodrigues AE (2010) Challenges of electric swing adsorption for CO_2 capture. Chemsuschem 3(8):892
67. Regufe et al (2019) Electrical conductive 3D-printed monolith adsorbent for CO_2 capture. Micropor Mesopor Mater 278:403
68. Regufe et al (2020) Development of hybrid materials with activated carbon and zeolite 13X for CO_2 capture from flue gases by electric swing adsorption. Ind Eng Chem Res 59(26):12197

69. Ribeiro RPPL, Grande CA, Rodrigues AE (2013) Activated carbon honeycomb monolith—zeolite 13X hybrid system to capture CO_2 from flue gases employing electric swing adsorption. Chem Eng Sci 104:304
70. Plaza MG, Rubiera F, Pevida C (2017) Evaluating the feasibility of a TSA process based on steam stripping in combination with structured carbon adsorbents to capture CO_2 from a coal power plant. Energy Fuels 31(9):9760
71. Alper E, Orhan OY (2017) CO_2 utilization: developments in conversion processes. Petroleum 3(1):109
72. Cheng et al (2021) CO_2 capture from flue gas of a coal-fired power plant using three-bed PSA process. Energies 14(12):3582
73. Hou S, Chiang C, Lin T-H (2020) Oxy-fuel combustion characteristics of pulverized coal under O_2 atmospheres. Appl Sci 10:1362
74. Karim et al (2020) CFD simulation of biomass thermal conversion under air/oxy-fuel conditions in a reciprocating grate boiler. Renew Energy 146:1416–1428
75. Roeder V, Hasenbein C, Kather A (2013) Evaluation and comparison of the part load behaviour of the CO_2 capture technologies oxyfuel and post-combustion. Energy Procedia 37:2420
76. Li Y, Lukszo Z, Weijnen M (2015) The implications of CO_2 price for China's power sector decarbonization. Appl Energy 146:53
77. Wu et al (2022) Progress and trends of carbon capture utilization and storage technology. In: Ecology and environmental monitoring of three gorges
78. Gabrielli P, Gazzani M, Mazzotti M (2020) The role of carbon capture and utilization, carbon capture and storage, and biomass to enable a net-zero-CO_2 emissions chemical industry. Ind Eng Chem Res 59(15):7033
79. Godin et al (2021) Advances in recovery and utilization of carbon dioxide: a brief review. J Environ Chem Eng 9(4):105644
80. Kamkeng et al (2021) Transformation technologies for CO_2 utilisation: current status, challenges and future prospects. Chem Eng J 409:128138
81. Development status and application prospect of carbon capture, utilization and storage (CCUS) technology. https://www.craes.cn/xxgk/zhxw/202107/t20210715_847263.shtml. Accessed 15 Mar 2022
82. Zhang JL, Yang TH (2016) Chemical utilization of carbon dioxide as a resource. Coal Chem Ind 44(03):1
83. Sun et al (2013) Progress in the chemical utilization of carbon dioxide. Chem Ind Eng Prog 32(07):1666
84. CO_2 bio-utilization technology will help carbon neutrality, and the commercialization prospects of various breakthrough technologies are expected. https://c.m.163.com/news/a/GOT6MLKQ0532PL1J.html. Accessed 23 Mar 2022
85. Ma et al (2018) Achieving simultaneous CO_2 and H_2S conversion via a coupled solar-driven electrochemical approach on non-precious-metal catalysts. Angew Chem Int Ed 57(13):3473
86. Liu et al (2020) Squaramide-derived framework modified periodic mesoporous organosilica: a robust bifunctional platform for CO_2. Chem Eng J 125682
87. Cai et al (2021) Cell-free chemoenzymatic starch synthesis from carbon dioxide. Science 373(6562):1523
88. Cao et al (2018) Measuring China's carbon emissions based on final consumption. Energy Proc 152:853
89. Viebahn P, Vallentin D, Höller S (2014) Prospects of carbon capture and storage (CCS) in India's power sector—an integrated assessment. Appl Energy 117:62
90. Ramachandra TV, Aithal BH, Sreejith K (2015) GHG footprint of major cities in India. Renew Sustain Energy Rev 44:473

An Introduction and Industrialization of Crude Oil

Rajesh Kumar, Kahkashan Begum, Shahid Parvez Ansari, and Jamal Akhter Siddique

Abstract The chapter based on the study about the crude oil produced from fossil fuel, a naturally arising hydrocarbon mixture made-up of ancient marine life. The industrialization of crude oil talks about to its transformation from naturally occurring materials into a primary global commodity with the help of extraction, refining, and the development of supporting infrastructure. The chapter thoroughly discusses the basic concept and differences between the terms crude oil, petroleum and fossil fuel, historical background, development, chemical compositions and their exponential growth.

Keywords Crude oil · Fossile fuel · Petroleum · Industrilisation · Energy sources

1 Fossil Fuel Versus Petroleum Versus Crude Oil

1.1 Fossil Fuels

Fossil fuels are a collective group of natural gas, coal, and crude oil, considered nonrenewable energy sources formed over millions of years under inordinate pressure and temp. in the Earth's crust. Originally, "fossil fuel" was coined by Caspar Neumann in 1759, a well-known German chemist [1]. The source of fossil fuels is the anaerobic decay of buried deceased entities. The transformation from the organic constituents to high-carbon fossil fuels is characteristically the outcome of a geological process

R. Kumar
Department of Chemistry, Maharaj Singh College, Saharanpur, India

K. Begum · J. A. Siddique (✉)
Department of Chemistry, Gandhi Faiz-E-Aam College (M.J.P. Rohilkhand University), Shahjahanpur, India
e-mail: sidjamal@gmail.com

S. P. Ansari
Department of Applied Chemistry, Faculty of Engineering and Technology, Aligarh Muslim University, Aligarh, India

J. A. Siddique et al. (eds.), *Energy Sources*, Green Energy and Technology,
https://doi.org/10.1007/978-981-95-1209-6_6

over numerous years [2]. They are essential constituents of the global energy system, driving transportation, electricity generation, and industrial processes. Due to their origins, they have high carbon content.

1.2 Petroleum

Petroleum considered a refined products (gasoline, diesel, and jet fuel) of crude oil or "rock oil," is a naturally occurring, dark-colored, oily liquid that is mainly composed of hydrocarbons. It's a type of fossil fuel made up of the remains of ancient marine organisms, like diatoms and algae. The term "petroleum" is a grouping of Latin words "petra" (rock) and "oleum" (oil), literally meaning "rock oil." The term was first used in the sixteenth century by German mineralogist Georgius Agricola [3].

1.3 Crude Oil

Crude oil refers to unrefined petroleum that originates in the microscopic apertures of sedimentary rocks like sandstone and limestone [4]. The term "crude oil" was established to precisely describe the raw, unprocessed liquid form of petroleum drilled from the earth, distinct from refined petroleum products. It's a naturally occurring liquid, principally a combination of hydrocarbons and additional organic materials. Crude oil is a raw material that undergoes refining processes to generate numerous products like gasoline, diesel, and jet fuel. It involves primarily impulsive liquid hydrocarbon-based complexes of hydrogen and carbon, but also comprises small extents of nitrogen, sulfur, and oxygen. The elements generate diverse, complex molecular configurations, some unidentified complexes. Typically, crude oil contains 82–87% carbon and 12–15% hydrogen by weight.

2 Historical Aspect of Crude Oil

Present crude oil production traces its roots to Baku, Azerbaijan, in 1837 [5]. The city is recognized as the world's first "oil town." The expansion of the global oil commerce, where the start of oil-refining was commercially recognized to produce paraffin (involved in lamp and central heating) by refining oil. A Russian engineer, "Vasily Semyonov," established modern-day oil well in the Peninsula named Absheron, situated in the northeast of Baku, by applying the thoughts of Nikolay Voskoboynikov in 1846 that reached a depth of 21 m [6]. These establishments in Baku are recognized for more than 90% of the global oil restoration, supplied to Persia (present Iran). Sustainable oil wells rapidly emerged in Bobrka, Poland (1854),

Bucharest, Romania (1857), Ontario, Canada (1858), and Pennsylvania, USA (1859), generating a "black diamond" flash in numerous territory.

3 Chemical Composition of Crude Oils

Crude oils consist of liquid paraffin hydrocarbons, hydrocarbons have five to fifteen carbon chains (C5–C15). These hydrocarbons contain unbranched, branched paraffins (isoparaffins), alkyl, benzyl, and cyclic paraffins (e.g., naphthenes and cycloparaffins), along with nuclear aromatics [7]. Generally, paraffins are considered low molecular weight, saturated hydrocarbons, frequently associated with gaseous phases. Naphthenes, or cycloparaffin, contain high molecular weight, tightly bonded hydrocarbons, and shares about 10% of crude oil composition. The elemental composition typically ranges from 83 to 87% carbon (Fig. 1) [8].

Crude oils also have heteroatomic compounds, like sulfur, carbon dioxide, oxygen, nitrogen, and trace metals, 10–14% hydrogen, and 0.1–6.0% sulfur [8]. Nitrogen gets involved as a dissolved gas from 0.01 to 2.0% [9], while the oxygen is present in viscous form as oxygen-bearing compounds.

Crude oil develops in geological "setups," which may be stratigraphic, paleogeomorphic [10], or a grouping. Paleogeomorphic setups contain structural folds, while stratigraphic traps result from lateral deviations in reservoir rock properties within a layer.

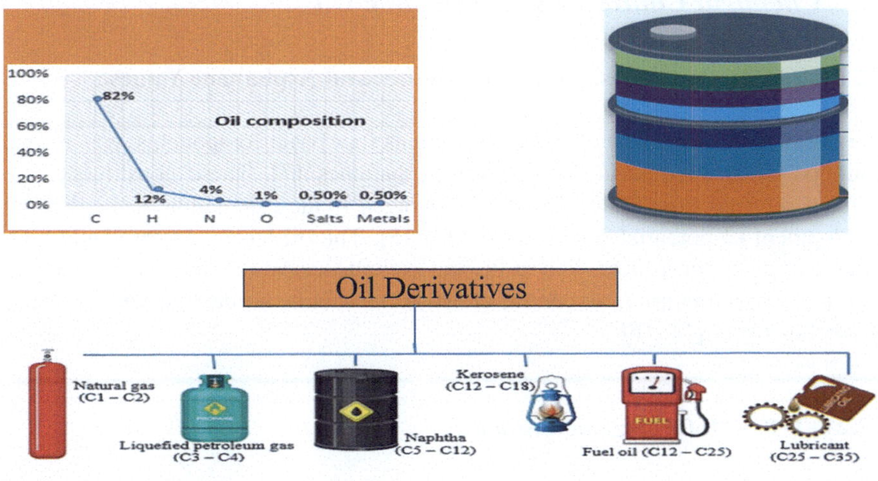

Fig. 1 Chemical structure of the petroleum, showing the number of carbons present in some oil derivatives

4 Characterization of Crude Oil

The physical characteristics of crude oils determine their quality and market evaluation. They contrast as per the configuration of the crude oil, the relative wealth of the clusters of hydrocarbons, and the presence of impurities to be governed by reservoir origin, temperatures, and pressures. The American Petroleum Institute [11], or "API," created the API gravity as a technique of evaluating the density of petroleum liquids compared to water. The API gravity gauge is measured in degrees. As water API gravity is 10°.

4.1 Physical Characteristics of Crude Oil

The American Petroleum Institute (API) calculates the density of crude oil [12]. Generally, crude oil has an average gravity between 20 and 30°. The API gravity of crude oil is subject to its configuration; however, lesser density of crude oils is lighter and greater API gravity. Density is suggestive of the configuration of the crude oil that relates to its processing methodology. Crude oils are ordered into three basic groups on the basis of API gravity: lighter (API > 35), intermediate (API: 25–35), and heavier (API < 25) [13].

4.2 Chemical Characteristics of Crude Oil

4.2.1 SARA—Saturates, Aromatics, Resins, and Asphaltene Values

The chemical arrangement of crude oil contains four basic components, viz., resins, saturated hydrocarbons, aromatics, and asphaltenes [7]. These constituents are together referred to as SARA. SARA is involved in the calculation of the thermal development of potential source rocks [14, 15]. The oil has a comparatively high volume of polar constituents like resins and asphaltenes, which indicates the sample rock has lesser maturity, as central constituents incline to decline with growing thermal maturity [14–16].

4.2.2 TOC—Total Organic Carbon

Total organic carbon (TOC) refers to the quantity of carbon existing in organic compounds within a section (rock), like crude oil. It's a measure of the organic richness and can be considered to assess the potential for generating hydrocarbons. The greater the organic carbon (OC) numbers, the healthier the sample rock and the maximum the possibility of hydrocarbon establishment [17]. The TOC found between 1 and 10% or sometimes higher across the same basins. TOC can be considered to regulate hydrocarbon development in oil, gas, or kerogen.

4.2.3 Sulfur Value

The sulfur (S) value in crude oil is a sign of accumulation situations and thermal development gauze [18]. Crude oil is labelled sweet, once its sulfur value is less than 1%, and labeled as sour after its sulfur value is higher than 1%.

4.2.4 CII—Colloidal Instability Index

The colloidal instability index (CII) is denoted by the accumulation of the collective amounts of saturates and asphaltenes, divided by the collective quantities of resins and aromatics. CII generally ranged between 0.97 and 2.86. If the CII figures of oil are less than 0.70, considered stable; equally, if the values of CII surpass 0.90, described as unstable [19, 20].

4.3 Biomarker Parameters of Crude Oil

4.3.1 Molecular Marker Ratios Pristane/Phytane (Oil Environmental Condition)

The ratio of pristane/phytane (pr/ph) is a broadly active gauge to presume the redox nature of the sample deposited in the environment. The ratio of pr/ph higher than 03.0 indicates an oxic accumulation situation, whereas lesser figures (< 0.8) indicate that high reducing (anoxic) situations [21]. Transitional values are agreed to indicate a suboxic state [21, 22]. The values of pr/ph ratio are within the range of less than 3 and higher than 0.8, which indicates that the oils were drilled in a suboxic atmosphere.

4.3.2 CPI—Carbon Preference Index (Oil Maturation Level)

The CPI is known as the relation of the long chain odd carbon number n-alkanes and the long chain even carbon number n-alkanes [14, 16]. The Carbon Preference Index (CPI) was originally presented by Bray and Evans in year 1961 and is well accepted as an important technique for distinguishing the undeveloped and developed n-paraffin dispersals in crude oils and cradle rocks [23]. It is generally involved to evaluate the thermal maturity stage of oil cradle rocks, particularly when nC25–nC33 n-alkanes exist [16].

4.3.3 TAR—Terrigenous to Aquatic Index (Source Rock Organic Matter Input)

The terrigenous to aquatic (TAR) ratio is a significant tool for estimating the amount of OC derived from land plants alongside to aquatic sources [14, 24]. A TAR figure higher than > 1 indicates a greater occurrence of land-plant derivative constituents, demonstrating a leading terrestrial involvement. On the contrary, crude oil specimens with TAR figures less than one indicate that the origin organic matter cradle is not from substantial land-plant participation. The lesser figures indicate that the origin is from aquatic sources [24]. The rocks can be categorized on the basis of their potential values like lesser potential (< 0.5% OC), modest potential (0.5–1.0% OC), satisfactory potential (1.0–2.0% OC), and extremely promising potential (> 2.0% OC) [25–27].

5 Classification

Crude oil is categorized on the basis of its quality, geographical origin, involvement of refining techniques, market values, chemical and physical parameters (like viscosity, density (API gravity; the density of crude oil is conventionally calculated by using the API gravity [11]), and presence of sulfur, content in it that plays a crucial role in purification and end use applications). These properties are also valuable in the oil market in the selection of a qualitative and technically sound product. There are different types of possible classifications listed below.

5.1 Classification Based on Density (API Gravity) and Sulfur Content

Chemical characteristics like sulfur content and API gravity play an important role in classification. For example, oils with low sulfur content are classified as sweet, greater sulfur content are classified as sour. The category last on the following five types of crude oil: name and characteristic properties with key points are as below.

5.1.1 Light Crude Oil

- Key features: flows easily with high API gravity (> 31.1°) and low density.
- Sulfur content: can be sweet or sour, ranging from low sulfur (< 0.5%) to high sulfur (> 0.5%).
- Overall: quite easy to refine into high-value end products like gasoline and diesel, having lower viscosity and higher yields of light portions.

- Common examples: West Texas Intermediate (WTI, sweet), Brent (sweet).

5.1.2 Heavy Crude Oil

- Key features: viscous flow with high density having low API gravity (< 22.3°).
- Sulfur content: frequently sour, demanding more complex refining processes.
- Overall: difficult to extract and refine, low yield of high-value products.
- Common examples: Canadian oil sands bitumen, Venezuelan Orinoco crude.

5.1.3 Medium Crude Oil

- Key features: API gravity between 22.3 and 31.1°, moderate density and viscosity.
- Sulfur content: ensuring both character sweet or sour (ranges from 0.5 to 2%).
- Overall: quite easier to refine with resourcefulness for various products.
- Common examples: Arabian Medium, some North Sea crudes.

5.1.4 Sweet Crude Oil

- Key features: low sulfur content (< 0.5%), regardless of density.
- Sulfur content: less than 0.5% by weight. This level is helpful and easier with less costly to refine, generating cleaner fuels with minimal environmental effect.
- Overall: it requires less refining to remove sulfur and is less corrosive, making it more desirable and expensive.
- Common examples: Nigerian bonny light, Libyan crude.

5.1.5 Sour Crude Oil

- Key features: the presence of sulfur also leads to higher levels of hydrogen sulfide (H_2S), a toxic and potentially fatal gas, density that ranges from approximately 0.7 to 0.99 g/cm^3.
- Sulfur content: high sulfur content (> 0.5%), regardless of density.
- Overall: it requires additional desulfurization during refining, increasing costs.
- Common examples: Dubai crude, Mexican Maya.

5.2 International Naming Conventions Classification

Crude oils are additionally characterized based on their geographic origin and chemical characteristics. Geographic origin can divulge to understandings the crude's worth and its typical uses.

5.2.1 Brent Crude

This is a chief trading category of crude oil coming from the North Sea. It is involved as a global benchmark for oil prices and is considered for its comparatively light and sweet qualities.

5.2.2 West Texas Intermediate (WTI)

The oil originates from Texas and plays a key benchmark in determining oil pricing in North America. WTI is light and sweet, these properties are highly required for refining into gasoline and other products.

5.2.3 Dubai Crude

This oil is responsible for crude oil pricing in the Asian markets and is known for being sourer and heavier compared to Brent and WTI.

5.3 Number of Known Crude Oil Classification

Universally, the variety of crude oil types is enormous, showing the variable geological developments and extraction approaches. Whereas the principal kinds of crude oils are light, heavy, sweet, and sour, which include a broad spectrum of classifications, there are several other variants beyond these primary classifications. Some of them are as follows:

5.3.1 Extra-Heavy Oils

Such type of oil has an API gravity of up to 10° and is viscous in nature. These oils are frequently derived from oil sands and involve improved extraction methods like thermal recovery. Athabasca oil sands in Alberta, Canada, is one of the prime examples.

5.3.2 Bitumens

Bitumen is an exceptionally viscous type of crude oil found in oil sands. It is drilled out along with surface mining or in-situ approaches and promoted to a lighter type of crude. The Canadian oil sands are a major source of bitumen.

5.3.3 Orinoco Belt Crude

Obtained from Venezuela, such crude is recognized for its heavy and sour features.

5.3.4 Kashagan Crude

The oil originated from Kazakhstan is one more example of heavy crude oil having great sulfur content, which has an impact on its quality.

There are different kinds of crude oil present globally, as the classification is based on their specific characteristics of oil from each reservoir or field. Globally, these are different distinct crude oil grades (more than 160 types are widely recognized), numerous variations in chemical composition (reservoir can yield crude with exclusive characteristics), and major organizations (like the EIA, OPEC, and Platts) retaining databases of key crudes. For example, Platts considers over 100 crude oil grades in its price valuations. This complexity highlights the significance of tailored refining practices and notifies global energy markets.

6 Industrialization

First commercial oil field, Bradford established in the 1880s in Pennsylvania [28], where they drilled 77% of the overall oil resource. By the late nineteenth century, Russian Kingdom established a company, namely "Branobel Company" in Azerbaijan, and had engaged as a leader in drilling-out [29]. Till the late nineteenth century, new oil fields were revealed near North America and the zone industrialized into a huge petrochemical refining center and exchange [30].

The contemporary US petroleum manufacturing becomes more mature due to the need for kerosene and oil lamps. It became a prime nationwide idea in the early part of the twentieth century. Invention of the internal combustion engine has propagated demands that have mainly sustained the industry to the present day. Ontario, along with Pennsylvania both were rapidly surpassed by demand; taking over to "oil-booms" in developed Galician oilfields as the globally 3rd largest oil generating Ohio, Texas, Oklahoma, and California.

Austro-Hungary country following the United States and the Russian Kingdom and became 5% shareholder of the worldwide oil production in 1908 [31]. The substantial oil fields were also introduced by Peru (1863, in Zorritos District), Dutch East Indies (1885, in Sumatra), Persia (1908, in Masjed Soleiman), Venezuela (1914, in Maracaibo Basin), and Mexico, and well established at an industrial level in the early nineteenth century.

Austro-Hungary empire lost its control in production of oil, which had been at the basic cause of the 1910 Petroleum War [31]. The accessibility of oil and control over it became of "fundamental significance" in armed supremacy earlier and afterwards World War I (WW-I), especially for naval forces that not only transformed from coal to oil but also with the establishment of motor transport, tanks, and airplanes. The philosophy would carry on in later clashes of the twentieth century, as well as World War II (WW-II), where oil facilities were led as the foremost tactical advantage and were massively accelerated [32]. In year 1938, giant reserves of oil were drilled out in the Al Ahsa territory in the Eastern region of the Kingdom of Saudi Arabia (KSA) alongside the coastline of the Arabian Gulf.

Today, roughly 90% of vehicular fuel requirements are met by oil. Petroleum's wealth is a movable, dense energy source operating the vast majority of automobiles, and as the need for many industrial chemicals has established, it is one of the world's most significant commodities.

Till 2023 leading oil-producing countries are the United States, Russia, and KSA. Around 80% of the globally willingly manageable reserves are positioned in the middle-east, with 62.5% coming from the Arab-five; Saudi Arabia (12.5%), UAE, Iraq, Qatar, and Kuwait.

6.1 *Rise of the Seven Sisters*

There are a few leading petroleum firms that are very well established today, easily find their roots, including occasions and current involvement.

- In 1901, Gulf oil and Texaco were recognized, known for the discovery of oil at Spindletop, Texas.
- Year 1907 is remembered as the merger year of Royal Dutch and Shell to create Royal Dutch/Shell, against the competitive higher pricing struggle from US firms.
- Arise of BP: the joint venture of the Anglo-Persian oil firm, established in 1908, working on search and drilling for oil in Iran.
- Chevron, Exxon, and Mobil (also known as Exxon Mobil) were initiated in 1911 when standard oil was split up by the United States of America's Supreme Court as a result of antitrust violations.

The international oil companies (IOCs), namely BP, Chevron, Exxon, Gulf Oil, Mobil, Royal Dutch/Shell, and Texaco, established as the "seven-sisters" and control approx. 85% of the global oil reserves were at their peak in the early 1970s.

6.2 *The Formation of OPEC*

In 1960, representatives of Iran, Kuwait, KSA, Iraq, and Venezuela met in Baghdad city to dialogue about handing the rate control enforced by the IOCs. They approved to establishment of OPEC, that is, Organization of Petroleum Exporting Countries (OPEC), aiming toward minimize competition between their nations and regulate the oil rates. Lately, in the next two decades, OPEC includes Indonesia, Qatar, Libya, United Arab Emirates (UAE), Nigeria, Algeria, Ecuador, and Gabon for expansion.

They lead to took control of oil reserves in their territories during the era of 1960 and 1976, by buying out or effectively taking shares from the IOCs [33].

7 Conclusion

Crude oil has been a dynamic entity to lead industrialization, driving economic growth and empowering technological innovations for over a century. Its role in the extraction of fuels, chemicals, and materials has molded modern civilization; however, the environmental and social costs of its production and use have initiated global debates. As the world changeovers in the direction of sustainable energy, the crude oil industry must take the initiative toward cleaner technologies and bargain reliance on fossil fuels. Balancing energy desires with environmental stewardship will express the future of crude oil in a progressing global landscape.

References

1. Bolin B et al (eds) (1979) The global carbon cycle. SCOPE report No. 13. Wiley, New York
2. Mann P, Gahagan L, Gordon MB (2003) Tectonic setting of the world's giant oil and gas fields. In: Halbouty MT (ed) Giant oil and gas fields of the decade, 1990–1999. American Association of Petroleum Geologists, Tulsa, Okla, p 50
3. Prescher H (1994) Dr. Georgius Agricola 1494–1555: a European scientist and humanist from saxony. GeoJournal 32(2):85–89. http://www.jstor.org/stable/41146714
4. Hsu CS, Robinson PR (2024) Petroleum science and technology. https://doi.org/10.1007/978-3-031-46641-0
5. Haiko H, Biletsky V (2019) The history of industrial oil extraction in Azerbaijan (the 19th—beginning of the 20th centuries). Skhid 4(162):35–40. https://doi.org/10.21847/1728-9343.2019.4(162).76938
6. Mir-Babayev NF, Houghton PJ (2002) Plants of the Republic of Azerbaijan with potential medicinal applications. Part III. Pharm Biol 40(1):16–22
7. Sayed K, Baloo L, Sharma NK (2021) Bioremediation of total petroleum hydrocarbons (TPH) by bioaugmentation and biostimulation in water with floating oil spill containment booms as bioreactor basin. Int J Environ Res Public Health 18(5):2226. https://doi.org/10.3390/ijerph18052226
8. Aitani AM (2004) Oil refining and products. In: Cleveland CJ (ed) Encyclopedia of energy. Elsevier, pp 715–729. ISBN: 9780121764807. https://doi.org/10.1016/B0-12-176480-X/00259-X
9. Ostle D (1975) (A. A.) Levinson. Introduction to exploration geochemistry. Calgary (Appl. Publishing Ltd.), 1974. 612 pp., 198 figs. Mineral Mag 40(311):323–324. https://doi.org/10.1180/minmag.1975.040.311.21
10. Balázs A et al (2021) Towards stratigraphic-thermo-mechanical numerical modelling: integrated analysis of asymmetric extensional basins. Glob Planet Change 196:103386. ISSN: 0921-8181. https://doi.org/10.1016/j.gloplacha.2020.103386
11. Aitani AM (2004) Oil refining and products. In: Cleveland CJ (ed) Encyclopedia of energy. Elsevier, pp 715–729. ISBN: 9780121764807. https://doi.org/10.1016/B0-12-176480-X/00259-X

12. Chinenyeze U, Madu AJ, Richard E (2015) Physical and chemical properties of crude oils and their geologic significances. Int J Sci Res (IJSR) 6(6):1514–1521. https://doi.org/10.21275/ART20174603
13. Wang Y, Feng J, Lin Q et al (2013) Effects of crude oil contamination on soil physical and chemical properties in Momoge wetland of China. Chin Geogr Sci 23:708–715. https://doi.org/10.1007/s11769-013-0641-6
14. Peters KE, Walters CC, Moldowan JM (2005) The biomarker guide vol. 1: biomarkers and isotopes in the environment and human history, 2nd edn. Cambridge University Press, New York. https://doi.org/10.1017/CBO9780511524868
15. Hakimi et al (2022) The influence of green manufacturing practices on the corporate sustainable performance of SMEs under the effect of green organizational culture: a moderated mediation analysis. J Clean Prod 376:134346. ISSN: 0959-6526. https://doi.org/10.1016/j.jclepro.2022.134346
16. Egbo OK, Adeigbe OC, Esegbue O (2020) Geochemical characterization of some biodegraded oils from the Niger Delta Basin, Nigeria. Environ Earth Sci Res J 7(4):127–133. https://doi.org/10.18280/eesrj.070401
17. Manurung PL, Dewanto O (2021) Total organic carbon (TOC) value prediction in source rock potential at North East Java Basin, Indonesia. JGEET 6(3):147–156. https://doi.org/10.25299/jgeet.2021.6.3.6644
18. Egbo et al (2024) Geochemistry of onshore oils within oil mining lease (OML A) of the Southern Niger Delta, Nigeria. Int J Innov Res Adv Stud (IJIRAS) 11(7)
19. Ashoori S, Sharifi M, Masoumi M, Salehi MM (2016) The relationship between SARA fractions and crude oil stability. Egypt J Pet 26(4):209–213. https://doi.org/10.1016/j.ejpe.2016.04.002
20. Bambinek K, Przyjazny A, Boczkaj G (2023) Compatibility of crude oil blends-processing issues related to asphaltene precipitation, methods of instability prediction—a review. Ind Eng Chem Res 62(1):2–15. https://doi.org/10.1021/acs.iecr.2c02532
21. Su S, Jiang Y (2022) Biomarker geochemistry of paleogene crude oils and source rocks from the Raoyang Sag, Bohai Bay Basin, NE China: an oil-source rock correlation study. J Pet Geol 45:5777. https://doi.org/10.1111/jpg.12807
22. Hoa F, Zhou X, Zhu Y, Zou H, Bao X, Kong Q (2009) Mechanisms of petroleum accumulation in the Bozhong sub-basin, Bohai Bay Basin, China. Part 1: origin and occurrence of crude oils. Mar Pet Geol 26(8):1528–1542. https://doi.org/10.1016/j.marpetgeo.2008.09.005
23. Dembicki H (2020) Reducing the risk of finding a working petroleum system using SAR imaging, sea surface slick sampling, and geophysical seafloor characterization: an example from the eastern Black Sea basin, offshore Georgia. Mar Pet Geol 115:104276. ISSN: 0264-8172. https://doi.org/10.1016/j.marpetgeo.2020.104276
24. El Sabagh A, Omar A, Saneoka H, Barutçular C (2015) Comparative physiological study of soybean (*Glycine max* L.) cultivars under salt stress. Yuzuncu Yıl Univ J Agric Sci 25(3):269–284
25. Thomas BM (1979) Geochemical analysis of hydrocarbon occurrences in Northern Perth Basin, Australia. AAPG Bull 63:1092–1107
26. Peters KE, Cassa MR (1994) Applied source rock geochemistry. In: Magoon LB, Dow WG (eds) The petroleum system—from source to trap, 60th edn. American Association of Petroleum Geologists, USA, pp 93–120
27. Al-Areeq NM et al (2018) Petroleum source rocks characterization and hydrocarbon generation of the Upper Jurassic succession in Jabal Ayban field, Sabatayn Basin, Yemen. Egypt J Pet 27(4):835–851. https://doi.org/10.1016/j.ejpe.2017.12.005
28. W. B. H. (1972) M. T. Halbouty (ed.) 1970. Geology of giant petroleum fields: a symposium. (53rd annual meeting of the American Association of Petroleum Geologists in Oklahoma City, 23rd–25th April 1968.) AAPG memoir 14, 575 pp. Geol Mag 109(1):74–74. https://doi.org/10.1017/S0016756800042400
29. Akiner S (ed) (2004) The Caspian: politics, energy and security, 1st edn. Routledge. https://doi.org/10.4324/9780203641675

30. Gow S (2005) End notes. In: Roughnecks, rock bits, and rigs: the evolution of oil well drilling technology in Alberta, 1883–1970. University of Calgary Press, Calgary, pp 395–432. https://doi.org/10.1515/9781552384442-017
31. Frank A (2009) The Petroleum War of 1910: standard oil, Austria, and the limits of the multinational corporation. Am Hist Rev 114(1):16–41. https://www.jstor.org/stable/30223641
32. Fromkin D (1989) A peace to end all peace: creating the modern Middle East, 1914–1922, 1st American edn. H. Holt
33. Baldwin H (1959) Oil strategy in WW-II archived Aug-15, year 2009 at the Wayback Machine. American Petroleum Institute quarterly centennial issue. American Petroleum Institute, pp 10–11

A Fundamental Approach to Renewable Energy: Classification and Development

Shaukat Khan

Abstract Renewable energy (RE) is essential in fostering sustainable, environmentally responsible, and economically viable electricity generation systems. The International Energy Agency (IEA) has reported a notable shift in the global energy landscape since 2019, marked by a steady decline in dependence on nonrenewable energy sources for the generation of electricity. This transition coincided with an increased integration of RE sources to address the ever-growing global energy needs. Advancements in RE technologies continue to evolve, with ongoing research focusing on enhancing energy conversion efficiencies and optimizing system performance to meet contemporary challenges. This chapter will provide a deep insight into the classification and development of various RE technologies, including hydro, solar, wind, bioenergy, geothermal, and hydrogen energy. Finally, the technical, economic, and energy conversion efficiency challenges associated with RE technologies are discussed, offering comprehensive recommendations to overcome these obstacles and facilitate the widespread adoption of clean energy solutions.

Keywords Renewable energy · Solar energy · Wind energy · Geothermal energy · Bioenergy · Hydrogen energy

1 Introduction

The escalating global energy demand, driven primarily by both developed and emerging economies, underscores the urgent need for the transition to more sustainable energy technologies capable of replacing traditional methods of electricity generation, such as those reliant on fossil fuels [1]. Fossil fuel-based energy systems are responsible for substantial environmental harm, with their contribution to global warming and climate change becoming increasingly significant over time [2]. The rise in greenhouse gas emissions from power generation has accelerated in recent

S. Khan (✉)
Department of Chemical Engineering, College of Engineering, Dhofar University, Salalah, Sultanate of Oman
e-mail: shkhan@du.edu.om

J. A. Siddique et al. (eds.), *Energy Sources*, Green Energy and Technology,
https://doi.org/10.1007/978-981-95-1209-6_7

decades, exacerbating these environmental crises [3]. In response to these challenges, renewable energy (RE) technologies have emerged as viable, sustainable alternatives [4]. These technologies—spanning solar, wind, hydroelectric, biomass, geothermal, and hydrogen energy—offer promising solutions for reducing carbon footprints and mitigating the detrimental effects of conventional energy production [5]. By harnessing natural and abundant resources, RE systems provide a pathway to cleaner electricity generation, thereby contributing to the global effort to combat environmental degradation and climate change [6].

RE technologies are gaining prominence due to their environmentally friendly attributes, including minimal or zero greenhouse gas emissions, and the growing societal awareness of the importance of a clean environment. In addition to promoting sustainability, RE also holds considerable economic value. It reduces electricity generation costs by utilizing natural, renewable resources and provides opportunities for economic benefit, as surplus electricity generated by consumers can be sold [7, 8].

Despite the increasing acceptance of RE technologies, fossil fuels still dominate global electricity generation due to challenges such as the variability of renewable resources and high investment costs. For instance, photovoltaic (PV) systems only operate during daylight, wind turbines require adequate wind flow, and hydro turbines depend on water flow for potential energy. Consequently, extensive research is being conducted worldwide to increase the productivity of RE technologies and address their inherent limitations [9].

This chapter aims to examine various RE technologies and subsequent power production, while introducing and promoting modern RE technologies. Additionally, it addresses the limitations of existing RE technologies and explores potential answers. This article is organized to initially provide a comprehensive overview of the present state of the energy sector, with a particular focus on the role of RE. It then examines the development and implementation of various RE sources. Subsequently, the discussion transitions to energy storage systems, analyzing their importance and practical applications. Finally, the article explores the challenges associated with these technologies and presents recommendations to address them.

2 Renewable Energy Technologies

RE is increasingly utilized for power generation worldwide, driven by growing societal concerns over the environmental impacts of conventional energy generation methods. RE sources, including solar, hydro, wind, geothermal energy, and bioenergy, have distinct operating conditions and energy conversion efficiencies, making their adoption dependent on specific locations and conditions.

2.1 Solar Energy

Solar energy (SE) refers to Sun's radiation energy and is among the most widely harnessed forms of RE globally [10]. To enhance the efficiency of SE, scientists continue to investigate advanced technologies for optimizing SE system designs [11]. Efforts are also focused on reducing costs and minimizing environmental impacts [12]. SE systems are broadly classified into two types: PV and solar thermal energy, both universally employed and continually refined to improve efficiency and effectiveness.

2.1.1 Photovoltaic Energy

PV technology is a leading SE innovation and is considered a promising energy source for the future [13]. Currently, PV energy accounts for almost 2% of global electricity demand [14]. However, the traditional solar cells are limited to a conversion efficiency of 31%, a threshold recognized as the Schockley–Queisser (SQ) limit [15]. Scientists have developed various advanced PV technologies to increase efficiency, as discussed in the subsequent subsections here. These innovations aim to significantly enhance the performance and scalability of PV technologies for future energy needs.

(a) **Concentrated PV (CPV)**

CPV technology uses focusing optics to converge light onto PV cells. This method captures both beam radiation and scattered radiation, directing them onto the solar cells for energy generation [16]. CPV systems have three types based on the concentration factor: low (1–40x), medium (40–300x), and high concentration (HCPV) (300–2000x) [17]. Among these, HCPV demonstrates the highest efficiency, reaching at least 40%, as reported by the International Energy Agency (IEA) [18]. Recent research efforts focus on employing various types of concentrators, each with distinct advantages and limitations. CPV systems hold significant potential for large-scale SE applications, offering promising performance and efficiency enhancements.

(b) **Hot Carrier Converter**

Hot carrier converters are specialized solar cells designed to harness excess photon energy to increase electricity generation [19]. In standard quantum-utilizing SE converters, absorbed photon energy facilitates electronic transitions between conduction and valence bands [20]. However, surplus photon energy—beyond the bandgap threshold—is typically lost as heat. Hot carrier converters address this limitation by reducing the cooling rate of photoexcited carriers, allowing more extended utilization of photon energy at elevated temperatures. These converters achieve conversion efficiencies of up to 65%, significantly outperforming conventional solar cells [21].

(c) **Multi-junction Solar Cells (MJSCs)**

MJSCs contain several stacked p-n junctions of unique semiconductors (subcells), connected in series to enhance SE conversion efficiency [22]. To prevent current blockage, interconnectors are employed. These interconnectors must possess high transparency and low electrical resistance [23].

The highest recorded MJSC efficiency for PV systems is 40.7%, achieved using InGaP/InGaAs/Ge subcell stacks [24]. The selection of topcell materials is crucial for achieving high efficiency. InGaP is favored for its superior properties, including minimal oxygen issues, low recombination velocities, and improved window layer performance compared to AlGaAs [25]. Similarly, InGaAs is ideal for lattice-matching with Ge substrates, enhancing open-circuit voltage (V_{oc}) and short-circuit current density (J_{sc}) due to its reduced bandgap energy and favorable lattice adaptation [26]. Ge subcells serve as the bottom cells, with the lowest bandgap energy to absorb longer wavelengths of solar radiation [27].

Spectral response analysis of Ge, GaAs, InGaP, and InGaP/InGaAs/Ge MJSCs demonstrates that each subcell performs optimally at specific wavelengths between 300 and 1800 nm [27]. The combined response of the InGaP/InGaAs/Ge MJSC shows enhanced performance within the 400–1600 nm range, with higher J_{sc} values. However, due to the high implementation costs, MJSCs are predominantly used in PV systems for satellites and space vehicles [28].

(d) **Floating Photovoltaic (FPV) Power System**

FPV systems are PV systems installed on water bodies [29]. This technology is extensively utilized in countries such as Japan, the United States, Brazil, Italy, and China [30]. FPV systems use conventional PV arrays, but with the added advantage of a cooling effect from the underlying water, which helps prevent solar cell overheating. This cooling benefit has been reported to slightly improve PV energy harvesting efficiency [31].

The FPV system offers reciprocal benefits between the solar cells and the water body. It reduces algal growth and limits water evaporation from reservoirs [32, 33], while the water helps cool the system, further enhancing the efficiency of solar cells [34]. Additionally, FPV systems contribute to sustainable land management by avoiding the use of land for energy generation. To date, no significant environmental concerns have been reported by regulatory authorities [35]. For example, the world's biggest FPV system in Korea spans over 30 km^2, incorporates around 5 million solar cells, and generates power of 2.1 GW [36].

(e) **Down Conversion of Photons**

The down-conversion method optimizes SE utilization by splitting high-energy photons into multiple low-energy photons, allowing solar cells to harness energy more effectively [37]. Conventional solar cells release the extra energy of high-energy photons as heat [38]. Down conversion addresses this issue by employing a photoluminescent converter fabricated on top of the solar cells to facilitate photon division [39, 40]. This process enhances radiative transitions and efficiently converts high-energy photons ($> 2E_g$) [41]. This leads to a 39.63% increase in efficiency [42].

(f) **Photovoltaic Thermal (PV/T) Systems**

The PV/T systems put together solar collectors with PV cells to transform SE into both electrical and thermal energy simultaneously [43]. This dual-function system enhances overall energy utilization by harvesting a broader spectrum of solar radiation [44].

The conventional PV systems convert only 20% of SE to electricity, with the remainder lost as heat [45]. Since the temperature of solar cell inversely affects the electrical efficiency [46], higher cell temperatures reduce performance. PV/T systems address this limitation by absorbing excess heat through a solar collector [47]. In this system, solar radiation first passes through a glazing cover and is concentrated onto solar cells, where wavelengths between 0.6 and 0.7 μm are converted into electrical energy [48]. Remaining radiation is transformed into heat, which is captured by a collector and transferred via fluid in flow channels to heating applications.

The layered structure of a typical PV/T module maximizes solar radiation utilization, achieving significantly higher SE conversion rates compared to separate PV cells and solar thermal collectors [49]. PV/T systems are an efficient, cost-effective solution for generating both thermal and electrical energy, offering enhanced performance and sustainability[50]. Table 1 indicates the key materials used in PV applications and their respective efficiencies.

2.1.2 Solar Thermal Energy (STE)

STE involves heat energy generation from solar radiation [52], widely used in manufacturing and household applications such as heat for industrial processes [53], electricity generation [54], and space or water heating [55]. Researchers have shown significant interest in STE for its potential to directly generate electricity from heat [56]. The process involves thermal energy generation from solar radiation through a fluid circulating beneath the solar receiver [57].

STE collectors are categorized based on temperature into low, medium, and high types [58]. Low and medium-temperature collectors, typically flat plates, are primarily used for heating space, water, or swimming pools. High-temperature collectors use lenses or mirrors to concentrate sunlight, possess high efficiency, rendering them suitable for electricity generation. While PV technology converts about 20% of solar radiation into electricity, STE systems can absorb over 90%, achieving much higher efficiency [59]. Research initiatives aimed at reducing the implementation costs of STE systems have primarily concentrated on the development of alternative materials to substitute high-cost components, including optical systems and vacuum tubes [60].

(a) **Porous Structure Receiver**

Porous structure receivers (PSRs) offer a promising substitute to traditional vacuum enclosures in solar collectors by reducing heat loss and improving STE system efficiency [61]. These receivers incorporate a porous layer within the solar thermal

Table 1 Key materials for solar photovoltaic applications and their efficiencies

Solar photovoltaic technologies	Materials	Advantages	Major challenge	Efficiency (%)
First-generation	Crystalline silicon (Si) technologies: monocrystalline Si (m-Si), polycrystalline Si (p-Si), and gallium arsenide (GaAs)	High availability, stability, nontoxic, resistant to radiation degradation (i.e., ideal for space applications), high absorptivity compatible with Si-based microelectronics industry	More at risk of losing some of their effectiveness at higher temperatures (hot sunny days)	m-Si: 24.4%, p-Si 19.9%, GaAs: 18.4–28.8%
Second-generation (thin-film solar cells)	Amorphous silicon (a-Si), microcrystalline (μc-Si), CdTe/CdS, CIGS solar cells	Using less material and lower-cost manufacturing processes	Lower efficiency, higher degradation in outdoor uses, and limitation of manufacturing materials	CIGS: 20.3%
Third-generation	Nanocrystalline films, active quantum dots, tandem/stacked multilayers of inorganics based on III–V materials (GaAs/GaInP), organic (polymer)-based solar cells, DSSC	Solution-processable, appropriate for huge scale, mechanical robustness, high efficiencies at high temperatures	To ease the cost, watt of delivered solar electricity	21.1–21.6%
Fourth-generation	Combines the low cost and flexibility of polymer thin films with the stability of novel inorganic nanostructures (such as metal nanoparticles and metal oxides) or organic nanomaterials (e.g., CNT, graphene, and its derivatives)	Inexpensive manufacturing and excellent material improve the cell operation, economic competitiveness	High upfront cost	~ 40%

Reproduced with permission from Jaiswal et al. [51], Copyright Elsevier, 2022 collector, mitigating conductive heat loss while accommodating higher flux levels [62]. PSRs are capable of storing conductive heat efficiently, making them a compelling solution for enhancing STE performance [63].

Researchers have explored various high-porosity materials with distinct thermal insulation properties to enhance the absorption of a broader spectrum of solar radiation wavelengths [64]. A typical STE collector consists of three layers: a glass cover (top), a porous transmitter layer, and a black absorber layer (bottom) [57]. The glass cover collects and transmits solar radiation to the porous transmitter layer, whose pore size is crucial for minimizing heat loss by managing heat penetration, radiation, and release. The black absorber layer retains the heat, with heat transfer fluids (HTFs) flowing through it to deliver thermal energy to domestic, industrial applications, or steam turbine-based power generation systems.

To enhance heat transfer efficiency in STE systems, nanofluids are increasingly utilized as HTFs owing to their superior thermophysical features compared to regular liquids such as water, oil, and ethylene glycol [65]. Nanofluids consist of nanoparticles (< 100 nm) suspended in a base fluid, significantly enhancing thermal conductivity [66]. These nanoparticles include metals, carbon-based materials, and nanocomposites, with nanocomposites demonstrating the highest efficiency and thermal conductivity [67].

2.2 *Wind Energy*

Wind energy is the second most utilized source of RE, following hydropower, due to its cost-effectiveness, simple infrastructure, and technological maturity [68, 69]. Electricity is generated in wind energy systems through wind turbines, which are deployed in power plants. These power plants are categorized into two types: **onshore farms** (on land) and **offshore farms** (situated in seas or freshwater bodies) [70, 71]. The selection of a wind farm's location depends on the annual median wind velocity. Optimal sites include elevated terrains or coastal regions, which are well-suited for onshore farms.

2.2.1 Classification

Wind energy systems are classified by **axis orientation** and the **type of generator** employed in the turbines.

(a) **Axis Types of Turbines**

Based on rotational axis orientation, wind turbines are classified into the following types:

- **Horizontal Axis Wind Turbines (HAWTs)**

HAWTs, which have a horizontally aligned rotation axis, are the most common wind turbine design. Their advanced technology, historical reliability, and high energy efficiency have made them the preferred option for both offshore and onshore applications over the past three decades [72]. HAWTs typically consist of steel or concrete towers, rotor blades, a yaw system for directing the turbine toward the wind, and gearboxes and converters [73].

The recent trend of increasing HAWT sizes aims to meet global energy demands by improving power output [74]. However, scaling up turbine components, especially rotor blades, increases system weight, leading to challenges such as fatigue-induced mechanical stress and higher maintenance needs [75]. These factors must be managed to ensure the cost-efficiency of large-scale HAWTs.

- **Vertical Axis Wind Turbines (VAWTs)**

VAWTs have a vertically oriented rotation axis and are primarily used in urban and remote applications. They can operate at lower speeds, harness wind from multiple directions without yawing mechanisms, and generate less noise. They require minimal structural support but have lower aerodynamic efficiency than HAWTs [76]. VAWTs are particularly suited for urban areas, locations distant from main grids, and environments where large-scale wind farms are not feasible due to environmental constraints [77]. Research continues to focus on optimizing VAWT designs to improve their efficiency and performance [78].

(b) **Generator Types for Wind Energy Conversion**

Wind turbines utilize various types of generators, which differ in operational mechanisms and efficiency:

- **Squirrel Cage Induction Generators (SCIGs)**

SCIGs are cost-effective, robust, and simple to implement, making them suitable for small-scale, fixed-speed applications. However, their limited speed control and high reactive power consumption pose challenges for larger or variable-speed systems [79].

- **Doubly Fed Induction Generators (DFIGs)**

DFIGs, introduced in the 1990s, enable variable-speed operation and ensure grid stability through converters that regulate power flow. Despite their benefits, DFIGs require complex control strategies and frequent maintenance due to brush and gearbox wear [80].

- **Electrically Excited Synchronous Generators (EESGs)**

EESGs are commonly employed in large-scale direct-drive systems due to their cost-effectiveness. However, their large rotor size complicates transportation and installation. EESGs rely on advanced converters to regulate output frequency and voltage [81].

- **Permanent Magnet Synchronous Generators (PMSGs)**

PMSGs utilize permanent magnets for excitation, ensuring high efficiency and reduced maintenance needs. They are ideal for variable-speed operations and exhibit superior energy conversion performance, despite their higher initial costs [82]. PMSG systems include controllers and converters to optimize power output and meet grid integration standards [83].

2.2.2 Modern Technological Advances

To meet increasing energy demands, the focus of wind energy technology has shifted toward larger generation units, taller towers, advanced generators, and extended rotor blades. Utility-scale turbines now employ pitch-regulated designs and speed control systems to minimize mechanical stress while maximizing energy output, even in low wind conditions. Additionally, partially or fully-rated power converters are widely adopted to ensure grid compatibility and enhance operational efficiency. These innovations underscore the ongoing progress in wind energy systems, aiming to optimize performance and cost-efficiency in response to global energy needs.

2.2.3 Maximum Power Point Tracking (MPPT)

MPPT is a widely employed technique used to optimize the rotational speed of wind turbine generators, thereby maximizing energy extraction in moderate wind conditions [84]. MPPT is implemented to enhance efficiency by regulating the turbine's rotary motion through a variable-speed system [85]. This approach results in enhanced efficiency compared to conventional systems [86].

MPPT controllers have four types: power signal feedback control, tip speed ratio (TSR), optimal torque control, and perturb and observe method.

- **Power Signal Feedback Control**

This system utilizes one sensor to control the rotary motion of the turbine and adjusts the output power to match a reference value. The controller receives error signals between the calculated and the reference power to make necessary adjustments [87].

- **Tip Speed Ratio Method**

The TSR method utilizes two sensors to measure both the rotary speed and the rotor radius. A tachometer and an anemometer are used to collect feedback signals, which are then processed to determine the turbine's optimal speed in relation to the wind speed [88].

- **Optimal Torque Control**

This method relies on a mechanical torque equation to calculate the optimal torque based on angular speed, ensuring maximum power extraction by adjusting the turbine's operating conditions accordingly [89].

- **Perturb and Observe Method**

This method utilizes electric measurement procedures in lieu of additional sensors. Though highly reliable, it is considered less efficient compared to the other MPPT methods, with a potential error rate of approximately 5% [90].

Conventional MPPT control strategies often result in energy losses of 1%–3% due to the difficulty in accurately predicting wind speed, which exhibits high variability both spatially and temporally [91]. To mitigate this issue, intelligent algorithms have been integrated with traditional MPPT techniques to reduce error rates and enhance system efficiency [92]. These combined methods rely on data collection for instant wind speed calculation and dynamic turbine performance analysis [93]. Intelligent MPPT methods employ artificial intelligence and prediction methods, enabling improved precision and reduced computational costs through effective use of historical data. As a result, estimation errors are minimized and prediction horizons are shortened [94].

2.3 Hydropower

Hydropower is a well-established and economical RE source that generates electricity through the kinetic energy of water streams. Among all RE resources, hydropower achieves the greatest conversion efficacy, approximately 90% [95], and accounts for 20% of global electricity production [96]. Hydropower systems are highly adaptable, enabling capacity utilization across varying load demands, and include various types, including pumped hydro storage, small hydropower plants, and cascaded reservoirs [97].

2.3.1 Pumped Hydro Storage (PHS)

PHS energy storage system uses two water pools made at different heights. Electricity is generated by running turbines using water and stores energy through pumping water to the higher reservoir [98]. PHS systems are categorized into open-looped, which are connected to natural water sources, and closed-looped, which operate independently [99, 100]. While PHS systems help balance electricity supply and demand, they are typically limited to daily energy storage cycles and require suitable geological conditions for large water volumes and fluctuating reservoir levels [101].

2.3.2 Small Hydropower Plants (SHPs)

SHPs are cost-effective, eco-friendly energy solutions, particularly suited for rural areas in developing countries [102]. They generally operate on a "run-of-river" basis, with minimal or no water storage [103]. SHP capacities vary by region according to local regulations [104]. A typical SHP collects and diverts river water through an intake at a weir, filtered through a settling tank before reaching the turbine. These plants offer reliable and environmentally sustainable energy, even in areas with low water flow [105].

2.3.3 Cascaded Reservoir Hydropower Plants (CRHP)

CRHP involves interconnected reservoirs at different elevations, generating electricity as water flows through multiple reservoirs [106]. These systems are complex, with varying hydraulic heads, flow rates, and nonlinear dynamics [107]. CRHP setups can range from simple two-reservoir systems to intricate networks, depending on natural topography [108]. In addition to power generation, CRHPs provide benefits, including irrigation, flood control, and water reserve management, contributing to overall ecosystem stability and resilience [109].

2.4 *Tidal Energy (TE)*

TE is generated through the conversion of the kinetic energy of water currents into electricity [110]. It arises from the difference in water levels between two bodies of water, with tidal currents and tides following predictable semidiurnal and spring-neap cycles [111]. Historically, tidal energy has been used along the Atlantic coast of North America and Europe, notably with tide mills [112]. Currently, two primary technologies are used for TE generation: hydrokinetic systems and oscillating water columns [113]. Technological advances are expected to enhance the productivity and cost-efficiency of tidal power [114].

2.4.1 Hydrokinetic System (HKS)

HKS transforms water's kinetic energy into electricity through a turbine, power converter, and generator [115]. The turbine, driven by the velocity of water, is coupled with a synchronous generator, which rotates without gears [116]. The generated electricity is either stored in batteries or fed into the grid. This system, which does not require reservoirs, is suitable for installation in fast-flowing water bodies, either on floating platforms or stationary structures [117]. It is currently used in industries like paper, textiles, and food industry [118].

2.4.2 Oscillating Water Column (OWC)

The OWC is a leading first-generation wave power technology. The design of each OWC plant is customized to local environmental factors, such as wave climate, geology, and hydrodynamics, which are analyzed through wave basin testing and computational models [119]. A typical OWC system consists of a concrete chamber installed on the seabed, with a horizontal-axis turbine attached to a generator over the water [120, 121]. This turbine captures air movement generated by wave interactions, converting it into electricity [122]. Key parameters like rotational speed and power output are optimized through programmable control systems [123]. Several OWC plants are currently operational around the world [124, 125].

2.5 Bioenergy

Bioenergy has become a prominent RE source, contributing to heating, transportation, and environmentally friendly electricity generation [126]. Derived from biomass, which encompasses both traditional and modern sources, bioenergy utilizes biological materials for various energy applications [127]. Traditional bioenergy uses agricultural materials like crop residues, fuelwood, charcoal, and animal waste, while modern bioenergy generates heat and electricity through biodiesel, biochar, and biogas using thermal conversion processes such as combustion, torrefaction, carbonization, gasification, and pyrolysis [128–130]. Biomass-derived products improve sanitation, reduce landfill waste, promote waste management, and support green development goals [131]. Bioenergy also holds importance in policies, especially for rural development and regional growth [132].

2.5.1 Biomass

Biomass energy extraction primarily occurs via combustion, converting chemical energy into heat for electricity generation. The energy conversion efficiency depends on factors like combustion method, biomass type, and application scale [133]. Biomass is categorized into four groups: dedicated energy crops, agricultural and wood residue, and municipal waste [134]. Unlike coal, biomass emits fewer nitrogen oxides (NO_x) and sulfur oxides (SO_x), reducing acid rain and ozone depletion. Using biomass instead of fossil fuels helps decrease greenhouse gas emissions, contributing to global warming mitigation [135].

2.5.2 Biodiesel

Biodiesel is a fuel made from organic oils and fats, capable of replacing gasoline and diesel in transportation and electricity generation applications [136]. It is engine-compatible and improves lubricity, while producing lower greenhouse gas emissions [137]. Biodiesel supports rural development by replacing gasoline in agricultural sectors [138]. However, engine problems may be caused by vegetable oils, such as carbon deposits and coking [139]. Transesterification of vegetable oils reduces viscosity and meets biodiesel engine requirements, enhancing efficiency and sustainability [140].

2.5.3 Biogas

Biogas, derived from waste-based biomass through anaerobic digestion, offers a promising alternative to fossil fuels like natural gas. Composed mainly of methane (65%) and carbon dioxide (35%), it can be processed for electricity generation and helps reduce organic pollutants by 50%–90% compared to traditional fuels [141, 142]. Key feedstocks include crop and agricultural residues and cattle waste [143]. Biogas reduces greenhouse gas emissions, mitigates pathogens and odors, and serves as an eco-friendly fuel for heat and electricity generation, benefiting both public grid systems and waste management [144]. Table 2 shows the different sources of biomass feedstock, their respective biofuels and applications.

Table 2 Different sources of biomass feedstock for renewable bioenergy and biofuels

Sources of biomass	Feedstocks	Bioenergy and biofuels	Applications
Wood by-products	Firewood, wood chips, wood pellets, lumber, bark, furniture mill sawdust, wood processing residues	Heat energy, light energy, combined heat and power, cellulosic ethanol, electricity production, compost	Domestic use for cooking/heating, natural fertilizers, soil conditioners, agriculture mulches
Agricultural by-products	Corn cobs, fruit pits, straw, leaves, grass trimmings, lumber, bark, branches, agricultural crop residues, algae, forestry residues	Heat energy, light energy, combined heat and power, cellulosic ethanol, electricity production, compost	Domestic use for cooking/heating, natural fertilizers, soil conditioners, agriculture mulches
Solid waste	Residential and commercial garbage, paper/cardboard, vegetable/fruit peels, leftover/waste food	Heat energy, light energy, electricity production, compost	Domestic use for cooking/heating, natural fertilizers, soil conditioners, agriculture mulches
Non-crop food plants, corn, sugarcane, agricultural residues	Cellulosic components	Bio-alcohol fuels, cellulosic ethanol, methanol, ethanol-blended petrol	Transportation vehicle usage
Landfills, leachate, manure, municipal wastewater, industrial wastewater	Biogas	Methane, liquid fuel, electricity generation, methanol production	Transportation and domestic fuel for cooking

Reproduced with permission from Jaiswal et al. [51], Copyright Elsevier, 2022

2.6 Geothermal Energy (GE)

GE is obtained from the heat generated by the radioactive decomposition of minerals and the primitive structure of the Earth itself [145]. Typically, the geothermal gradient increases at a rate of 0.03 °C per meter of depth, meaning that 99% of the Earth's temperature exceeds 1000 °C [146]. Unlike other intermittent RE sources such as wind, solar, and hydro, geothermal energy remains rich and unlimited within the Earth's core [147]. Moreover, geothermal energy is inherently stable and does not contribute to CO_2 emissions [148]. Geothermal energy possesses substantial economic potential, especially in regions rich in hydrothermal resources, particularly in countries with active volcanic activity [149]. Currently, approximately 26 nations are generating electricity using geothermal energy [150]. Geothermal power plants are categorized into two main types: steam- and binary cycle-based plants [151]. To maximize heat insulation in the pipes, most plants are constructed near geothermal resources [152]. There has been significant advancement in the development and execution of geothermal energy in recent years through the modernization of traditional concepts and techniques.

2.6.1 Organic Rankine Cycle (ORC)

The ORC, also known as a binary cycle, is used in geothermal energy applications where resources contain chemical contaminations that prevent efficient flashing, or where the geothermal fails to produce steam due to its low heat content [153]. The ORC systems are advantageous as they operate mainly at lower temperatures, leading to reduced mechanical stress on turbines. Additionally, the current ORC systems are erosion-free due to the lack of humidity from vapor expansion in turbines. Furthermore, ORC systems, operating under vacuum conditions, do not experience leakage issues [154]. In comparison, ORC plants generate zero CO_2 emissions, whereas flash steam plants emit approximately 27 kg/MWh of CO_2 [155]. The binary cycle process involves several stages: the hot fluid enters the plant at its inlet temperature, with gases that are non-condensable remaining in the fluid unless the pressure is high enough to eliminate them [156]. The fluid is then vaporized in the vaporizer before entering the turbine. The vapor from the turbine gets into a regenerator where heat exchange takes place between the vapors and the condenser fluid. The vapors are subsequently transferred back to the vaporizer, which cools the extra fluid and injects it back into the source well. The vapor is subsequently cooled, condensed into a liquid, and circulated back through the system using a pump, increasing the pressure for the next cycle.

2.6.2 Steam Power Cycle (SPC)

The SPC is generally classified into two systems: the single and the double-flash systems. In the single-flash systems, the geothermal fluid undergoes flashing, resulting in the separation of water and steam and subsequent sending to the steam turbine. In the double-flash systems, the water is further flashed into steam at low pressure, followed by sending it to a second turbine for additional power production.

(a) **Single Flash Cycle (SFC)**

During this process, geothermal fluid is extracted and transferred to a separator, where it is boiled to produce steam [157]. The brine is then sent for reinjection into the well, and the steam goes to the steam turbine. After expanding through the turbine, the steam is cooled in the condenser, with cooling air entering and exiting through designated paths. The cycle is completed through reinjection of the condensed fluid into the well.

(b) **Double Flash Cycle (DFC)**

The DFC begins by collecting geothermal fluid from the well and passing it to a high-pressure separator [157]. The brine is transferred to a low-pressure separator through a regulating valve, while the steam from both separators is sent to the turbines. After expanding through the turbines, the steam is cooled in the condenser, while the fluid is sent back into the well after condensation. The process generates additional electricity by capturing more thermal energy from the same geothermal resource. The brine is reinjected into the well for sustainable resource management.

2.7 Hydrogen Energy

Hydrogen energy is a clean energy with a high energy density by mass; 9.5 g of hydrogen holds the same energy as 25 g of gasoline [157]. Hydrogen can be stored through direct or electrolytic procedures before undergoing a chemical reaction to release electricity [158]. Various production methods for hydrogen include water electrolysis, coal gasification, thermochemical decomposition, and solar-driven processes [159, 160]. Hydrogen is sustainable owing to its nonrenewable and renewable resources [161], and has applications across multiple sectors, including transportation, commercial, institutional, and residential uses [162]. As a key energy carrier, hydrogen has the potential to transform energy systems by integrating electricity and hydrogen [163]. Its by-product in fuel cells (FCs)—water vapor—is harmless, making it an appealing alternative fuel for power generation and hydrogen plants [164].

2.7.1 Hydrogen Fuel Cell

A hydrogen FC is an electrochemical device that generates electricity by converting chemical energy. FCs operate by continuously requiring oxygen and hydrogen (or other fuels like methane or alcohol) to produce electricity [165, 166]. A typical polymer electrolyte membrane fuel cell (PEMFC) has three components: the anode, cathode, and catalytic electrolyte [167, 168]. These components facilitate the conversion of hydrogen and oxygen into water, generating electricity. FCs can achieve 60% efficiency in converting energy into electricity, and 80% when both thermal and electrical energy are co-generated. FCs produce up to 90% fewer pollutants compared to conventional electricity generation methods [169]. Table 3 shows the properties of major FC technologies.

2.7.2 Hydrogen Gas Turbine

Hydrogen gas turbines offer a highly efficient method for large-scale electricity generation, building on traditional gas turbine technology [171]. These turbines generate electricity without emitting CO_2 or NO_x, making them an environmentally friendly option [172]. Current applications include integrated gasification combined cycle power plants, power plants with precombustion CO_2 capture through carbon capture and sequestration [173], and systems where hydrogen is stored as a secondary energy source or integrated into RE systems. A recent thermal cycle concept combining hydrogen and steam turbines has been proposed, where hydrogen generated through solar-driven water photolysis powers a combined turbine system [174]. This approach significantly enhances energy conversion efficiency, utilizing both hydrogen and steam turbines.

3 Challenges and Suggestions

RE is progressively becoming the main source of electricity production. To facilitate this shift, it is necessary to integrate alternative RE sources into the existing conventional energy systems. However, the transition from non-sustainable to RE sources is undeniably challenging due to significant technological barriers, despite the growing application of RE technologies [175, 176]. The limitations of RE sources primarily stem from their intrinsic characteristics, which are yet to be fully addressed.

3.1 Challenges

At present, sponsorships and support for the implementation of RE are evolving across industries, governmental bodies, and nongovernmental organizations. Experts

Table 3 Typical characteristics of major FC technologies

Type of hydrogen fuel cells	Solid oxide FC (SOFC)	Polymer electrolyte membrane FC (PEMFC)	Phosphoric acid FC (PAFC)	Molten carbonate FC (MCFC)	Alkaline FC (AFC)
Electrolyte type (ion/media)	O^{2-} or ceramic matrix with free oxide ions	H^+ or ion-conducting polymer membrane	H^+/H_3PO_4	$CO_3{}^{2-}$ or mixture of molten alkaline carbonates	OH^- or KOH aqueous solution
Typical materials used	Ceramic, high-temperature metals	Metal, carbon, plastic	Ceramic, carbon	Ceramic, high-temperature metals	Metal, plastic
Fuel	H_2 or reformate gas	H_2 or methanol	H_2 or reformate gas	H_2 or reformate gas	H_2
Internal reforming oxidant	Yes	No	No	Yes	No
Working temperature (°C)	750–1000	25–85	190–210	650–700	90–260
System electric efficiency, % of HHV$^+$	45–55	25–45	35–45	40–50	30–40
Total system efficiency, % (electrical + thermal) of HHV$^+$	68–77	60–75	69–80	60–65	65–68
Main sensitivities to contaminants	Sulfur	Sulfur, CO, NH_3	Sulfur, CO if over 1%	Sulfur	Sulfur, CO, CO_2

Reproduced with permission from Ang et al. [170], Copyright Elsevier, 2022

are focusing on environmental, energy, and developmental agendas at local and global levels to develop policies, economies, and RE production in both developed and developing nations [177]. In many developing countries, hydropower remains the dominant RE source due to its cost-effectiveness and efficiency, while nations endowed with abundant RE resources are making strides toward sustainability measures [178]. However, various challenges persist within the RE markets, industries, and policies. A key obstacle to RE implementation is the continuing reliance on fossil fuels, which detracts attention from the significance of RE adoption. Many authorities still prefer to subsidize fossil fuel projects over RE, owing to the advancement of fossil fuel technologies, whereas RE technologies are generally in earlier stages of development in terms of technical maturity, market readiness, and policy. No financial support from governments is another obstacle to the development, deployment, and promotion of RE technologies. Furthermore, developing countries not only lack mechanisms to address the economic disparity between fossil fuels and RE, but also a comprehensive RE policy framework. Many underdeveloped nations also lack experience in managing the potential hostile environmental impacts of RE power plants, especially when subjected to external economic or environmental shocks [178]. Consequently, despite their abundant RE resources, fossil fuels continue to dominate electricity generation in these regions. On the other hand, high-GDP nations with greater energy sustainability awareness typically possess the financial capacity and support for researching and implementing RE technologies, further boosting their economy and energy sector [179]. Moreover, developed countries often take the lead in establishing policies, while developing countries tend to adopt these policies with additional regulations for adaptation [178]. The selection of which RE technologies to implement in a country largely depends on the development of special support policies within a given population.

3.2 Suggestions

Ongoing research is essential for enhancing the economic, technical, and conversion efficiencies of RE systems. Technological advancement plays an important part in increasing the global adoption of RE technologies, as the continuous evolution of these technologies requires sustained global efforts. Understanding global market trends and patterns in RE is crucial for fostering technological capacity across various industries, enabling progress that extends beyond domestic boundaries [180]. Developing technological expertise and knowledge is integral not only for poverty alleviation and socioeconomic development but also for enhancing local production capacities [180]. Such expertise is critical for improving system performance, ensuring cost-effectiveness, and addressing the pressing challenges posed by climate change [181].

Equally important, comprehensive and effective policy frameworks are necessary to overcome the obstacles hindering the widespread implementation of RE. Policies should be carefully crafted to align with each country's unique socioeconomic objectives, environmental concerns, and financial limitations [182]. Effective policy tools are required to navigate the complexities of renewable resource accessibility, market dynamics, and policy objectives, ensuring that certain sectors do not bear disproportionate impacts [183]. The development of RE production in developing nations, in particular, is closely tied to regulatory policies and economic instruments [184]. Thus, governments must implement strong and forward-thinking RE policies that consider the interconnectedness of various factors influencing the sustainability and availability of RE [185].

The successful deployment of advanced RE technologies largely depends on the ability to build, manage, and uphold a robust energy setup. This requires active collaboration among decision-makers, scientists, and companies nationally and internationally [186]. Moreover, it is essential to balance economic growth with environmental preservation, ensuring that energy development strategies lead to mutually beneficial outcomes that foster long-term sustainability and prosperity [187].

4 Conclusion

Environmental challenges, such as global warming and ozone layer depletion, are escalating in severity and urgency. To mitigate these critical issues, both developed and developing nations are increasingly joining forces to advance the adoption of RE technologies. This collective effort calls for enhanced research and innovation, focusing on the exploration of new methods and components to elevate the efficiency, affordability, and scalability of RE-based power generation.

This chapter provides an outline of the different types of RE technologies, identifies the inherent limitations within these technologies, and proposes strategies to address these challenges. It is argued that many of the obstacles hindering the broader deployment of RE can be overcome soon through the adoption of the suggested measures. In addition to technological innovations, the successful integration of RE will require affordable pricing structures, enabling regulatory frameworks, and heightened public awareness and acceptance of these technologies.

Governments, along with nongovernmental organizations, play vital roles in promoting these foundational elements. By implementing supportive policies and encouraging broad societal engagement, they can help drive the widespread adoption of RE technologies. Future research must prioritize these socioeconomic and policy-driven aspects, ensuring that RE resources are chosen and installed effectively, reflecting the diverse requirements and contexts of various areas [100].

References

1. Rizzi F, van Eck NJ, Frey M (2014) The production of scientific knowledge on renewable energies: worldwide trends, dynamics can challenges and implications for management. Renew Energy 62:657–671
2. Vine E (2008) Breaking down the silos: the integration of energy efficiency, renewable energy, demand respond and climate change. Energy Effic 1:49–63
3. Manish S, Pillai IR, Banerjee R (2006) Sustainability analysis of renewables for climate change mitigation. Energy Sustain Dev 10:25–36
4. Santika WG, Anisuzzaman A, Bahri PA, Shafiullah G, Rupf GV, Urmee T (2019) From goals to joules: a quantitative approach of interlinkages between energy and the sustainable development goals. Energy Res Soc Sci 50:201–214
5. Raheem A, Abbasi SA, Memon A, Samo SR, Taufiq-Yap YH, Danquah MK, Harun R (2016) Energy Suatain Soc 6:16

6. Ludin NA, Mustafa NI, Hanafiah MM, Ibrahim MA, Mat Teridi MA, Sepeai S, Zaharim A, Sopian K (2018) Prospects of life cycle assessment of renewable energy from solarphotovoltaic technologies: a review. Renew Sustain Energy Rev 96:11–28
7. Baños R, Manzano-Agugliaro F, Montoya FG, Gil C, Alcayde A, Gómez J (2011) Optimization methods applied to renewable and sustainable energy: a review. Renew Sustain Energy Rev 15:1753–1766
8. Qazi A, Hussain F, Rahim NA, Hardaker G, Alghazzawi D, Shaban K, Haruna K (2019) Towards sustainable energy: a systematic review of renewable energy sources technologies, and public opinions. IEEE Access 7:63837–63851
9. Kardooni R, Yusoff SB, Kari FB (2016) Renewable energy technology acceptance in Peninsular Malaysia. Energy Policy 88:1–10
10. Alhamrouni I, Danial M, Salem M, Awalin LJ, Ismail B (2020) Design of 2LC-Y DC-DC converter for high voltage/low current renewable energy application. Test Eng Manag 83:2111–2117
11. Alhuyi Nazari M, Salem M, Mahariq I, Younes K, Maqableh BB (2021) Utilization of data-driven methods in solar desalination systems: a comprehensive review. Front Energy Res 9:541
12. Diakaki C, Grigoroudis E, Kabelis N, Kolokotsa D, Kalaitzekis K, Stavrakakis G (2010) A multi-objective decision model for the improvement of energy efficiency in buildings. Energy 35:5483–5496
13. Alhafadhi L, Teh J, Lai C-M, Salem M (2020) Predictive adaptive filter for reducing total harmonics distortion in PV systems. Energies 13:3286
14. Bughneda A, Salem M, Richelli A, Ishak D, Alatai S (2021) Review of multilevel inverters for PV energy system applications. Energies 14:1585
15. Shockley W, Queisskr HJ (1961) Detailed balance limit of efficiency of p-n junction solar cells. J Appl Phys 32
16. Khamooshi M, Salati H, Egelioglu F, Hooshyar Faghiri A, Tarabishi J, Babadi S (2014) A review of solar photovoltaic concentrators. Int J Photoenergy 1–17
17. Bassam NE (2021) Distributed renewable energies for off-grid communities, 2nd edn. Oxford
18. Bahar H (2020) Concentrating solar power (CSP). In: IEA (ed) Paris
19. Conibeer G, Shrestha S, Huang S, Patterson R, Xia H, Feng Y, Zhang P, Gupta N, Tayebjee M, Smyth S, Liao Y, Zhang Z, Chung S, Lin S, Wang P, Dai X (2014) Hot carrier solar cell absorbers: materials, mechanisms and nanostructures. Next Gener Technol Solar Energy Convers 9178:917802
20. Ross RT, Nozik AJ (1982) Efficiency of hot-carrier solar energy converters. J Appl Phys 53:3813–3818
21. Conibeer G (2007) Third-generation photovoltaics. Mater Today 10:42–50
22. Palik E (1991) Handbook of optical constants of solids II. Academic Press, San Diego
23. Gilman JM, O'Neill AG (1993) Device applications of interband tunneling structures with one, two, and three dimensions. J Appl Phys 74
24. King RR, Law DC, Edmondson KM, Fetzer CM, Kinsey GS, Yoon H, Sherif R, Karam NH (2007) 40% efficient metamorphic GaInP/GaInAs/Ge multijunction solar cells. Appl Phys Lett 90:183516
25. Olson JM, Kurtz SR, Kibbler AE, Faine P (1990) A 27.3% efficient $Ga_{0.5}In_{0.5}P$/GaAs tandem solar cell. Appl Phys Lett 56:623–625
26. Takamoto T, Ikeda E, Agui T, Kurita H (1999) Characteristics of GaAs based concentrator cell. Technical digest PVSEC-11
27. Soley SS, Dwivedi ADD (2021) Numerical simulation and performance analysis of InGaP, GaAs, Ge single junction and InGaP/GaAs/Ge triple junction solar cells. Mater Today Proc 5:39
28. Bailey S, Raffaelle R (2011) Space solar cells and arrays. In: Handbook of phoyovoltaic science and engineering. Wiley, Chichester, pp 365–401
29. Cazzaniga R, Cicu M, Rosa-Clot P, Tina GM, Ventura C (2018) Floating photovoltaic plants: performance analysis and design solutions. Renew Sustain Energy Rev 81:1730–1741

30. IEA (2015) IEA-PVPS annual report 2015. International Energy Agency
31. Santafé MR, Trapani K (2014) Progress in photovoltaics, research and applications
32. Alam MZB, Ohgaki S (2001) Evaluation of UV-radiation and its residual effect for algal growth control. In: Advances in water and wastewater treatment technology. Elsevier, Oxford, pp 109–117
33. Ferrer-Gisbert C, Ferrán-Gozálvez JJ, Redón-Santafé M, Ferrer-Gisbert P, Sánchez-Romero FJ, Torregrosa-Soler JB (2013) A new photovoltaic floating cover system for water reservoirs. Renew Energy 60:63–70
34. Bahaidarah H, Subhan A, Gandhidasan P, Rehman S (2013) Performance evaluation of a PV (photovoltaic) module by back surface water cooling for hot climatic conditions. Energy 59:445–453
35. Rosa-Clot M, Tina GM (2017) Submerged and floating photovoltaic systems: modelling, design and case studies. Elsevier Inc.
36. Ministry of Trade, Industry and Energy (2019) The world's largest floating photovoltaic business in Saemangeum, 2019. [Online]. Available: https://www.motie.go.kr/motie/ne/presse/press2/bbs/bbsView.do?bbs_seq_n=161921&bbs_cd_n=81
37. Wegh RT, Donker H, Oskam KD, Meijerink A (1999) Visible quantum cutting in $LiGdF_4:Eu^{3+}$ through downconversion. Science 283:663–666
38. Wolf M (1971) A new look at silicon solar cell performance. Energy Convers 11:63–73
39. Trupke T, Green MA, Würfel P (2002) Improving solar cell efficiencies by downconversion of high-energy. J Appl Phys 92:1668–1674
40. Richards BS (2006) Enhancing the performance of silicon solar cells via the application of passive luminescence conversion layers. Sol Energy Mater Sol Cell 90:2329–2337
41. Vos AD, Szymanska A, Badescu V (2009) Modelling of solar cells with downconversion of high energy photons, anti-reflection coatings and light trapping. Energy Convers Manag 50:328–336
42. Wang Y, Wang S, Zhang Y, Mao Q, Su S, Chen Z (2021) Enhancing efficiencies of solar thermophotovoltaic cells by downconversion of high-energy photons. J Renew Sustain Energy 13
43. Khanjari Y, Pourfayaz F, Kasaeian AB (2016) Numerical investigation on using of nanofluid in a water-cooled photovoltaic thermal system. Energy Convers Manag 122:263–278
44. Khanjari Y, Kasaeian AB, Pourfayaz F (2017) Evaluating the environmental parameters affecting the performance of photovoltaic thermal system using nanofluid. Appl Therm Eng 115:178–187
45. Suresh AK, Khurana S, Nandan G, Dwivedi G, Kumar S (2018) Role on nanofluids in cooling solar photovoltaic cell to enhance overall efficiency. Mater Today Proc 5:20614–20620
46. Good C, Andresen I, Hestnes AG (2015) Solar energy for net zero energy buildings—a comparison between solar thermal, PV and photovoltaic–thermal (PV/T) systems. Sol Energy Mater Sol Cell 122:986–996
47. Zhang X, Zhao X, Smith S, Xu J, Yu X (2012) Review of R&D progress and practical application of the solar photovoltaic/thermal (PV/T) technologies. Renew Sustain Energy Rev 16:559–617
48. Kern EC, Russell MC Jr (1978) Combined photovoltaic and thermal hybrid collector systems. In: Proceedings of the 13th IEEE PV specialist conference, Washington, DC
49. Zondag HA, de Vries DW, van Steenhoven AA, van Helden WG, van Zolingen RJ (1999) Thermal and electrical yield of a combi-panel. In: Proceedings of ISES bi-annual conference on CD-ROM, Jerusalem
50. Zhao X, Zhang X, Riffat SB, Su Y (2011) Theoretical study of the performance of a novel PV/e roof module for heat pump operation. Energy Convers Manag 52:603–614
51. Jaiswal KK, Chowdhury CR, Yadav D, Verma R, Dutta S, Jaiswal KS, Karuppasamy KSK (2022) Renewable and sustainable clean energy development and impact on social, economic, and environmental health. Energy Nexus 7:100118
52. Cleveland CJ, Morris C (2015) Dictionary of energy, 2nd edn. Elsevier Limited, Oxford

53. Karanikola V, Moore SE, Deshmukh A, Arnold RG, Elimelech M, Sáez AE (2019) Economic performance of membrane distillation configurations in optimal solar thermal desalination systems. Desalination 472:114164
54. Huang J, He Y, Hu Y, Wang X (2018) Steam generation enabled by a high efficiency solar absorber with thermal concentration. Energy 165:1282–1291
55. Ayadi O, Al-Dahidi S (2019) Comparison of solar thermal and solar electric space heating and cooling systems for buildings in different climatic regions. Sol Energy Mater Sol Cell 188:545–560
56. Yahya K, Salem M, Iqteit N, Khan SA (2020) A thermoelectric energy harvesting system. IntechOpen
57. Yashim MM, Sainorudin MH, Mohammad M, Fudholi A, Asim N, Razali H, Sopian K (2021) Recent advances on lightweight aerogel as a porous receiver layer for solar thermal. Sol Energy Mater Sol Cell 228:111131
58. Bhatia SC (2014) Solar thermal energy. In: Advanced renewable energy systems. Woodhead Publishing India, pp 94–143
59. Chang C, Wang Z, Fu B, Ji Y (2020) High-efficiency solar thermoelectric conversion enabled by movable charging of molten salts. Sci Rep 10
60. Avila-Marin AL, Fernandez-Reche J, Martinez-Tarifa A (2019) Modelling strategies for porous structures as solar receivers in central receiver systems: a review. Renew Sustain Energy Rev 111:15–33
61. Li Q, Zhang Y, Wen ZX, Qiu Y (2020) An evacuated receiver partially insulated by a solar transparent aerogel for parabolic trough collector. Energy Convers Manag 214:112911
62. Liu W, Wei G, Huang P, Xu C, Du X (2018) Design and performance analysis of volumetric solar receiver based on porous foam ceramics. AIP Conf Proc 040022
63. Fend T, Hoffschmidt B, Pitz-Paal R, Reutter O, Rietbrock P (2004) Porous materials as open volumetric solar receivers: experimental determination of thermophysical and heat transfer properties. Energy 29:823–833
64. Wang Z, Quan X, Zhang Z, Cheng P (2017) Optical absorption of carbon-gold coreshell Nanoparticles. J Quant Spectros Radiat Trans 205
65. Gupta SK, Gupta G (2021) The role of nanofluids in solar thermal energy: a review of recent. Mater Today Proc 44:401–412
66. Ghalandari M, Maleki A, Haghighi A, Shadloo MS, Nazari MA, Tlili I (2020) Applications of nanofluids containing carbon nanotubes in solar energy systems: a review. J Mol Liq 313:113476
67. Nagarajan PK, Subramani J, Suyambazhahan S, Sathyamurthy R (2014) Nanofluids for solar collector applications: a review. Energy Proc 61:2416–2434
68. Herbert GJ, Iniyan S, Sreevalsan E, Rajapandian S (2007) A review of wind energy technologies. Renew Sustain Energy Rev 11:1117–1145
69. Breeze P (2014) Wind power generation technologies, chap 11. In: Power generation technologies, 2nd edn. Newnes, London
70. Rashad A, Kamel S, Jurado F (2017) The basic principles of wind farms. In: Distributed generation systems. Butterworth-Heinemann, Oxford, pp 21–67
71. Breeze P (2019) Wind power. In: Power generation technologies, 3rd edn. Newnes, London, pp 251–273
72. Twidell J (2015) Renewable energy resources. Routledge, Oxfordshire
73. Liu J, Lin H, Zhang J (2019) Review on the technical perspectives and commercial viability of vertical axis wind turbines. Ocean Eng 182:608–626
74. Vestas Wind System A/S (2021) Vestas launches the V236-15.0 MW to set new industry benchmark and take next step towards leadership in offshore wind. Vestas Wind System A/S, Aarhus
75. Shires A (2013) Design optimisation of an offshore vertical axis wind turbine. Proc Inst Civ Eng Energy 166:7–18
76. Tjiu W, Marnoto T, Mat S, Ruslan MH, Sopian K (2015) Darrieus vertical axis wind turbine for power generation I: assessment of Darrieus VAWT configurations. Renew Energy 75:50–67

77. Bishop JDK, Amaratunga GAJ (2008) Evaluation of small wind turbines in distributed arrangement as sustainable wind energy option for Barbados. Energy Convers Manag 49:1652–1661
78. Bhutta MM, Hayat N, Farooq AU, Ali Z, Jamil SR, Hussain Z (2012) Vertical axis wind turbine—a review of various configurations and design techniques. Renew Sustain Energy Rev 16:1926–1939
79. Hansen LH, Helle L, Blaabjerg F, Ritchie E, Munk-Nielsen S, Bindner H, Soerensen P, Bak-Jensen B (2001) Conceptual survey of generators and power electronics for wind turbines. Riso National Laboratory, Roskilde
80. Carlin PW, Laxson AS, Muljadi EB (2001) The history and state of the art of variable-speed wind turbine technology. National Renewable Energy Laboratory, Golden, Colorado
81. Eckels PW, Snitchler G (2008) 5 MW high temperature superconductor ship propulsion motor design and test results. Nav Eng J 117:31–36
82. Zhou D, Blaabjerg F, Franke T, Tønnes M, Lau M (2015) Comparison of wind power converter reliability with low-speed and medium-speed permanent-magnet synchronous generators. IEEE Trans Ind Electron 62:6575–6584
83. Anaya-Lara O (2009) Wind energy generation: modelling and control. Wiley, Oxford
84. Abdullah MA, Yatim AH, Tan CW, Saidur R (2012) A review of maximum power point tracking algorithms for wind energy systems. Renew Sustain Energy Rev 16:3220–3227
85. Muyeen SM, Tamura J, Murata T (2008) Stability augmentation of a grid-connected wind farm. Springer, London
86. Carrasco JM, Franquelo LG, Bialasiewicz JT, Galván E, PortilloGuisado RC, Prats MM, León JI, Moreno-Alfonso N (2006) Power-electronic systems for the grid integration of renewable energy sources: a survey. IEEE Trans Ind Electron 53:1002–1016
87. Ardjal A, Mansouri R, Bettayeb M (2017) Nonlinear synergetic control of wind turbine for maximum power point tracking. Presented at the 5th international conference on electrical engineering, Boumerdes, Algeria
88. Delfino F, Pampararo F, Procopio R, Rossi M (2012) A feedback linearization control scheme for the integration of wind energy conversion systems into distribution grids. IEEE Syst J 6:85–93
89. Oh K-Y, Park J-Y, Lee J-S, Lee J (2015) Implementation of a torque and a collective pitch controller in a wind turbine simulator to characterize the dynamics at three control regions. Renew Energy 79:150–160
90. Zhang Y, Zhang L, Liu Y (2019) Implementation of maximum power point tracking. Processes 7:158
91. Johnson KE, Pao LY, Balas MJ, Fingersh LJ (2006) Control of variable-speed wind turbines: standard and adaptive techniques for maximizing energy capture. IEEE Control Syst Mag 26:70–81
92. Bekakra Y, Attous DB (2014) Optimal tuning of PI controller using PSO optimization for indirect power control for DFIG based wind turbine with MPPT. Int J Syst Assur Eng Manag 5:219–229
93. Huang C (2015) Maximum power point tracking strategy for large-scale wind generation systems considering wind turbine dynamics. IEEE Trans Ind Electron 62:2530–2539
94. Zeng J, Qiao W (2012) Short-term wind power prediction using a wavelet support vector machine. IEEE Trans Sustain Energy 3:255–264
95. Killingtveit Å (2020) Future energy: improved, sustainable and clean options for our planet. Elsevier Ltd., Oxford
96. Alternate Energy (2021) Hydroelectric power. Available: https://www.altenergy.org/renewables/hydroelectric.html
97. IRENA (2012) Renewable energy cost analysis: hydropower. IRENA, United Arab Emirates
98. W. P. T. Office (2021) Pumped-storage hydropower, 13 June 2021. Available: https://www.energy.gov/eere/water/pumped-storage-hydropower
99. Hockenberger J, Dudenhoeffer B (2012) Pump storage arrangement, method of operating a pump storage arrangement and pump storage hydropower plant. Germany patent

100. Hunt JD, Zakeri B, Lopes R, Barbosa PS, Nascimento A, De Castro NJ, Brandão R, Schneider PS, Wada Y (2020) Existing and new arrangements of pumped hydro storage plants. Renew Sustain Energy Rev 129:109914
101. Kougias I, Szabó S (2017) Pumped hydroelectric storage utilization assessment: forerunner of renewable energy integration or Trojan horse? Energy 140:318–329
102. Haidar AMA, Senan MFM, Noman A, Radman T (2012) Utilization of pico hydro generation in domestic and commercial loads. Renew Sustain Energy Rev 16:518–524
103. Paish O (2002) Small hydro power: technology and current status. Renew Sustain Energy Rev 6:537–556
104. Edenhofer O, Pichs-Madruga R, Sokona Y, Seyboth K, Kadner S, Zwickel T, Eickemeier P, Hansen G, Schlömer S, von Stechow C, Matschoss P (2011) Renewable energy sources and climate change mitigation. IPCC, Cambridge
105. Ohunakin OS, Ojolo SJ, Ajayi OO (2011) Small hydropower (SHP) development in Nigeria: an assessment. Renew Sustain Energy Rev 15:2006–2013
106. Piekutowski MR, Litwinowicz T, Frowd RJ (1994) Optimal short-term scheduling for a large-scale cascaded hydro system. IEEE Trans Power Syst 9:805–811
107. Moreno SR, Kaviski E (2015) Daily scheduling of small hydro power plants dispatch with modified particles swarm optimization. Pesqui Oper 35:25–37
108. Mahmoud M, Dutton K, Denman M (2004) Dynamical modelling and simulation of a cascaded reservoirs hydropower plant. Electr Power Syst Res 70:129–139
109. Zhang L, Huang Q, Liu D, Deng M, Zhang H, Pan B, Zhang H (2021) Long-term and mid-term ecological operation of cascade hydropower plants considering ecological water demands in arid region. J Clean Prod 279:123599
110. Neill SP, Hashemi MR, Lewis MJ (2016) Tidal energy leasing and tidal phasing. Renew Energy 85:580–587
111. Waters S, Aggidis G (2016) Tidal range technologies and state of the art in review. Renew Sustain Energy Rev 59:514–529
112. Vlachogiannis JG (2014) Marine-current power generation model for smart grids. J Power Sources 249:172–174
113. Greaves D, Iglesias G (2018) Wave and tidal energy. Wiley, New Jersey
114. Altawell N (2021) Energy technologies and energy storage systems for sustainable development. In: Rural electrification. Elsevier, Oxford, pp 231–248
115. Khan MJ, Iqbal MT, Quaicoe JE (2008) River current energy conversion systems: progress, prospects and challenges. Renew Sustain Energy Rev 12:2177–2193
116. Ibrahim WI, Mohamed MR, Ismail RMTR, Leung PK, Xing WW, Shah AA (2021) Hydrokinetic energy harnessing technologies: a review. Energy Rep 7:2021–2042
117. Anyi M, Kirke B (2010) Evaluation of small axial flow hydrokinetic turbines for remote communities. Energy Sustain Dev 14:110–116
118. Tanier-Gesner F, Stillinger C, Bond A, Egan P, Perry J (2014) Design, build and testing of a hydrokinetic H-Darrieus turbine for developing countries. In: IEEE general meeting power & energy society. Washington, DC
119. Falcão AFO (2004) First-generation wave power plants: current status and R&D requirements. J Offshore Mech Arctic Eng 126:384–388
120. Moretti G, Rosati Papini GP, Daniele L, Forehand D, Ingram D, Vertechy R, Fontana M (2019) Modelling and testing of a wave energy converter based on dielectric elastomer generators. Proc R Soc A 475:20180566
121. Falcão AFO, Sarmento AJ, Gato LM, Brito-Melo A (2020) The pico OWC wave power plant: its lifetime from conception to closure 1986–2018. Appl Ocean Res 98:102104
122. Sarmento AJNA, e Melo AB, Pontes MT (2020) The influence of the wave climate on the design and annual production of electricity by OWC wave power plants. J Offshore Mech Arctic Eng 125:139144
123. Falcão AFO (2002) Control of an oscillating water column wave power plant for maximum energy production. Appl Ocean Res 24:73–82

124. Heath T, Whittaker TJT, Boake CB (2001) The design, construction and operation of the LIMPET wave energy converter (Islay, Scotland) [land installed marine powered energy transformer]. In: 4th European wave energy conference. Aalborg, Denmark
125. Falcão AFO (2000) The shoreline OWC wave power plant at the Azores. In: 4th European wave energy conference. Aalborg, Denmark
126. Belyakov BN (2020) Sustainable power generation. Academic Press, Cambridge, United States, pp 461–474
127. Pishvaee MS, Mohseni S, Bairamzadeh S (2021) An overview of biomass feedstocks for biofuel production. In: Biomass to biofuel supply chain design and planning under uncertainty. Academic Press, Cambridge, United States, pp 1–20
128. Sharmina M, McGlade C, Gilbert P, Larkin A (2017) Global energy scenarios and their implications for future shipped trade. Mar Pol 84:12–21
129. Sun Y, Gao B, Yao Y, Fang J, Zhang M, Zhou Y, Chen H, Yang L (2014) Effects of feedstock type, production method, and pyrolysis temperature on biochar and hydrochar properties. Chem Eng J 240:574–578
130. Tripathi M, Sahu J, Ganesan P (2016) Effect of process parameters on production of biochar from biomass waste through pyrolysis: a review. Renew Sustain Energy Rev 55:467–481
131. Kumar A, Bhattacharya T, Hasnain SMM, Nayak AK (2020) Applications of biomass derived materials for energy production, conversion and storage. Mater Sci Energy Technol 3:905–920
132. Lago NCEC, Lechon Y (2019) Key challenges and opportunities. In: The role of bioenergy in the emerging bioeconomy. Academic Press, Cambridge, United States, pp 297–378
133. Wolf DJ (2013) Biomass combustion for power generation: an introduction. In: Biomass combustion science, technology and engineering. Woodhead Publishing Limited, Sawton, United Kingdom, pp 3–8
134. Easterly JL, Burnham M (1996) Overview of biomass and waste fuel resources for power production. Biomass Bioenergy 10:79–92
135. Chupka M, Howarth D, Zoi C (1992) Renewable electric generation: an assessment of air pollution prevention potential: final report. Environmental Protection Agency, Washington, DC
136. Atadashi IM, Aroua MK, Aziz AA, Sulaiman NM (2012) Production of biodiesel using high free fatty acid feedstocks. Renew Sustain Energy Rev 16:3275–3285
137. Hassan MH, Kalam MA (2013) An overview of biofuel as a renewable energy source: development and challenges. Procedia Eng 56:39–53
138. Pizarro AVL, Park EY (2003) Lipase-catalyzed production of biodiesel fuel from vegetable oils contained in waste activated bleaching earth. Process Biochem 38:1077–1082
139. Silitonga AS, Atabani AE, Mahlia TMI, Masjuki HH, Badruddin IA, Mekhilef S (2011) A review on prospect of *Jatropha curcas* for biodiesel in Indonesia. Renew Sustain Energy Rev 15:3733–3756
140. Huang D, Zhou H, Lin L (2012) Biodiesel: an alternative to conventional fuel. Energy Procedia 16:1874–1885
141. Appels L, Lauwers J, Degrève J, Helsen L, Lievens B, Willems K, Van Impe J, Dewil R (2011) Anaerobic digestion in global bio-energy production: potential and research challenges. Renew Sustain Energy Rev 15:4295–4301
142. Lansing S, Botero RB, Martin JF (2008) Waste treatment and biogas quality in small scale agricultural digesters. Bioresour Technol 99:5881–5890
143. Yu Q, Liu R, Li K, Ma R (2019) A review of crop straw pretreatment methods for biogas production by anaerobic digestion in China. Renew Sustain Energy Rev 107:51–58
144. Ciotola RJ, Lansing S, Martin JF (2011) Emergy analysis of biogas production and electricity generation from small-scale. Ecol Eng 37:1681–1691
145. Turcotte DL, Schubert G (2002) Geodynamics, 2nd edn. Cambridge University Press, Cambridge
146. Thamer M (2021) Geothermal energy. In: Distributed renewable energies for off-grid communities, 2nd edn. Elsevier, Amsterdam, Netherlands, pp 247–261

147. Barbier E (2002) Geothermal energy technology and current status: an overview. Renew Sustain Energy Rev 6:3–65
148. Li K, Bian H, Liu C, Zhang D, Yang Y (2015) Comparison of geothermal with solar and wind power generation systems. Renew Sustain Energy Rev 42:1464–1474
149. Badescu V (2007) Economic aspects of using ground thermal energy for passive house heating. Renew Energy 32:895–903
150. Lund JW, Toth AN (2021) Direct utilization of geothermal energy 2020 worldwide review. Geothermics 90:101915
151. El Haj Assad M, Bani-Hani E, Khalil MJ (2017) Performance of geothermal power plants (single, dual, and binary) to compensate for LHC-CERN power consumption: comparative study. Geotherm Energy 5:17
152. IEA (2009) World energy balances. Paris
153. Salameh Z (2014) Emerging renewable energy sources. In: Renewable energy system design. Academic Press, Massachusetts, United States, pp 299–371
154. DiPippo R (2020) Geothermal energy as a source of electricity: a worldwide survey of the design and operation of geothermal power plants. U.S. Department of Energy, Washington, DC, p 19880
155. Salem M, Fahim Alavi M, Mahariq I, Accouche O, El Haj Assad M (2021) Applications of thermal energy storage in solar organic rankine cycles: a comprehensive review. Front Energy Res 9:638
156. Valdimarsson P (2011) Geothermal power plant ctcles and main components. In: Short course on geothermal drilling,resource development and power plants. Santa Tecla, El Salvador
157. Ultanir MO (1997) Hidrojenin yakit olarak kullanimi ve özellikleri. In: Evre-Enerji Kongresi C. TMMOB Makina Mühendisleri Odasi
158. Koohi-Fayegh S, Rosen MA (2020) A review of energy storage types, applications and recent developments. J. Energy Storage 27:101047
159. Rosen MA, Scott DS (1998) Comparative efficiency assessments for a range of hydrogen production processes. Int J Hydrogen Energy 23:653–659
160. Marchetti C (2006) Long-term global vision of nuclear-produced hydrogen. Int J Nucl Hydrogen Prod Appl 1:13–19
161. Dincer I, Rosen MA (2009) Exergy as a driver for achieving sustainability. Int J Green Energy 1:1–19
162. Zulkiflee MSNM, Razak ASA, Ahmad A, Zakaria Z, Wahid ZA, Zakaria (2011) Hydrogen as the future sustainable energy—a review. Int J Civ Eng Geo-Environ 2:47–57
163. Midilli A, Rzayev P, Olgun H, Ayhan T (2000) Solar hydrogen production from hazelnut shells. Int J Hydrogen Energy 25:723–732
164. González A, McKeogh E, Gallachoir BO (2004) The role of hydrogen in high wind energy penetration electricity systems: the Irish case. Renew Energy 29:471–489
165. Hamad KB, Salem M, Luta DN (2020) Harmonic mitigation for a megawatt grid-tied fuel. In: 2020 IEEE international conference on power and energy (PECon). Penang
166. Hamad KA, Salem M (2020) Harmonics mitigation in a large-scale grid connected fuel cell stack. In: Africa and international energy congress, p 2020
167. Cheng X, Shi Z, Glass N, Zhang L, Zhang J, Song D, Liu Z-S, Wang H, Shen J (2007) A review of PEM hydrogen fuel cell contamination: impacts, mechanisms, and mitigation. J Power Sources 165:739–756
168. Gamburzev S, Appleby AJ (2002) Recent progress in performance improvement of the proton exchange membrane fuel cell (PEMFC). J Power Sources 107:5–12
169. Basile A, Iulianelli A (2014) Advances in hydrogen production, storage and distribution. Woodhead Publishing, Cambridge
170. Ang T-Z, Salem M, Kamarol M, Das HS, Nazari MA, Prabaharan N (2022) A comprehensive study of renewable energy sources: classifications, challenges and suggestions. Energy Strat Rev 43:100939
171. Juste GL (2006) Hydrogen injection as additional fuel in gas turbine combustor. Evaluation of effects. Int J Hydrogen Energy 31:2112–2121

172. Qin L et al (2012) Effect of recooling cycle on performance of hydrogen fueled scramjet. Int J Hydrogen Energy 37:18528–18536
173. Ditaranto M, Heggset T, Berstad D (2020) Concept of hydrogen fired gas turbine cycle with exhaust gas recirculation: assessment of process performance. Energy 192:116646
174. Wu W, Zang S, Zhong C (2015) Study of the hydrogen-steam turbine composite cycle. Procedia CIRP 26:735–739
175. Verbruggen A, Fischedick M, Moomaw W, Weir T, Nadaï A, Nilsson LJ, Nyboer J, Sathaye J (2010) Renewable energy costs, potentials, barriers: conceptual issues. Energy Policy 38:850–861
176. Gielen D, Boshell F, Saygin G, Bazilian MD, Wagner N, Gorini R (2019) The role of renewable energy in the global energy transformation. Energy Strategy Rev 24:38–50
177. Bayer P, Dolan L, Urpelainen J (2013) Global patterns of renewable energy innovation, 1990–2009. Energy Sustain Dev 17:288–295
178. Kochtcheeva LV (2016) Renewable energy: global challenges. E-International Relations
179. Müller S, Brown A, Ölz S (2011) Renewable energy: policy considerations for deploying renewables. International Energy Agency, Paris
180. Huenteler J, Niebuhr C, Schmidt TS (2016) The effect of local and global learning on the cost of renewable energy in developing countries. J Clean Prod 128:6–21
181. Lema A, Lema R (2013) Technology transfer in the clean development mechanism: insights from wind power. Global Environ Change 23:301–313
182. Djiby RT (2011) An energy pricing scheme for the diffusion of decentralized renewable technology investment in developing countries. Energy Policy 39:4284–4297
183. Waissbein O, Glemarec Y, Bayraktar H, Schmidt TS (2013) Derisking renewable energy investment. United Nations Developmenet Programme, New York
184. Pfeiffer B, Mulder P (2013) Explaining the diffusion of renewable energy technology in developing countries. Energy Econ 40:285–296
185. Owusu PA, Asumadu-Sarkodie S (2016) A review of renewable energy sources, sustainability issues and climate change mitigation. Cogent Eng 3:1667990
186. MacLeod JM, Rosei F (2015) Supporting the development and deployment of sustainable energy technologies through targeted scientific training. In: Sustainable access to energy in the global south. Springer International Publishing, London, pp 231–233
187. Mohuiddin S (2016) Expanding the role of microfinance in promoting renewable energy access in developing countries. Georgetown Public Policy Rev 2:119–124

Solar Renewable Energy Systems: Study and Applications with Present and Future Potential

Ranjan Dey, Eshita R. Kakodkar, Praseeda P. Nair, and Suvanjan Bhattacharyya

Abstract Climate change and sustainability mandates an urgent global transition from fossil fuels to solar energy. This chapter provides a comprehensive overview of the solar energy systems emphasizing on its current state, technological advances and diverse applications. Solar power stands out because it is abundant, scalable and affordable. We then discuss the evolution of photovoltaic materials across three generations from traditional silicon to innovative perovskite and quantum dot technologies. Several exciting developments like tandem solar cells, flexible printed panels and hybrid systems using nanomaterials have been highlighted. The discussion then moves on to solar still technology and on how integration of phase change materials and magnetic nanoparticles can enhance the efficiency. In the current world artificial intelligence and machine learning have become an integral aspect of research, technologies and everyday life. AI/ML play an important role in solar energy systems by improving maintenance predictions, boosting efficiency and discovery of new materials. In conclusion, the key to unlock a greener, more decentralized and resilient energy systems lies in the synergy of solar energy, advanced materials and artificial intelligence.

Keywords Solar energy · Sustainable · PCM-MNP · AI-ML · Desalination

1 Introduction

The era of fossil fuels, which has reigned the energy production worldwide for over a century, is heading towards a significant transformation. Due to the detrimental environmental effects, such as greenhouse gas emissions which results in climate

R. Dey (✉) · E. R. Kakodkar · P. P. Nair
Thermophysical Properties Lab, Department of Chemistry, Birla Institute of Technology and Science, Sancoale, Goa, India
e-mail: ranjandey@goa.bits-pilani.ac.in

S. Bhattacharyya
Department of Mechanical Engineering, Birla Institute of Technology and Science, Pilani Campus, Pilani, Rajasthan, India

J. A. Siddique et al. (eds.), *Energy Sources*, Green Energy and Technology,
https://doi.org/10.1007/978-981-95-1209-6_8

change and the exhaustible nature of fossil fuels, there arises a global urge to shift towards a cleaner, sustainable and renewable source of energy. The various source of renewable energy like wind, geothermal, biomass, hydro and solar have emerged as the most sought-after viable alternatives to overcome the limitations associated with the fossil fuels. Out of all these, the solar energy surpasses the others due to its scalability, abundance and the rapidly advancing research and technology in this sector. Solar energy is undoubtedly the cleanest and 'greenest' amongst all the Renewable energy options available globally. It is estimated that the energy from the Sun available in a year is more than the cumulative sum of the energy generated from the fossil fuel reserves in India. India as a country is blessed naturally with abundance of Sunlight with a major part of the country having good amount of solar insolation throughout the year barring a month or two of the monsoons. Solar power utilizes the energy from the sun in the form of radiations, that are absorbed by the solar power systems, thus converting the sunlight into a usable form of energy without the emission of any greenhouse gases during the conversion process. This underscored the importance of solar energy as a key contributor in the transition towards a sustainable and greener energy future.

The recent advances in solar technology have further solidified its importance in the energy transition landscape. The development of hybrid perovskite tandem cells has improved the performance of the solar panels for energy production. For instance, Hanwha Qcells developed a large silicon-perovskite tandem solar cell which increased the efficiency by 28.6% indicating their potential to gain an increased energy output from panels with smaller areas [1]. The addition of artificial intelligence to the solar-powered systems provides real time energy management and predictive maintenance which further optimizes their performance. An innovative tool to improve the land utility is the development of floatovoltaics or floating solar farms. In these, the solar devices are installed in the water bodies thus cooling the devices to enhance its efficiency while simultaneously being an effective approach to overcome land scarcity. These innovations along with reduced cost and supportive government policies, have boosted solar energy to surpass the 2000 TWh mark in energy generation thus achieving a global energy transition milestone of contributing approximately 7% of global electricity generation and surpassing in 2024.

The rapid growth of the adoption of solar energy is not only due to its technological feasibility but also due to its economic resilience. The solar-powered devices have globally emerged as one of the cheapest electricity sources with the Levelized Cost of Electricity (LCOE) for utility-scale solar PV dropping to about 90%since 2010 [2]. In contrast to other sources like wind or hydro which are restricted by their location requirements and broad scale infrastructure, the solar PV offer flexibility and modularity since they can be employed in a variety of environments ranging from rural rooftops to urban solar farms [3]. Furthermore, solar energy is in line with decentralized energy strategies, thus making it a reliable force for providing energy access in remote areas of sub-Saharan African and Southeast Asian regions which lack robust grid infrastructure [4]. Being the world's most energy deficient continent, Africa accommodates 75% of the global population who are energy deprived with 600 million people devoid of energy access and around 98% of them are located in the sub-Saharan African region. Innovative solutions based on solar power such as mini grids, solar house systems, etc., can play a pivotal role in transforming the situation in these regions [5]. The environmental impact of solar technology is gradually improving the global situation with the use of recycling techniques and use of efficient and sustainable materials, thus highlighting the importance of solar energy as an eco-friendly source of renewable energy (Fig. 1).

Fuelled by the aim to reduce the greenhouse gas emissions and achieve net-zero carbon emission targets, the sector of solar energy is geared up to undergo o significant transformation in the near future. The International Energy Agency (IEA) has predicted that by 2050 the solar power will contribute to around 33% of global energy generation thus playing a leading role to move towards a sustainable and greener development scenario. Many government and private agencies in countries like China, the U.S., and India have increasing invested in solar-powered setups to carry out large scale installations and energy production [6].

Presently, India stands next to China and United States as the third largest solar energy globally [7]. The Government has set a target of achieving 500 GW of installed capacity from non-fossil fuel sources by 2030, in line with the Prime Minister's announcement at COP-26, held in Glasgow [8]. India's solar power installed capacity

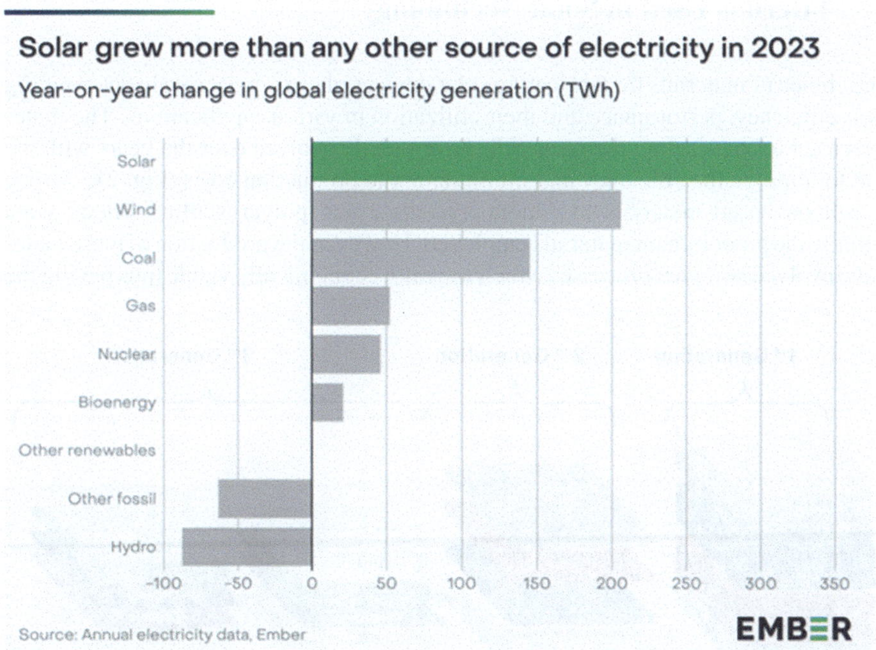

Fig. 1 Bar chart showing the year-on-year change in global electricity generation in terawatt-hours (TWh) for 2023. Solar energy experienced the highest growth, followed by wind and coal. Other sources like gas, nuclear, and bioenergy showed smaller increases. Hydro and other fossil sources decreased. The chart highlights solar as the leading growth source

was 100.33 GW as of 31st January 2025 [9]. This comprises of Ground Mounted Solar Plant: 76.95 GW; Grid Connected Solar Rooftop: 16.28 GW; Hybrid Projects (Solar Component): 2.85 GW and Off-Grid Solar: 4.25 GW. These are crucial targets as they are the key to India's commitment for net zero emission by 2070.

As per the International Energy Forum (IEF), Solar energy remains the third largest renewable electricity technology behind hydropower and wind—but it accounted for just 4.5% of total global electricity generation in 2022 [10]. With the advent of data-driven technology and modelling playing a key role in all fields of technology and industrial applications, datasets provided by Global Energy Monitor (GEM) through the Global Solar Power Tracker, a worldwide dataset of utility-scale solar photovoltaic (PV) and solar thermal facilities [11] are gaining significance.

Recent innovations in solar technology include the integration of photovoltaics in building, development of quantum dot solar cells that offers high efficiency through the use of advanced nanomaterials [12, 13]. In addition to these, solar-powered systems are also being integrated with existing energy storage devices like lithium ion and solid-state batteries to overcome irregular energy production and enabling unhindered power supply thus providing a viable solution for energy storage [14, 15].

2 Materials Used in Solar Technology

The choice of materials for fabricating solar devices play a crucial role in determining their efficiency, performance and their utilization in various applications. The materials used for fabricating the solar cells have greatly evolved over the years with the aim to improve the efficiency and minimizing the production costs (Fig. 2).

Silicon based materials like monocrystalline and polycrystalline silicon were initially the predominant materials employed. However, the production of these materials involved complex processes which were not economically viable thus paving the

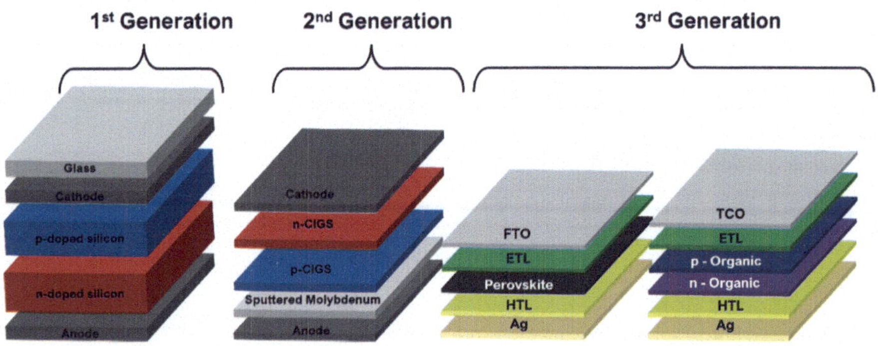

Fig. 2 Diagram illustrating three generations of solar cell structures. The first generation includes layers labelled as glass, cathode, p-doped silicon, -doped silicon, and anode. The second-generation features layers labelled as cathode, -CdS, p-CIGS, sputtered molybdenum, and anode. The third generation shows two structures: one with layers labelled FTO, ETL, perovskite, HTL, and Ag; the other with TCO, ETL, p-organic, -organic, HTL, and Ag. Each generation is visually separated and labelled accordingly

way for the emergence of second-generation alternative based on thin film technologies. These materials included amorphous silicon (a-Si), cadmium telluride (CdTe), and copper indium gallium selenide (CIGS), etc. which gave rise to a new class of lightweight and flexible solar panels, thus making them a viable option for utilization in various applications [16]. The drawbacks of these materials could be associated to their low efficiency, toxicity and scarcity which thus impeded their wide scale use. These drawbacks gave rise to a third-generation of solar technology based on perovskites which exhibited high efficiencies and could be fabricated using low-cost processes. However, the stability and toxicity of these materials pose a challenge for their utilization in large scale processes. Research in solar technology is constantly evolving and the recent focus have shifted to the development of organic photovoltaics and quantum dot solar cells which can be utilized to create lightweight, flexible and semi-transparent devices which could be integrated with different surfaces and structures, thus extending their use in a wide range of applications [17]. Recently, the efforts are now diverted towards the development of hybrid and tandem perovskite--silicon solar cells that combine the properties of both the components and yield higher efficiencies in comparison to the individual constituents [18].

Material	Key characteristics	Applications	References
Perovskite--silicon tandem cells	Perovskite's light absorption is combined with silicon's stability ensuring high stability	Solar installations for residential and commercial purposes	Hanwha Qcells [1]
Graphene-based conductive ink	Replaces rare materials like silver and exhibits high conductivity	Cost-effective solar panels; potential use in batteries	CSIRO [19]
Ultra-thin perovskite solar cells	Lightweight and flexible, suitable for diverse surfaces	Building-integrated photovoltaics; portable electronics	Urry [20]
Printed solar cells	Lightweight and flexible, fabricated via printing methods	Used in disaster relief, mining, construction and agriculture	CSIRO [21]
$MoSi_2N_4/WSi_2N_4$ heterostructures	Have strong optical absorption and exceptional excitonic solar cell conversion efficiency	Used in photonics devices and high-performance ultrathin excitonic solar cells	Chen et al. [22]
Solution-processed 2D materials	Exhibit tunable optoelectronic properties and scalable production	Organic solar cells; dye-sensitized solar cells; perovskite solar cells	Bellani et al. [23]
Graphene-enhanced thermal interface materials	Possess high thermal conductivity which in turn improves heat dissipation	Thermal management in photovoltaic solar cells; enhancing efficiency	Saadah et al. [24]

(continued)

(continued)

Material	Key characteristics	Applications	References
Cadmium telluride (CdTe)	Has a high absorption coefficient and low-cost production	Thin-film solar panels; utility-scale solar installations	Machín and Márquez [17]
Copper indium gallium selenide (CIGS)	High efficiency, flexible and lightweight	Portable electronics; building-integrated photovoltaics; flexible solar panels	Machín and Márquez [17]
Copper zinc tin sulphide (CZTS)	Earth-abundant, non-toxic, with tunable bandgap	Sustainable thin-film solar cells; alternative to CdTe and CIGS	Dwivedi et al. [25]
Transition metal dichalcogenides (TMDs)	Ultrahigh light absorption, suitable for ultrathin applications	Wearable devices; sensors; lightweight aerospace applications	Dyatkin [26]
Graphene-based Schottky junctions	High electrical conductivity and optical transparency	Inexpensive solar cells; potential for high conversion efficiency	Pu et al. [27]
Perovskite quantum dots	High photoluminescence with tunable bandgap	High-efficiency solar cells; potential for power conversion efficiency	Liu et al. [28]
Solar paint (Photovoltaic paint)	Converts solar energy into electricity and can be applied like paint	Charging electric vehicles; building facades; car parks	Urry [20]
Quantum dot solar cells	Tunable absorption spectrum and potential for high efficiency	Next-generation solar panels; flexible electronics	Huang et al. [12]
Organic photovoltaics (OPVs)	Flexible, lightweight and low-cost production	Wearable electronics; portable chargers; building-integrated photovoltaics	Machín and Márquez [17]
$BaCd_2P_2$ (Zintl-phosphide)	Long carrier lifetime, stable and high performance	Advanced photovoltaic applications	Yuan et al. [29]
Dye-sensitized solar cells (DSSCs)	Low-cost, flexible, semi-transparent and utilizes dye molecules for light absorption	Building-integrated photovoltaics; portable electronics; low-light environments	Gong et al. [30]
Flexible perovskite solar cells	High efficiency, lightweight, bendable and solution-processable	Wearable electronics; portable chargers; flexible displays	Machín and Márquez [17]

3 Advances in Solar Technology

Latest developments in solar technology have greatly improved their performance, adaptability and their utilization in various applications. In context to solar stills, the addition of phase change materials (PCM) like paraffin wax and nano enhances PCMs have greatly enhanced the thermal energy storage resulting in higher yield of freshwater. A study by Rashid et al. [31] proved that the integration of PCMs into the solar stills led to an increase in productivity when compared to the conventional still design. Furthermore, the addition of metal oxide-based nanofluids such as Al_2O_3 and CuO increased the overall efficiency by improving the thermal conductivity and evaporation rates [32]. Another approach by Nazari et al. [33] to increase the still performance employed the integration of thermoelectric module which converted the waste heat generated to electricity.

In the field of photovoltaics, the perovskite--silicon tandem cells have emerged as highly efficient fabrication materials which surpass the convention silicon-based cells with an efficiency of 28.6% [1]. Another significant progress in this field is the development of flexible, printed solar cells which have attained efficiency of 15.5% are lightweight, bendable and can be a viable solution for different applications due to their characteristics which enhances their integration into multifarious surfaces and structures [21]. The fabrication of Semi-transparent perovskite solar cells (ST-PSCs) have proved to be suitable alternatives for building-integrated photovoltaics (BIPV) due to their high conversion efficiency and moderate visible transmittance [34].

The incorporation of artificial intelligence (AI) and machine learning (ML) into solar-powered systems has helped in improving the output and predictive maintenance. In recent times, AI algorithms have been used to increase the energy performance of the system, facilitate predictive maintenance and enhance the efficiency in solar farms [35]. AI based systems have been designed to optimize the conditions of the systems to achieve accurate forecasting of solar power, enabling the maximum utilization of energy and to enhance the reliability of the solar panel through predictive maintenance [36].

Solar energy has wide applications in various sectors highlighting its reliability and prominence towards attaining sustainable and greener energy goals. Solar stills embedded with PCMs and nanofluids have been extensively used due to their enhanced performance for the purpose of desalination and wastewater treatment, thus presenting environmentally sustainable strategies in regions facing issues of water-scarcity [32]. Advances in the production of semi-transparent and flexible PV modules has greatly improved the use of these materials in building-integrated photovoltaics (BIPV), since they can be easily incorporated in building elements to offer efficient architectural designs [34]. The development of lightweight, printed solar cells are efficient in providing power solutions which are portable and can be highly beneficial in emergency situations and remote locations [21]. Solar energy has also been utilized for hydrogen production by utilizing it to power the photoelectrochemical processes, thus providing a sustainable and greener alternative to fossil fuels.

Finally, solar power also finds its way in agriculture wherein solar energy is now being used to power many agricultural systems such as embedding the semi-transparent PV modules on rooftops of greenhouses, thus contributing towards sustainable farming goals [37].

3.1 Integration of Latent Heat Energy Storage with Solar Energy

The intermittent nature of sunlight poses a huge challenge to solar energy systems, Latent heat thermal energy storage (LHTES) using phase change materials (PCMs) emerges as a solution to this challenge [38]. Excess thermal energy is stored during peak sunlight and released during low-insolation periods thereby enhancing the reliability and efficiency of solar technologies [39]. PCMs absorb and release large amounts of energy during phase transitions. Sensible heat storage relies on temperature changes in materials like water or rocks wherein latent heat storage offers higher energy density [40]. PCMs can be categorized into organic, inorganic and eutectic mixtures. Organic PCMs are predominantly used for solar applications due to their low corrosion risk while inorganic PCMs offer higher thermal conductivity. Paraffin wax is a widely used organic PCM, its isothermal nature makes it ideal for applications such as solar desalination or cooling of buildings [41]. Recent advancements in PCMs focus on hybrid composites, where PCMs are combined with nanomaterials to mitigate low thermal conductivity. Paraffin when infused with copper oxide nanoparticles enhance thermal conductivity which significantly improves the charge and discharge rates [42]. Integrating PCMs into solar thermal systems requires strategic engineering to maximize heat transfer and storage efficiency. Encapsulation helps to prevent PCM leakage and increase the surface area for heat exchange [43].

Solar stills, which utilize solar energy to evaporate and condense freshwater from saline water, face inherent limitations due to intermittent sunlight. Integrating PCMs addresses these challenges by stabilizing basin temperatures and extending distillation cycles into non-sunlit hours. During daytime, PCMs absorb excess heat, delaying water temperature drops and maintaining evaporation rates. At night, the stored heat is released, prolonging freshwater production (Fig. 3).

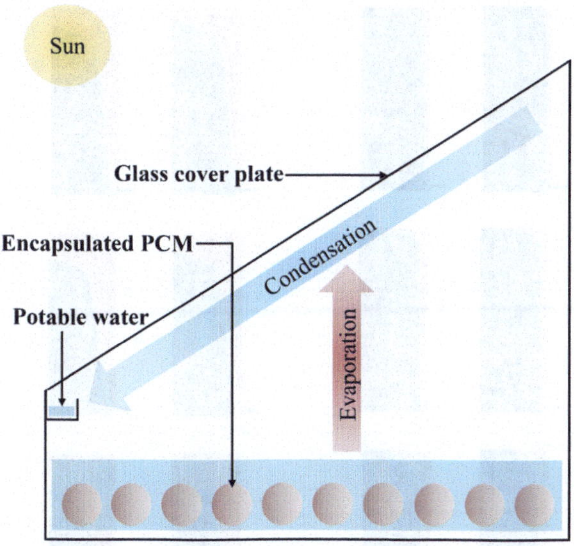

Fig. 3 Schematics of solar still with encapsulated PCM

3.2 CFD Validation of PCM-MNP Systems in Solar Desalination

Computational fluid dynamics (CFD) studies have provided critical insights into the thermal performance of RT-35 paraffin enhanced with Fe_3O_4 magnetic nanoparticles (MNPs) for latent heat storage applications. These simulations, validated against experimental prototypes of solar stills, reveal how nanoparticle integration optimizes phase change dynamics and energy storage efficiency (Fig. 4).

The inclusion of Fe_3O_4 MNPs significantly accelerates the melting process of RT-35 paraffin. At 100 s, the MNP-PCM composite exhibited a 13.32% higher liquid fraction (0.064) compared to pure PCM (0.056), with this improvement peaking at 16.59% by 750 s. This acceleration is attributed to the nanoparticles' role in disrupting thermal boundary layers and promoting convective heat transfer. By 5000 s, the MNP-PCM achieved near-complete melting (97% liquid fraction), outperforming pure PCM (91%), which underscores its superior thermal storage capacity. CFD temperature contours demonstrated that nanoparticles reduce interfacial thermal resistance, enabling faster heat diffusion across the PCM domain (Fig. 5).

Fig. 4 Liquid fraction contours at different time points from 100 to 5000 s

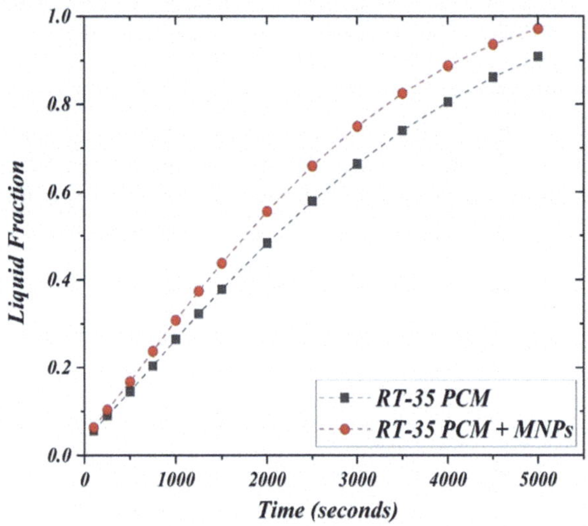

Fig. 5 Comparison of liquid fraction of RT-35 and RT-35 + MNP at different time intervals

3.3 Integration with AI-ML

The integration of AI-ML in solar-powered systems play a key role in the advancement of solar technology. The algorithms optimize the solar-powered systems by reducing the material discovery time, forecasting the system's energy output, detecting faults, etc. which in turn improves the performance of the system [44]. The most common AI-ML models include linear regression models, Support vector machines, Artificial neural networks, Decision trees, Long short-term memory networks, etc. Linear Regression models utilize the linear relationship between solar irradiance and power output and is a simple linear model but it faces challenges with non-linear patterns. This limitation is overcome by support vector machines which are suitable for moderate forecasting and uses flexible hyperplanes for non-linear modelling. A slightly more advanced algorithm is the Artificial Neural Networks which replicates the human brain for highly complex, non-linear forecasting. Long Short-Term Memory networks are effective for forecasting the solar output over time as they excel at capturing time-series data dependencies [45].

A new approach by Shetty et al. [46] used natural language processing to extract over 861 polymer solar cell systems from around 3300 research papers and included their electron-donor pairs and performance metrics. This data was then used to train machine learning models to identify 10 new potential photovoltaic materials with a predicted power conversion efficiency of above 10%. This approach greatly reduced the discovery time by 75% which is equivalent to an acceleration of 15 years. Ledmaoui et al. [47] comparatively studied the predictive accuracy of six different machine learning models namely-, Random Forest (RF), Artificial Neural Networks (ANN), Decision Tree (DT), XGBoost, Generalized Additive Models (GAM), and Support Vector Regression (SVR) using the previously collected daily production, irradiance, and temperature data from a 24 kW solar power plant in Morocco. The predictive accuracy of the models for forecasting the solar output was determined on the basis of their calculated Mean Absolute Error (MAE), Mean Absolute Scaled Error (MASE), Root Mean Square Error (RMSE) and the coefficient of determination (R^2). It was found out that the artificial neural network outperformed others achieving a near perfect prediction accuracy with RMSE $= 2.6 \times 10^{-8}$, MAE $= 0.00013$, MASE $= 0.29$, and $R^2 = 0.99$ followed by Random Forest and XGBoost, thus highlighting the potential of these models for predictive maintenance and solar energy utilization to support sustainable energy management strategies. Machine learning models like Artificial Neural Networks were used to detect faults in a solar based water pumping system. The model allowed fault recognition with an accuracy of 95 and 92% accuracy for detection of fault type which greatly reduces system downtime and maintenance cost [48]. At a greater magnitude, the NEOM city project utilized AI to optimize energy systems by integrating solar, wind, and battery storage. AI algorithms predicted energy supply and demand with over 85% accuracy thus enhancing system efficiency by 15% and reducing waste by 10% [49].

4 Applications

4.1 Solar Energy in Water Treatment

Water is the most precious life sustaining natural resource. However, approximately 97% of the water on Earth is attributed to seawater with only 3% available as freshwater. With the current global scenario, water scarcity continues to be a threat with the growing global population. Currently, attempts are being made to desalinise the seawater and the integration of solar energy to power these desalination and distillation systems has proved to be an efficient route to carry out such energy intensive processes.

A stepped solar still equipped with a corrugated copper absorber plate was designed by Elsheikh et al. [50] for freshwater production through solar-powered distillation process. The absorption of the solar radiation was greatly enhanced due to the increased surface area and improved thermal conductivity of the absorber plate which aids the distillation process. The corrugated absorber plate also improved the evaporation and condensation rates within the solar still by improving the heat transfer. The solar still was tested under real climatic conditions and a pronounced improvement in the water yield was observed. The stepped design of the solar still further contributed to this improved yield by providing an increased surface area for evaporation. Thus, the solar still was able to produce a maximum daily output of 5.87 L/m^2 which was significantly higher than the conventional flat plate solar stills. The water production of the stepped solar still and the conventional solar still was predicted by employing a long short term memory neural network, which is a deep learning model effective for time-series predictions. The various climatic variables such as solar radiation, ambient temperature, wind speed, humidity, etc. were used as the input data to train the model and it achieved a highly accurate correlation of 0.99 and 0.97 for the stepped solar still and the conventional one respectively. Thus, the stepped solar still can efficiently be used for sustainable water production in remote areas with abundant solar radiations while the integration of AI can greatly help in forecasting the yield.

In a study by Parsa et al. [51], different configurations of identical solar stills were tested for increasing the efficiency and yield, wherein, the solar still was equipped with thermoelectric heating, cooling and nanofluid. The modification was carried out such that the first system comprised of thermoelectric heating (TEH), while the second system was loaded with silver nanofluid in addition to TEH and the third system consisted of TEH and nanofluid and in addition also incorporated a double slope symmetric condenser that one side has thermoelectric cooling while the other side is cooled by a water film. The silver nanoparticles were used due to their remarkable properties which facilitates the water disinfection. The solar still equipped with the external condenser and nanofluid achieved the highest productivity of 7760 cc/m^2/day, which had a productivity increase of 47.6% as compared to the SS without condenser/nanofluid and about 16.5% with the one loaded with nanofluid without condenser. The external condenser improved the productivity by 26.3% with the thermoelectric side showing 5 times better productivity than the side having the water film. The efficiency and daily yield of SS with only nanofluid was 30.6% and 50.8% respectively in contrast to the conventional SS without nanofluid and condenser. The efficiency and daily yield of SS with condenser/nanofluid was 26.7% and 100.5% respectively in contrast to the conventional SS without nanofluid and condenser.

In a similar study by Shalaby et al. [52], a Photovoltaic powered reverse osmosis system was introduced with cooling systems to improve its efficiency in the desalination process. The cooling system helped in cooling the photovoltaic while simultaneously heating the feed water prior to the RO treatment. Two different cooling system designs were tested for the system wherein one design utilized PVC tubes having half-circular cross section to allow direct contact of the cooling fluid with the back surface of PV while the second design employed copper tubes with square cross section attached to the PV back surface. The first design showed better performance which could be attributed to the effective heat transfer due to the direct contact cooling of the nanofluid thus showing the importance of heat transfer area on the performance of the PV-RO system. H_2O/Al_2O_3 and H_2O/CuO were used as the cooling nanofluid using distilled water as the reference and the electric efficiency observed with the first design along with the fluids was found to be 34.8, 39.5 and 27.3% respectively in comparison to the uncooled PV. The concentration variation of the nanofluid in the cooling system was carried out to determine the optimum concentration for desalination of brackish water. The findings revealed that when brackish water having a salinity of 3000 ppm was used, the PV-RO system managed to yield 366 L/day of fresh water.

Another study by Prasanna and Deshmukh [53] explored and assessed an advanced modification to enhance the performance of the traditional solar stills by integrating the use of an air cavity below the absorber plate which greatly improved the heat transfer thus yielding higher water temperatures and boosting the distillation output. The influence of two different basin materials—aluminium and polycarbonate were evaluated under varying water depths to assess the energy and exergy efficiency as well as the economic viability of the solar still. Temperatures as high as 89 °C was observed which indicated a good thermal absorption. The depth variation analysis proved that shallow water depths allowed for quicker heating and higher distillate yield thus having higher efficiencies. The results demonstrated that the aluminium basin exhibited better performance than the polycarbonate basin. The aluminium basin solar still was able to achieve energy and exergy efficiency of 48.8% and 3.44% respectively, whereas for the polycarbonate basin still, they were observed to be 42.4% and 2.3% respectively. The aluminium based still was able to produce about 2.8–3.8 L of distilled water per square metre per day in comparison to the 1.8–3.4 L of the polycarbonate based solar still. The economic analysis revealed that the cost of water per litre was Rs. 0.723 for aluminium and Rs. 1.361 for polycarbonate based solar still with a capital recover time of 10 and 11 months respectively.

4.2 Solar Energy in Agricultural Sector

Agrivoltaics is an emerging field that integrates the solar energy with the existing agricultural setups to simultaneously produce electricity as well as food. In the agricultural sector, solar energy is often used to power energy intensive greenhouse systems, for solar irrigation, etc. the use of these devices is often found to affect the growth of the plants, however with the correct balance of the changes in the microclimate, these PV devices have shown positive or no impact on the plants and have even improved the yield and food quality.

A study by Carreño-Ortega et al. [54] integrated photovoltaic with agriculture to develop agrivoltaics system that produced both food and renewable energy simultaneously. Mini photovoltaic modules were installed in various patterns to evaluate their performance in energy production and effect in the growth of the plants. The PV mini modules were designed and installed in three different patterns i.e. checkerboard, linear and diamond. The patterning of these solar modules played a vital role in the development and growth of the lettuce plant by creating a better microclimate. The PV modules produced a shading effect which in turn moderated the temperature and scaled down the light intensity received by the plant, thus protecting it from heat stress. Out of all the PV module configurations the checkerboard pattern produced lettuce yields that were equal or greater than the controls that was placed in full sunlight and the ones having concentrated shade due to traditional PV panels. The scattered shading effect of the patterned mini-PV modules enhanced the fresh weight, number of leaves, length, etc. of lettuce in spring and summer season. These results open up a new prospect of using to solar energy for dual purposes in agriculture and energy production while efficiently utilizing land which could greatly improve productivity in areas of land scarcity and high temperatures, thus paving the way for sustainable agricultural systems.

A similar result was observed by Kavga et al. [55] wherein the pepper crops were cultivated in a solar-powered greenhouse which was self-sufficient in generating the electricity required to run the greenhouse. The study integrated the traditional greenhouse design with semi-transparent solar panels made of polycrystalline silicone and covering about 20% of the greenhouse rooftop which not only produced electricity but also allowed partial light penetration for the plant growth. The performance of the solar-powered greenhouse was compared with that of a traditional greenhouse without any PV panels. The findings revealed that although a slight reduction was observed in the fruit size due to the filtered light, however the overall produce of the peppers remained economically viable with no significant effect on the quality of the pepper fruits. The energy performance of the solar-powered greenhouse revealed that the contribution of the solar energy greatly reduce the demand for additional external energy, thus achieving almost net-zero energy operation, which could be very efficient and economical in places with high sunlight. The life cycle analysis of the system showed that the when used for longer timespans the economic and environmental output greatly outweighed the initial installation expenditure.

A similar study was carried out by Tang et al. [56] which incorporated opaque photovoltaic modules and semi-transparent photovoltaic modules that were attached on the rooftops of two identical greenhouses and their performance was comparatively evaluated for energy production and strawberry cultivation. The PV modules were aligned such that they formed a tilt angle of about 30° with OPV covering about 25.9% of the roof area while the STPV accounted for 20% of the area and generated an annual energy output of 880 kWh and 388 kWh respectively when subjected to the same conditions. This greatly minimized the dependence of the greenhouse on external energy sources. Microclimate assessment was carried out which revealed a temperature drop of 2.9° and 1.1 °C for OPV and STPV shaded areas compared to unshaded control areas which could be due to the contribution of these systems in cooling thus minimizing the energy for climate control. The photosynthetically active radiation (PAR) decreased by 43.5% for OPV and 31.7% for STPV compared to unshaded areas, however the fruit quality and growth remained robust. The strawberries produced under OPV modules showed the highest soluble solids content at 16.4 and 15.7 mg/g for STPV modules—both higher than the unshaded control which indicate improved fruit quality and sweetness. The strawberries grown under OPV panels showed better overall results than STPV which shows that the microclimate plays a vital role in the enhancement of certain crop characteristics.

The performance of a stilt mounted photovoltaic panels was evaluated by Sekiyama and Nagashima [57] to generate electricity and for crop production. The experiment was carried out on a 100 m^2 corn farm wherein the panels were set up in three different ways. The first section comprised of a control land with no PV modules installed, while the second section consisted of low-density PV arrays arranged such that the distance between the two adjacent panels was 1.67 m and the third section had high density of PV panel arrays with each panel distanced 0.71 m from each other. The planting of crops was carried out such that each square metre had a density of nine corn stalks each. The results indicated that the low-density PV arrangement improved the crop production, with an increase of 4.9% in corn stover biomass and an increased yield of 5.6%/m^2 in comparison to the control. However, the high-density PV array showed a decrease in the biomass of 96.9% of the output produced by the control. The increased yield can be attributed to reduced heat stress, conserved soil moisture due to prevention of water evaporation, lowered leaf temperature, protection from harsh environment, etc. Economic analysis revealed that the high-density setup generated a revenue 8.3 times greater the control while the revenue of low-density setup was 4.7 times higher. Thus, these sustainable agrivoltaics systems could effectively improve the crop production, even for shade intolerant crops like corn.

A solar tree was designed by Kumar et al. [35] which was made up of vertically arranged arrays of photovoltaic panels for efficiently powering submersible water pump for irrigating the Indian agricultural lands efficiently with a space efficient design that can be highly beneficial for agriculture in places with limited land availability. Being a renewable and eco-friendly source of the system effectively harnesses solar energy and provides a sustainable irrigation source for agriculture. The solar tree had a total installed capacity of 1.8 kW and was able to generate electricity of about 6–7 kWh per day which could sufficiently power a 1 HP submersible pump

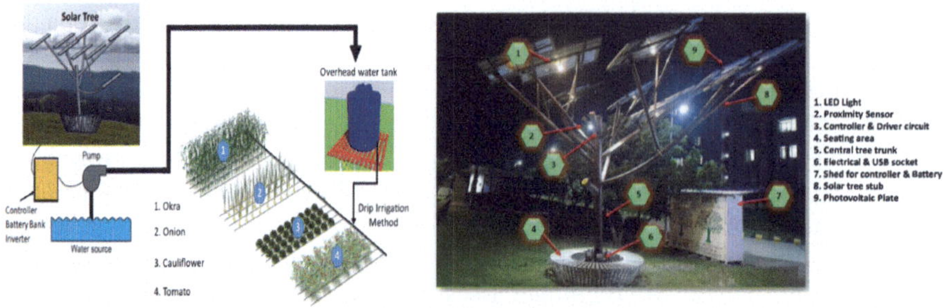

Fig. 6 Diagram illustrating a solar-powered irrigation system. On the left, a flow chart shows a solar tree connected to a pump, which draws water from a source to an overhead tank. The water is then distributed to a field using a drip irrigation method, watering crops like okra, onion, cauliflower, and tomato. On the right, a photograph of the solar tree at night highlights components such as LED lights, proximity sensors, a controller and driver circuit, seating area, central tree trunk, electrical and USB socket, shed for controller and battery, solar tree stub, and photovoltaic plates

of for approximately 5 h daily. This energy was sufficient to deliver nearly 18,000–20,000 L of water which could adequately irrigate around 1 acre of land. The solar tree although compact in design maintained stable performance under different sunlight conditions thus proving the system to be robust and reliable for agricultural use on a daily basis. The solar tree is low maintenance with the additional benefit of having lower operation cost and fuel-free operation as compared to the traditional diesel pumps. The solar tree also profited the farmers by increasing the crop yield and reducing the irrigation downtime due to consistent water availability. The solar tree system is highly beneficial for irrigating small and marginal lands, thus providing highly effective solutions for sustainable agriculture practices. This new approach could contribute greatly for India's sustainable goals to integrate renewable energy with agriculture (Fig. 6).

4.3 Solar Energy in Energy Sector

Due to the global need to shift towards cleaner and greener fuel alternatives, the environmental footprint of hydrogen as a fuel continues to grow. The hydrogen production sector is on the rise with many innovative technological advances and strategies aimed at producing clean hydrogen with high efficiencies that could be directly used to generate energy. At present, hydrogen is mainly generated via steam methane reforming which is a route heavily dependent on carbon thus leading to high carbon emissions that has several detrimental effects on the environment. Another route to harness hydrogen includes using renewable energies such as solar energy to aid the electrolysis process which is a greener approach with net-zero emissions [58]. It is estimated that the process of electrolysis requires 9 kg of high purity water to produce 1 kg of hydrogen. However, actual findings show that due to system inefficiencies this requirement rises to around 10–22.4 kg of water for per kg of hydrogen [59]. Nevertheless, the new and innovative technological advances in the areas of desalination and wastewater treatment have been essential in making these alternative sources of water viable specially in regions facing the water scarcity problems. The development of photovoltaic-electrolysis (PV-EC) systems coupled with a good and economical catalyst have greatly improved the efficiency of these systems while making the process economically feasible leading to a rise of the development of such solar-powered hydrogen production methods. In such systems, solar energy plays a crucial role due to its high availability, reduced cost and its capacity to power the energy production with minimal impact on the environment. The integration of such solar photovoltaic systems with electrolyzers can achieve solar-to-hydrogen (STH) efficiencies of 15–20% after optimization. Similarly, photoelectrochemical (PEC) water splitting methods exhibit efficiencies of 10–20% at the lab-scale and are emerging as promising hydrogen production solutions. These findings reveal the significant progress in solar hydrogen technology, thus proving their potential as sustainable solutions for energy production at a scalable level especially in regions with high solar irradiance. The integration of solar power technology and smart water management can further pave the way for the production of green hydrogen for achieving cleaner ang greener energies in the future [60]. The main approaches that leverage solar energy for hydrogen production include photovoltaic-electrolysis (PV-EC) systems, the photoelectrochemical (PEC) water splitting photocatalytic (PC) water splitting and solar thermochemical (STC) water splitting (Fig. 7).

The Photovoltaic-electrolysis systems utilize photovoltaic panels for the absorption of sunlight which is then converted to electricity to drive the electrolysis of water for hydrogen production. In such systems, the solar power acts as the indirect energy source for the water splitting reaction. These systems are well established, commercially deployable and display high efficiencies. In a study, a monolithic photoelectrochemical device incorporating photovoltaic tandem heterojunctions were able to achieve a solar to hydrogen efficiency of 19.3% under AM 1.5 G solar radiation which could be attributed to device design which reduces losses due to reflection and is equipped with a protective layer that improves the charge transfer at the cathode-electrolyte interface. However, the high installation cost, stability of PV materials over time and the complexity hinders the widespread use of such systems [61]. In the photoelectrochemical (PEC) water splitting systems the use of solar energy is more direct wherein the solar energy is converted to chemical energy by photoelectrodes. The semiconductor photoelectrodes absorb the solar radiation that leads to the generation of electron–hole pairs thus driving the oxidation reaction at the photoanode and reduction reaction at the photocathode and thus splits the water molecules into hydrogen and oxygen. A tandem PEC cell comprising of a $Pt/TiO_x/PIP/CuO_x$ photocathode and a $Co_4O_4/pGO/BiVO_4/SnO_x$ photoanode achieved a solar to hydrogen efficiency of 4.3% due to its enhanced spectral absorption and better charge separation, thus improving the overall solar energy utilization. These systems face the challenges of electrode degradation and stability, and a low hydrogen energy output. Research to overcome these challenges are ongoing to design systems yielding high

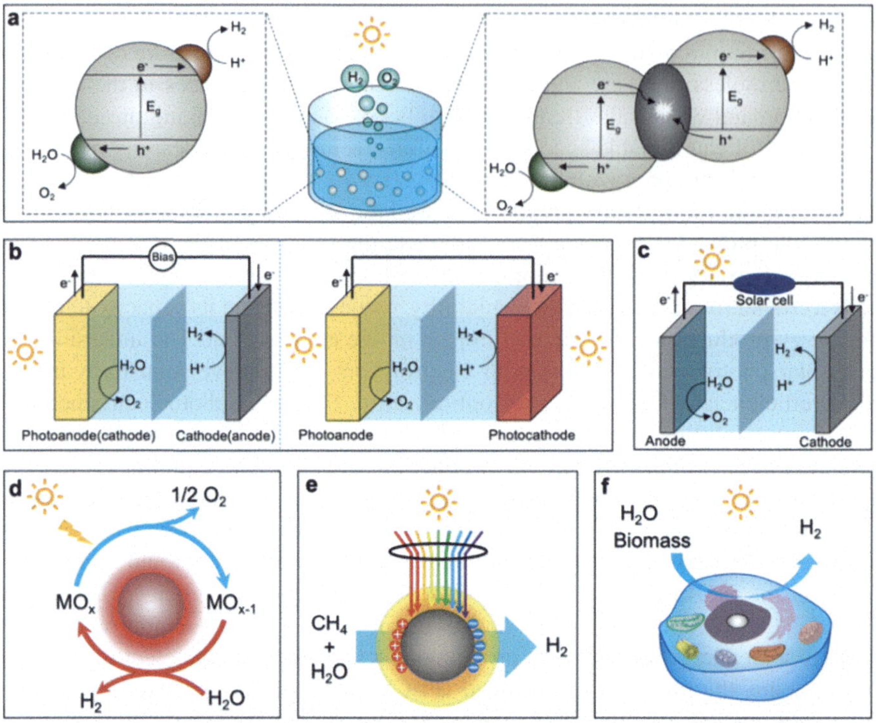

Fig. 7 Diagram illustrating various methods of hydrogen production using solar energy. Panel **a** shows a schematic of water splitting with solar energy, depicting electron and hole movement, and hydrogen and oxygen generation. Panel **b** illustrates a photoelectrochemical cell with a photoanode and photocathode, showing electron flow and gas evolution. Panel **c** depicts a solar cell-assisted electrolysis setup. Panel **d** represents a thermochemical cycle with a redox reaction involving metal oxides. Panel **e** shows a photothermal process using methane and water to produce hydrogen. Panel **f** illustrates biomass conversion to hydrogen using solar energy. Each panel highlights the role of sunlight in facilitating these processes

efficiencies and energy outputs with stable electrode materials [62]. In the case of photocatalytic (PC) water splitting, photocatalysts dispersed in water directly absorb the photons in the solar radiation which activates the catalyst and generates excited electrons and holes which initiates the redox reactions to split the water molecule into hydrogen and oxygen. These systems do not require any additional circuits or electrodes thus making them simpler and more affordable than the PEC systems. A nitrogen deficient (C_3N_4)–based Z-scheme photocatalysts doped with boron was able to achieve a solar to hydrogen efficiency of 1.16%. The reactivity of the photocatalyst increased due to improved charge separation, as a result of modifying the level of doping and the formation of structural defects in the material. These systems are eco-friendly, easy to fabricate and are affordable which make them a prime target for decentralized hydrogen production strategies in remote areas. Currently, these systems face issues of low efficiency and catalyst stability. However, these systems are still in their preliminary phase of development and research advances aimed at improving efficiencies are still underway [63]. In solar thermochemical (STC) water splitting, concentrated solar thermal energy is utilized to carry out the redox reaction in metal oxides at high temperatures. In these systems, the solar energy is harnessed in the form of heat which is used to thermally reduce the metal oxides, which are then re-oxidized in the presence of water, thus producing hydrogen. This thermochemical cycle consisting of two steps uses solar concentrators such as heliostat fields and parabolic mirrors to capture the solar heat and convert it to chemical energy. These systems can be used for large scale and continuous hydrogen production without requiring any expensive electrodes or electrical interfaces. However, the materials used for this process should be able to withstand high temperature cycling. Cerium-based metal oxides have been reported for hydrogen production using the solar thermochemical water splitting pathway. These systems under ideal conditions with optimized materials and reactor geometries were predicted to have a theoretical solar to hydrogen efficiency of 20% [64]. However, experimentally this efficiency of such cerium-based system with high fuel purity and selectivity was determined to be 5.25% [65].

Another innovative approach in this area is the design of a combined floating photovoltaic system by Temiz et al. [66] which could simultaneously produce electricity as well as hydrogen. The design of this FPV incorporates 1008 PV modules which were mounted and secured on a 3010 m^2 floating platform and could be oriented to capture maximum solar radiation. These FPV system were placed in the dam reservoirs which enhanced the efficiency of the panel due to the cooling effect provided by the water surface. It also reduced the water evaporation from the reservoir due to the shading effect of the panels while also providing a sustainable solution for areas facing land scarcity issues. The study also utilizes simulation tools like PvSyst and HOMER Pro to evaluate the performance of the system in producing electricity and hydrogen. The system was able to generate about 211.94 MWh/year of electricity, out of which 60.30 MWh/year was directly consumed to meet the energy needs of a nearby community and was able to fulfil 99.43% of their electricity demands. The remaining electricity was then used to power a proton exchange membrane (PEM) electrolyzer for hydrogen production. The hydrogen obtained is stored for future use, particularly to generate electricity during low solar irradiance periods. The levelized cost of electricity for this system was evaluated at 0.6124/kWh. Thus, this approach provides a sustainable route for energy production while also addressing land scarcity issues and can set a good example for many community-based energy generation systems in the future.

5 Conclusion

In the current global scenario with a push towards cleaner low-carbon energy technologies, solar energy is a game changer—its endless supply and easy scalability make it economically compelling. The chapter highlights multidimensional advancements in solar technologies spanning from photovoltaic materials, solar thermal systems to integrated energy solutions. Evolution of conventional silicon-based cells to advanced perovskite--silicon tandems underscores the continuous pursuit for high efficiency, lower material cost and adaptability across diverse applications. There is a key focus on the integration of latent heat thermal energy storage using phase change materials particularly in solar desalination systems. The incorporation of nanoparticle-enhanced PCMs demonstrated significant improvements in heat transfer thereby improving the overall system efficiency and effectively addressing the challenge of solar intermittency. The convergence of artificial intelligence and machine learning with solar energy systems represents a transformative paradigm wherein AI-driven models enable predictive maintenance, optimization and material discovery. The versatility of solar power is illustrated by its multifarious applications across water treatment, agrivoltaics and hydrogen production. Despite all the advancements several challenges persist pertaining to material stability, large-scale implementation and system integration. Continued interdisciplinary research that combines material science, thermal engineering and data-driven methodologies is henceforth essential for overcoming these barriers. In conclusion, the future of solar energy is in the synergistic integration of advanced materials, energy storage technologies and intelligent computational tools. This approach is pivotal for the achievement of resilient, decentralized and sustainable energy infrastructures to contribute towards global efforts for net-zero emissions and environmental sustainability.

Acknowledgements Authors Ranjan Dey, Praseeda P Nair and Suvanjan Bhattacharyya acknowledge the financial assistance from the CDRF Project Grant C1/23/152 from BITS Pilani, University and Ranjan Dey, and Suvanjan Bhattacharyya acknowledge the funding from the SPARC Project: SPARC/2024-2025/ENSU/P3314 from Govt. of India.

References

1. Hanwha. Perovskite tandem solar cells closer to reality. https://www.hanwha.com/newsroom/news/feature-stories/perovskite-silicon-tandem-cells-move-from-theory-to-reality-with-hanwha-qcells-breakthrough.do. Accessed 20 Apr 2025
2. IRENA (2024) Renewable power generation costs in 2023. International Renewable Energy Agency, Abu Dhabi
3. Gielen D, Boshell F, Saygin D, Bazilian MD, Wagner N, Gorini R (2019) The role of renewable energy in the global energy transformation. Energy Strat Rev 24:38–50. https://doi.org/10.1016/j.esr.2019.01.006
4. Lwakatare B, Vyavahare P, Mehta K, Zörner W (2024) Electrification planning for off-grid communities in Sub-Saharan Africa: advancing energy access. Energies 17:5994. https://doi.org/10.3390/en17235994

5. Baskaran G, Coste S (2024) Achieving universal energy access in Africa amid global decarbonization. CSIS. https://www.csis.org/analysis/achieving-universal-energy-access-africa-amid-global-decarbonization
6. IEA (2023) World energy outlook 2023—analysis. https://www.iea.org/reports/world-energy-outlook-2023. Accessed 20 Apr 2025
7. Ember (2025) Global electricity review 2024. https://ember-energy.org/latest-insights/global-electricity-review-2024/
8. India's solar energy potential is estimated to be 748 GWP. Union Power and New & Renewable Energy Minister. https://pib.gov.in/Pressreleaseshare.aspx?PRID=2003561
9. Physical achievements. Ministry of New and Renewable Energy, India. https://mnre.gov.in/en/physical-progress/
10. International Energy Forum (2024) The remarkable rise of solar power. https://www.ief.org/news/the-remarkable-rise-of-solar-power.
11. Global Energy Monitor (2025) Global solar power tracker. https://globalenergymonitor.org/projects/global-solar-power-tracker/
12. Huang J, Zhou J, Jungstedt E, Samanta A, Linnros J, Berglund LA, Sychugov I (2022) Large-area transparent "Quantum dot glass" for building-integrated photovoltaics. ACS Photonics 9:2499–2509. https://doi.org/10.1021/acsphotonics.2c00633
13. Jelle B (2015) Building integrated photovoltaics: a concise description of the current state of the art and possible research pathways. Energies 9:21. https://doi.org/10.3390/en9010021
14. Lozanova S (2025) Solid state battery technology: the future of energy storage. GreenLancer. https://www.greenlancer.com/post/solid-state-batteries. Accessed 20 Apr 2025
15. Monolith. Solid-state battery tech: 2024 energy storage advancements. https://www.monolithai.com/blog/solid-state-batteries-energy-storage. Accessed 20 Apr 2025
16. Maziviero FV, Melo DM, Medeiros RL, Oliveira ÂA, Macedo HP, Braga RM, Morgado E (2024) Advancements and prospects in perovskite solar cells: from hybrid to all-inorganic materials. Nanomaterials 14:332. https://doi.org/10.3390/nano14040332
17. Machín A, Márquez F (2024) Advancements in photovoltaic cell materials: silicon, organic, and perovskite solar cells. https://doi.org/10.20944/preprints202401.2160.v1
18. Correa-Baena J-P, Saliba M, Buonassisi T, Grätzel M, Abate A, Tress W, Hagfeldt A (2017) Promises and challenges of perovskite solar cells. Science 358:739–744. https://doi.org/10.1126/science.aam6323
19. CSIRO. CSIRO achieves record efficiency for next-gen roll-to-roll printed solar cells. https://www.csiro.au/en/news/All/News/2024/March/CSIRO-achieves-record-efficiency-for-next-gen-roll-to-roll-printed-solar-cells. Accessed 20 Apr 2025
20. Urry A (2018) Thin, flexible new solar cells could soon line your shirt. Wired. https://www.wired.com/story/thin-flexible-new-solar-cells-could-soon-line-your-shirt/. Accessed 20 Apr 2025
21. The Guardian (2024) CSIRO claims new record for energy efficiency in lightweight printed solar cells. https://www.theguardian.com/australia-news/2024/mar/12/csiro-claims-new-record-for-energy-efficiency-in-lightweight-printed-solar-cells
22. Chen Y, Jia B, Guan X, Han L, Wu L, Guan P, Lu P (2022) Design and analysis of III-V two-dimensional Van Der Waals heterostructures for ultra-thin solar cells. Appl Surf Sci 586:152799. https://doi.org/10.1016/j.apsusc.2022.152799
23. Bellani S, Bartolotta A, Agresti A, Calogero G, Grancini G, Di Carlo A, Kymakis E, Bonaccorso F (2021) Solution-processed two-dimensional materials for next-generation photovoltaics. Chem Soc Rev 50:11870–11965. https://doi.org/10.1039/d1cs00106j
24. Saadah M, Hernandez E, Balandin A (2017) Thermal management of concentrated multi-junction solar cells with graphene-enhanced thermal interface materials. Appl Sci 7:589. https://doi.org/10.3390/app7060589
25. Dwivedi SK, Tripathi SK, Tiwari DC, Chauhan AS, Dwivedi PK, Eswara Prasad N (2021) Low-cost copper zinc tin sulphide (CZTS) solar cells fabricated by sulphurizing sol-gel deposited precursor using 1,2-ethanedithiol (EDT). Sol Energy 224:210–217. https://doi.org/10.1016/j.solener.2021.04.046

26. Dyatkin B (2022) Transition-metal dichalcogenide solar cells possess new capabilities. MRS Bull 47:354–354. https://doi.org/10.1557/s43577-022-00328-z
27. Pu D, Anwar MA, Zhou J, Mao R, Pan X, Chai J, Tian F, Wang H, Hu H, Xu Y (2023) Enhanced photovoltaic effect in graphene–silicon Schottky junction under mechanical manipulation. Appl Phys Lett. https://doi.org/10.1063/5.0128962
28. Liu L, Najar A, Wang K, Du M, Liu S (Frank) (2022) Perovskite quantum dots in solar cells. Adv Sci. https://doi.org/10.1002/advs.202104577
29. Yuan Z, Dahliah D, Hasan MR, Kassa G, Pike A, Quadir S, Claes R, Chandler C, Xiong Y, Kyveryga V, Yox P, Rignanese G-M, Dabo I, Zakutayev A, Fenning DP, Reid OG, Bauers S, Liu J, Kovnir K, Hautier G (2024) Discovery of the Zintl-phosphide BaCd P as a long carrier lifetime and stable solar absorber. Joule 8:1412–1429. https://doi.org/10.1016/j.joule.2024.02.017
30. Gong J, Sumathy K, Qiao Q, Zhou Z (2017) Review on dye-sensitized solar cells (DSSCS): advanced techniques and research trends. Renew Sustain Energy Rev 68:234–246. https://doi.org/10.1016/j.rser.2016.09.097
31. Rashid FL, Al-Obaidi MA, Dulaimi A, Bahlol HY, Hasan A (2023) Recent advances, development, and impact of using phase change materials as thermal energy storage in different solar energy systems: a review. Designs 7:66. https://doi.org/10.3390/designs7030066
32. Pathak AK, Tyagi VV, Anand S, Pandey AK, Kothari R (2022) Advancement in solar still integration with phase change materials-based TES systems and nanofluid for water and wastewater treatment applications. J Therm Anal Calorim 147:9181–9227. https://doi.org/10.1007/s10973-021-11182-7
33. Nazari S, Safarzadeh H, Bahiraei M (2019) Performance improvement of a single slope solar still by employing thermoelectric cooling channel and copper oxide nanofluid: an experimental study. J Clean Prod 208:1041–1052. https://doi.org/10.1016/j.jclepro.2018.10.194
34. Tan X, Li Y (2024) Innovations and challenges in semi-transparent perovskite solar cells: a mini review of advancements toward sustainable energy solutions. J Compos Sci 8:458. https://doi.org/10.3390/jcs8110458
35. Kumar R, Kumar A, Gupta MK, Yadav J, Jain A (2022) Solar tree-based water pumping for assured irrigation in sustainable Indian agriculture environment. Sustain Prod Consum 33:15–27. https://doi.org/10.1016/j.spc.2022.06.013
36. Singh A, Verma S, Patel M (2025) AI-based solar energy optimization. Int J Prog Res Eng Manag Sci 2(2):38575. https://doi.org/10.56726/IRJMETS38575
37. Widmer J, Christ B, Grenz J, Norgrove L (2024) Agrivoltaics, a promising new tool for electricity and food production: a systematic review. Renew Sustain Energy Rev 192:114277. https://doi.org/10.1016/j.rser.2023.114277
38. Alva G, Lin Y, Fang G (2018) An overview of thermal energy storage systems. Energy 144:341–378. https://doi.org/10.1016/j.energy.2017.12.037
39. Sarbu I, Sebarchievici C (2018) A comprehensive review of thermal energy storage. Sustainability 10:191. https://doi.org/10.3390/su10010191
40. Zhang H, Baeyens J, Cáceres G et al (2016) Thermal energy storage: recent developments and practical aspects. Prog Energy Combust Sci 53:1–40. https://doi.org/10.1016/j.pecs.2015.10.003
41. Tyagi VV, Chopra K, Kalidasan B et al (2021) Phase change material based advance solar thermal energy storage systems for building heating and cooling applications: a prospective research approach. Sustain Energy Technol Assess 47:101318. https://doi.org/10.1016/j.seta.2021.101318
42. Singh SK, Kumar R, Verma SK (2021) CuO nanoparticle enhanced paraffin for latent heat storage applications. IOP Conf Ser Mater Sci Eng 1116:012061. https://doi.org/10.1088/1757-899X/1116/1/012061
43. Tan FL, Hosseinizadeh SF, Khodadadi JM, Fan L (2009) Experimental and computational study of constrained melting of phase change materials (PCM) inside a spherical capsule. Int J Heat Mass Transf 52:3464–3472. https://doi.org/10.1016/j.ijheatmasstransfer.2009.02.043

44. Mohammad A, Mahjabeen F (2023) Revolutionizing solar energy: the impact of artificial intelligence on photovoltaic systems. Int J Multidiscip Sci Arts. https://doi.org/10.47709/ijmdsa.v2i1.2599
45. Gupta A, Gupta K, Saroha S (2021) A review and evaluation of solar forecasting technologies. Mater Today Proc 47:2420–2425. https://doi.org/10.1016/j.matpr.2021.04.491
46. Shetty P, Adeboye A, Gupta S, Zhang C, Ramprasad R (2024) Accelerating materials discovery for polymer solar cells: data-driven insights enabled by natural language processing. Chem Mater. https://doi.org/10.1021/acs.chemmater.4c00709
47. Ledmaoui Y, El Maghraoui A, El Aroussi M, Saadane R, Chebak A, Chehri A (2023) Forecasting solar energy production: a comparative study of machine learning algorithms. Energy Rep 10:1004–1012. https://doi.org/10.1016/j.egyr.2023.07.042
48. Sowmmiya U, Abinaya K, Jhunjhunwala R, Satpathy P, Arul R (2024) Fault classification and identification through machine learning approaches for a solar PV—battery based water pumping system. Multimedia Tools Appl. https://doi.org/10.1007/s11042-024-19593-y
49. Boretti A (2021) Integration of solar thermal and photovoltaic, wind, and battery energy storage through AI in NEOM City. Energy AI 3:100038. https://doi.org/10.1016/j.egyai.2020.100038
50. Elsheikh AH, Katekar VP, Muskens OL, Deshmukh SS, Elaziz MA, Dabour SM (2021) Utilization of LSTM neural network for water production forecasting of a stepped solar still with a corrugated absorber plate. Process Saf Environ Prot 148:273–282. https://doi.org/10.1016/j.psep.2020.09.068
51. Parsa SM, Rahbar A, Koleini MH, Aberoumand S, Afrand M, Amidpour M (2020) A renewable energy-driven thermoelectric-utilized solar still with external condenser loaded by silver/nanofluid for simultaneously water disinfection and desalination. Desalination 480:114354. https://doi.org/10.1016/j.desal.2020.114354
52. Shalaby SM, Elfakharany MK, Mujtaba IM, Moharram BM, Abosheiasha HF (2022) Development of an efficient nano-fluid cooling/preheating system for PV-Ro water desalination pilot plant. Energy Convers Manage 268:115960. https://doi.org/10.1016/j.enconman.2022.115960
53. Prasanna YS, Deshmukh SS (2022) Energy, exergy and economic analysis of an air cavity appended passive solar still of different basin material at varying depth. Energy Sustain Dev 71:13–26. https://doi.org/10.1016/j.esd.2022.09.008
54. Carreño-Ortega A, Paço T, Diaz-Perez M, Gomez-Galan M (2025) Lettuce production under Mini-PV modules arranged in patterned designs. SSRN Electron J. https://doi.org/10.2139/ssrn.5170058
55. Kavga A, Strati IF, Sinanoglou VJ, Fotakis C, Sotiroudis G, Christodoulou P, Zoumpoulakis P (2018) Evaluating the experimental cultivation of peppers in low-energy-demand greenhouses. An interdisciplinary study. J Sci Food Agric 99:781–789. https://doi.org/10.1002/jsfa.9246
56. Tang Y, Li M, Ma X (2019) Study on photovoltaic modules on greenhouse roof for energy and strawberry production. E3S Web Conf 118:03049. https://doi.org/10.1051/e3sconf/201911803049
57. Sekiyama T, Nagashima A (2019) Solar sharing for both food and clean energy production: performance of agrivoltaic systems for corn, a typical shade-intolerant crop. Environments 6:65. https://doi.org/10.3390/environments6060065
58. Ishaq H, Dincer I (2021) Comparative assessment of renewable energy-based hydrogen production methods. Renew Sustain Energy Rev 135:110192. https://doi.org/10.1016/j.rser.2020.110192
59. Simoes SG, Catarino J, Picado A, Lopes TF, di Berardino S, Amorim F, Gírio F, Rangel CM, Ponce de Leão T (2021) Water availability and water usage solutions for electrolysis in hydrogen production. J Clean Prod 315:128124. https://doi.org/10.1016/j.jclepro.2021.128124
60. Song H, Luo S, Huang H, Deng B, Ye J (2022) Solar-driven hydrogen production: recent advances, challenges, and future perspectives. ACS Energy Lett 7:1043–1065. https://doi.org/10.1021/acsenergylett.1c02591
61. Cheng W-H, Richter MH, May MM, Ohlmann J, Lackner D, Dimroth F, Hannappel T, Atwater HA, Lewerenz H-J (2018) Monolithic photoelectrochemical device for direct water splitting with 19% efficiency. ACS Energy Lett 3:1795–1800. https://doi.org/10.1021/acsenergylett.8b00920

62. Ye S, Shi W, Liu Y, Li D, Yin H, Chi H, Luo Y, Ta N, Fan F, Wang X, Li C (2021) Unassisted photoelectrochemical cell with multimediator modulation for solar water splitting exceeding 4% solar-to-hydrogen efficiency. J Am Chem Soc 143:12499–12508. https://doi.org/10.1021/jacs.1c00802
63. Zhao D, Wang Y, Dong C-L, Huang Y-C, Chen J, Xue F, Shen S, Guo L (2021) Boron-doped nitrogen-deficient carbon nitride-based Z-scheme heterostructures for photocatalytic overall water splitting. Nat Energy 6:388–397. https://doi.org/10.1038/s41560-021-00795-9
64. Hoes M, Muhich CL, Jacot R, Patzke GR, Steinfeld A (2017) Thermodynamics of paired charge-compensating doped Ceria with superior redox performance for solar thermochemical splitting of H_2O and CO_2. J Mater Chem A 5:19476–19484. https://doi.org/10.1039/c7ta05824a
65. Marxer D, Furler P, Takacs M, Steinfeld A (2017) Solar thermochemical splitting of CO_2 into separate streams of CO and O_2 with high selectivity, stability, conversion, and efficiency. Energy Environ Sci 10:1142–1149. https://doi.org/10.1039/c6ee03776c
66. Temiz M, Javani N (2020) Design and analysis of a combined floating photovoltaic system for electricity and hydrogen production. Int J Hydrogen Energy 45:3457–3469. https://doi.org/10.1016/j.ijhydene.2018.12.226

Wind Energy Technology: Application and Impact

Riya Chugh, Indrani, and Rajeev Kumar

Abstract Energy plays a crucial role in both social and economic advancement. The growing impacts of climate change have prompted the use of renewable sources of energy, with wind being one of the main sources, to generate electricity. Wind energy technology captures the kinetic energy of wind and converts it into mechanical power or electricity. This technology is implemented primarily through wind turbines. This chapter wind energy technology primarily addresses the kinds of turbines or wind energy systems that are employed and their potential future applications. This chapter also highlights some of the forthcoming technologies for the production of wind energy (like multi-rotor, offshore floating, smart rotor, tip-rotors, composites, vortex nano, and airborne). Moreover, the applications of wind energy such as electricity generation, renewable hydrogen, water pumping, milling, industrial and agricultural uses, energy storage, transportation and mobility, and sailing. This chapter also gives an overview of the impacts of wind energy technology, including environmental, economic, and social impacts. As technology advances, wind energy continues to play an increasingly significant role in the global transition to renewable energy, contributing to sustainability, energy independence, and economic growth.

Keywords Wind energy technology · Types of systems · Applications · Impacts · Technologies for wind energy

R. Chugh (✉) · Indrani · R. Kumar
Department of Applied Sciences, Manav Rachna International Institute of Research and Studies, Faridabad, Haryana, India
e-mail: riyachugh.set@mriu.edu.in

Indrani
e-mail: indrani.set@mriu.edu.in

R. Kumar
e-mail: rajeevkumar.set@mriu.edu.in

J. A. Siddique et al. (eds.), *Energy Sources*, Green Energy and Technology,
https://doi.org/10.1007/978-981-95-1209-6_9

1 Introduction

Energy plays a vital role in the overall economic development of a country and also plays a vital role in human life. Due to the huge consumption of fossil fuels such as coal, petroleum, oil, and natural gas, various environmental and health issues can be produced. It is predicted by the researchers and scientists that the need and consumption of electric energy will grow tremendously in the future. Addressing the energy demands requires a mindful use of the available resources of energy, with a particular emphasis on renewable sources of energy, as conventional energy sources like fossil fuels will eventually fizzle away. However, it has become very important to switch to some renewable, clean, and green alternatives to conventional energy sources or fossil fuels such as wind energy, geothermal energy, biomass energy, tidal energy, hydropower, and solar energy (Fig. 1). One of the greatest sources of electrical energy is wind energy, which has grown too fast since 2000 [46]. It is basically the wind or air conversion to some useful form of energy. As an alternative to fossil fuels, wind energy or wind power is considered to be a green and clean energy source. The second most significant renewable energy source is wind energy, after hydroelectric energy, and thus, it is considered the most promising energy source across the world. Wind energy is an impactful, eco-friendly technology that might be able to address a lot of issues related to the environment and health such as greenhouse gas (GHG) emission, change in climatic conditions, sustainable development, cost, pollution-related issues, and reliability of electric energy [39]. Wind energy can be converted to other forms of energy by sails, wind turbines, wind mills, and wind pumps [11].

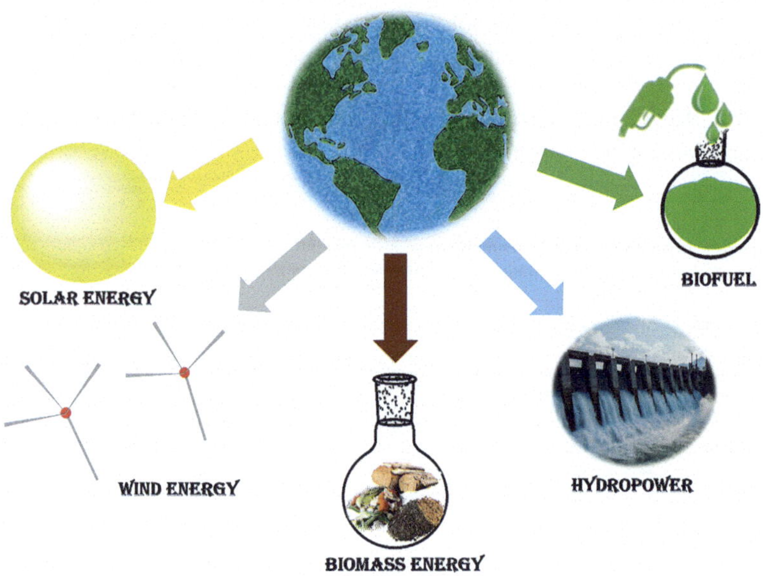

Fig. 1 Types of renewable energy sources

Earlier in the course of the twentieth century, the first-generation wind turbines were designed with the purpose of producing electrical energy. From the early beginning of 1970s, this technology has been progressively developed. After the 1990s, almost twenty years later, wind energy technology has made a significant comeback as the largest and most important renewable source of energy [31]. Among all the technologies, the most recent advancements in renewable energy sources are the wind energy source, as the wind turbine units have already expanded from a few hundred kilowatts (kW) to multipower generators during the past few years. By considering the continual ongoing rise in the power electrical output of the wind turbines and their growing connectivity into the electric grids, more upgraded generators, electrical power convertors, and control units' systems are required to improve the units of wind turbines, making them more suitable for incorporation into the power grid [43].

The explosive development of wind energy as a renewable energy source and wind energy technology has significant consequences in the field of research and education. Wind energy technology has also significantly impacted the professionals and experts employed by the wind energy sectors and electric power companies. It was observed and noted by the researchers and scientists that only five nations, including India, Germany, Spain, Denmark, and the United States, have more than 83% of the globe's wind turbines or wind generators installed, but, on the other hand, wind energy technology is growing rapidly into other parts of the world [33]. Thus, it is also very important to disseminate the information that is presently available worldwide. However, wind energy technology is highly complicated, since it has been used for more than 3000 years [29].

The transformation of the wind's kinetic energy into useful forms of energy is what drives wind energy as an important renewable energy source. The technique by which current wind turbines do this is as follows:

(i) Wind captured by blades: A turbine's blades rotate when wind passes across them. The blades can capture and maximize the wind's force because of their design, which is very similar to that of the wings of an airplane.
(ii) Interaction of rotor and generator: The rotor in the wind turbines, which is generally attached to the blades, transfers the rotational energy to a generator, usually through a gearbox.
(iii) Generation of electricity: The received energy by the generator transforms the rotational energy into electrical power that can be used further locally or sent to the grid.

As wind energy is a renewable source of energy that does not get depleted by consumption. With this, it is a greener alternative to other energy sources as it does not produce any greenhouse gases during production. The production cost of wind energy is also very less; thus, it has become one of the most important, reliable, and cost-effective renewable sources of energy [32].

2 History and Evolution of Wind Energy Technology

For centuries, humans have harnessed the kinetic energy of wind to generate mechanical power. The development of water wheel-driven hydropower stands as one of humanity's earliest power sources. Although it is known that wind power has been used in a variety of different ways over time, documented historical evidence tells that wind energy has been purposefully utilized for more than a thousand years. The use of wind energy is said to have originated in the Asian civilizations of China, Tibet, India, Afghanistan, and Persia [15]. The earliest known record about wind turbine usage comes from Hero of Alexandria in the second or third century BC. He documented a basic horizontal-axis wind turbine (HAWT) designed to power an organ, though it has been debated whether this device served any practical purpose beyond functioning as a novelty. More concrete evidence shows that by the seventh century BC, the Persians employed wind energy effectively with a vertical-axis wind machine. Over time, the use of wind power from different parts of Asia started spreading throughout European civilization. Historical data indicate that windmills were in use in England by the eleventh or twelfth century. Additionally, a firsthand report from 1190 suggests that German crusaders introduced windmill-building techniques to Syria. This implies that windmill technology was likely widespread throughout Europe by the Middle Ages. Windmills and water wheels have historically, and continue in some developing regions, to support basic, low-energy tasks like water pumping and grinding grain. Since windmills were first described, their design has evolved over centuries, adapting differently across various regions. The well-known Dutch windmills are a prime example, used extensively for water pumping, while numerous Mediterranean islands are celebrated for their charming, historic windmills [62]. With the introduction of the steam engine in the eighteenth century, the world gradually changed its demand for power to techniques and machines based on thermodynamic processes. The benefits of fossil fuels, such as coal, oil, and natural gas, became particularly evident with their introduction. Compared to traditional machinery, steam engines, steam and gas turbines, and engines powered by oil and gas are far more compact and can deliver significantly more power—well beyond what is required for water pumping or grain grinding. Additionally, these engines are not restricted by the need for flowing water or strong wind locations, allowing greater flexibility in placement. Finally, these newer machines offered a more dependable power source than wind turbines. Therefore, the importance of wind energy as a source of power decreased during the nineteenth century [37].

2.1 *Ancient and Medieval Wind Technology*

Humans have latched onto the wind as a source of power for a very long time, with civilizations understanding its use to reduce human effort, especially in areas with consistent wind patterns. Historical accounts suggest that the first wind-powered machines emerged around 200 B.C. in Persia during the time period when grains needed to be milled and excess water needed to be pumped out. These devices were also known as panemone windmills, which had vertical shafts with simple cloth or reed sails [64]. Such simple machines would have made the best advantage of the directional and predictable desert winds of the region.

Africa has allowed the spread of technology from Arab civilizations, which has aided civilizations like Egypt. This specific mix of cultures has aided the spread of technologies through trade routes. Vertical wind-powered machines evolved to horizontal wind-powered machines between the 1100 s and 1300 s. This also allowed for tower mills to be built, integrating a rotating top which allowed the sails to face the wind and, as such, increasing the device's efficiency by a margin [45]. Furthermore, historical sources underline the importance of the Crusades in providing ground for the expansion of this technology [42]. There are those who theorize that the returning crusaders brought back knowledge of windmills from the Middle East and modified the technology for the European terrain and climate. More complex uses, including land drainage in the low-developed countries, were made possible by the transition to the horizontal-axis windmills, which were more technologically advanced than the simple grain milling applications. Such innovations and adaptations provide evidence of active technological change, which was linked to the social setting. The historical context also explains the interplay of factors that caused such changes in the use of wind power. Wind technology efficiently penetrated the day-to-day life of Europe's rural population during the late medieval period. The construction of windmills in England and the Netherlands, for instance, greatly boosted agricultural productivity and nutrient management. This precolonial wind technology smoothened and helped to address intra-regional complex differences through the use of wind power across multiple regions [47].

2.2 *Wind Energy Through the Industrial Revolution*

Wind technology had rapid development during the Industrial Revolution (between the late eighteenth century and mid-nineteenth century). Such developments marked changes in agriculture, transportation, and most importantly, energy sources. Coal and steam engines are known to have been the rulers of the century; therefore, wind energy was not the main source of energy—it was still invaluable, though dominated largely by coal and steam engines—it still played an essential role in mechanical operations, especially in rural and agricultural settings [60]. Originally, windmills had been utilized in Europe for centuries for the purposes of grinding grain and water lifting, among other things. Following the rise of the Industrial Revolution, it was possible to see an increase in the use of windmill technology as there were vast improvements made in the discipline of engineering. For example, changes were made to the design of windmills in England and the Netherlands to perform new industrial functions such as the operation of saw mills, paper mills, and other mechanized equipment. During the period, engineers attempted to improve the windmill's efficiency by making structural changes to it, such as the application of multibladed rotor systems and fancy sails, which enabled the mill to turn automatically to face the wind [57]. One of the major inventions made long after the post mills was the tower mill, which had an unmovable structure but a moving cap, unlike its predecessor, the post mills. This enhanced the strength and stability of the structure, enabling it to bear larger sails and perform more sophisticated tasks. The increased knowledge of aerodynamics in this period increased the efficiency of blade design in wind technology to make the most energy while losing the least energy into mechanical form [52]. The changes caused wind technology to continue to be relevant when fossil fuel energy systems were developed, owing to the fact that wind technology was useful for tasks that could not be performed efficiently by steam engines in remote areas. The Industrial Revolution led to the transition in the field of wind energy, which basically laid the foundation for future developments in renewable energy; however, it was overshadowed by steam technology. Engineers maintained the usefulness of wind energy by modifying windmills to meet new industrial requirements and surroundings, laying the groundwork for later exploration and development in the twentieth and twenty-first centuries [24].

3 Wind Energy Technology: Its Components and Mechanisms

3.1 Twentieth-Century Innovations in Wind Turbines

The main advancements in the field of wind turbine technology during the twentieth century ultimately transformed the production of renewable energy. These innovations led to developments for present-day wind power, which is basically a combination of scientific research, as well as engineering breakthroughs, resulting in effective and large-scale power solutions. The construction of the first megawatt-scale wind turbine in the 1930s known as the "Smith-Putnam," which was in Vermont, USA, proved to be the biggest turning point. This wind turbine was designed by Palmer Cosslett Putnam, having 1.25 MW power turned out to be a major engineering advancement [1]. However, it had mechanical issues, which were the major stone in its path of success. In spite of these challenges, it provided a valuable understanding of large-scale turbine design and hence accentuated the need for more enduring materials and streamlined blade shapes. Another major innovation in the 1950s came from Denmark, which is now a global leader in wind energy. In contrast to previous vertical-axis models, engineers invented HAWTs, which aligned the electrical generator and rotor shaft on a horizontal plane. This led to greater proficiency and adjustability according to the changing direction of the wind, establishing the HAWT model as the leading design standard in the industry. The introduction of fiberglass and carbon fiber into turbine blade construction in 1980s, as a result of rapid advancements in materials science, allowed for larger and more aerodynamic blades. There were added advantages since materials were not only lighter but also resistant to environmental wear, which made turbines more durable and capable of harnessing higher energy outputs. By the end of the century, turbines could produce up to 1.5 MW, setting the stage for large wind farms [49]. Basically, the innovations that took place in twentieth century established a firm groundwork for the wind energy technology we observe today. Early wind turbine designs, developed through trial and error and continuous improvements, went from being experiments to reliable sources of clean energy. They solved engineering problems and helped meet environmental needs. These innovations have made it possible for wind energy to grow worldwide.

3.2 Brief History of Wind Power and Wind Energy Development in Denmark

So, when we consider the timeline of wind energy progress in Denmark, it is the story of a nation that has worked its way up into becoming one of the world leaders in the field of renewable energy sources. The Danish wind energy industry began with low-tech, small wind power generating systems at the close of the 1800s and developed through such significant historical events like World War I, World War II, and the oil crises of the 1970s.12 Such energy import dependencies were made visible through these crises; thus, the government made it a point to focus on developing increased energy independence and diversification. The first windmills in Denmark were built at the beginning of the twentieth century, which marked the start of the wind energy revolution in Denmark [27].

By the 1970s, the oil crisis spurred Denmark to establish energy policies focused on efficiency and sustainable sources, with wind power at the forefront. Through both grassroots efforts and government policies, Denmark built a robust wind energy infrastructure. The establishment of the Danish Wind Electricity Company in 1903, which trained rural electricians, and subsequent local initiatives underscored the value of community-driven energy projects. In the 1980s, Denmark's antinuclear movements pushed for renewable sources, further popularizing wind energy. Wind energy surpassed the limits of politeness in the early 2000s when Denmark galloped into the race by introducing several wind farms, which even included offshore facilities. Nevertheless, this surge in development did not come easily. One big hurdle came from the locals, who began opposing the establishment of several wind farms in the starting stages, raising environmental and sanctity matters, surveying the issue of renewable energy in a more detailed manner, and feeling helpless. Local co-ownership and recompensating for loss in property value were employment schemes that were used to recruit people for these projects. Historically, Denmark has taught us that as soon as you become aggressive technologically, neglect local people and the planet at your own risk [13].

3.3 American Wind Development (1930–1940)

The period between the 1930s and 1940s saw unprecedented use of wind mass energy in the USA, as there were start and mid-level attempts and initiatives targeted toward exploiting wind in rural areas, and in addition to other sources like coal, the construction industry to support massive wind turbine development and deployment, increasing usage during difficult economic situations. The electricity within the country was sought to be improved by the Rural Electrification Administration (REA) which was established in 1935, especially as not many rural farms had concrete connections to the electricity grids. Earlier in history, as well, wind had seen dominance with small generators, as they were mounted in the Midwest region of America, wind generators produced output below 1kw and helped in generating electrifying energy at pockets. In 1941, the construction of the Smith-Putnam wind turbine marked a major milestone. The team was led by Palmer Cosslett Putnam and funded by Vermont utilities. With a capacity of 1.25 MW and a rotor diameter of 175 feet, the Smith-Putnam turbine was the world's first megawatt-scale wind turbine. However, because the blade failed, it was only in operation for a few years. Even with time constraints, its creation showed that it was possible to produce wind energy at a vast scale. Various experiments were done to explore oceanic winds during World War II, as there was a noticeable tilt, but the resources were half-heartedly redirected to the world's war effort. Events like these turned into the building blocks for future prospects. The continuous demand for evolution, in turn, made way for the technological renaissance of wind energy in the 1970s [50].

3.4 Modern and Offshore Wind Technology

Part of the evolution to modern wind technology is seen in the efficient scaling of turbines and wind technology. With time, the initial wind-powered turbines, which were efficient on land, began giving way to larger turbines powered by wind on oceans. As per research conducted recently, the rate of electricity generation from offshore wind turbines is three times higher than that produced from onshore wind turbines. This is attributed to the size of the turbines used and the constant wind speeds that are constant at sea. Turbine size has greatly contributed to this major shift: a typical offshore wind turbine now weighs more than 10 MW, whereas an onshore turbine has a capacity of 2–3 MW. As far as expanding the offshore wind farms is concerned, Europe has been ahead of everyone else because of the adequate policies and funds set aside for offshore wind technology [8]. Significant nations in terms of offshore wind capacity include the United Kingdom, Denmark, and Germany, where the North Sea and the Baltic have become prime areas. Also, due to the EU's 2050 carbon neutrality targets, there is an increased demand for clean energy, hence the exponential growth of the offshore wind technology. This spurred the further development of more robust offshore turbines by European companies like Siemens

Gamesa and Vestas. These turbines have the capacity for renewal and adaptation to the harsh marine environments. The evolution of floating wind turbines is one of the latest trends in the offshore sector, intended to take advantage of deeper waters and locations previously classified as inappropriate for wind farms. In contrast to fixed-bottom turbines, floating turbines are attached to the seabed by means of cables, which allows for locations in water deeper than sixty meters. Companies such as Siemens Gamesa and Vestas are at the forefront of floating turbines development, with the focus of trying to decrease costs and increase stability. For example, Vestas's new floating tidal turbine project off Norway's coast aims to be completed by 2026, demonstrating the contours and potentials of floating structures. Floating turbines increase the scope of wind energy development, such as construction far from coastal areas with dense populations, where there is a lack of room for conventional power plants and energy infrastructure. As these technologies mature, the ability of floating turbines to increase renewable energy supply in the world is becoming more and more realistic. With new technologies and more resources being put into offshore wind layouts, including floating technologies, this energy source will be integral to the energy transition processes [48].

4 Applications of Wind Energy Technology

Wind energy technology has lots of applications in different applied fields across different sectors, owing to its ability to generate clean and sustainable power [5, 41]. Figure 2 represents the use of wind energy in different directions. This section explores the different applications of wind energy.

4.1 Electricity Generation

Utilizing the kinetic energy of the wind to rotate turbines, which generate electricity, is the process of generating electricity using wind energy. Because of its effectiveness, scalability, and environmental advantages, it is one of the renewable energy sources with the fastest rate of growth. Wind energy generators in homes, businesses, and industries are powered by electricity from large wind farms that are fed into the grid [6]. Smaller wind turbines lessen reliance on the grid by supplying energy directly to nearby consumers.

4.2 Rural and Remote Power Supply

Wind energy offers sustainable and renewable electricity, making it a potential option for isolated and rural power supplies. Since centralized electricity systems

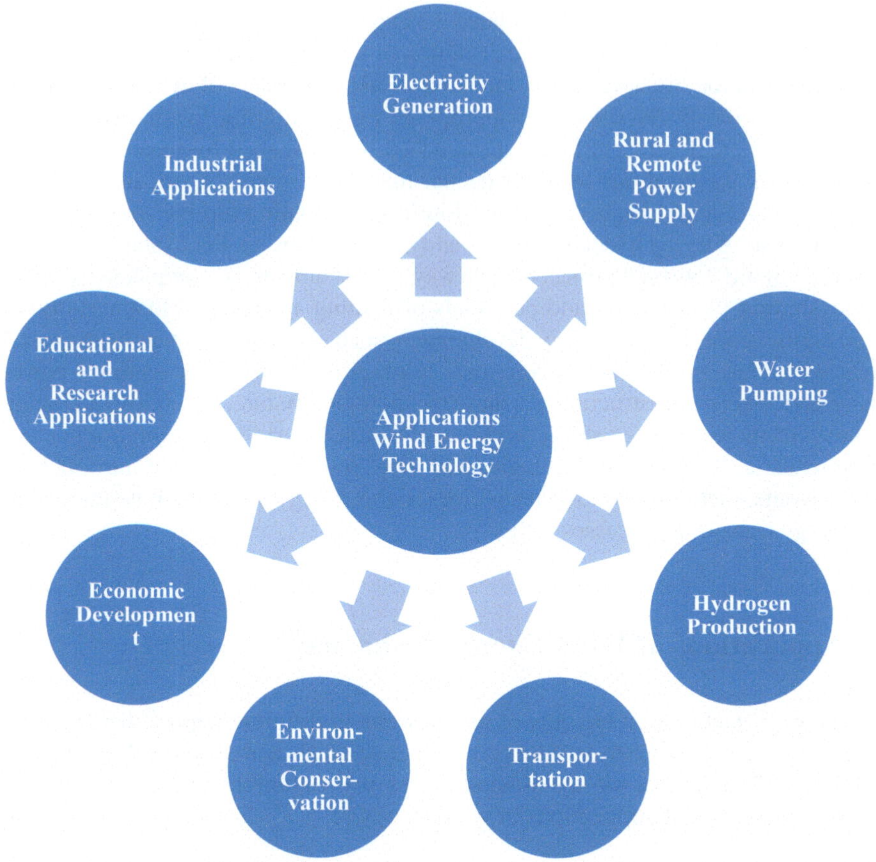

Fig. 2 Application of wind energy technology in different applied fields

are frequently unavailable in these places, wind energy is a viable substitute. Remote locations without access to a major electrical grid, including islands, rural settlements, or research outposts, are powered by wind turbines [35]. In some isolated locations, wind energy is combined with solar power or diesel generators to provide a steady source of electricity.

4.3 Water Pumping

Particularly in rural and isolated locations with limited access to power, using wind energy for water pumping is a practical and ecological alternative. Since ancient times, wind-powered water pumps have been in use, and they are still developing with new innovations. Water is pumped by windmills for human consumption, cattle, and agriculture [2]. Particularly in areas with limited water supplies, wind-powered systems are used to turn saltwater into freshwater.

4.4 Hydrogen Production

One viable strategy for developing a clean and sustainable energy system is the use of wind energy for hydrogen synthesis. Water is divided into hydrogen and oxygen using an electrolyzer powered by electricity produced by wind turbines. After that, the hydrogen can be used for a variety of purposes, including fuel cells, industrial operations, or as a clean energy carrier [7]. It can also be stored or transported. Through electrolysis, excess wind energy can be converted into hydrogen, which can then be used as a clean energy source.

4.5 Transportation

The possibility of using wind energy to power several kinds of transportation has been investigated. Despite not being a major energy source in the transportation industry just yet, a number of creative uses have shown potential. EV charging stations are supported by wind-generated electricity, which encourages environmentally friendly transportation [63]. Ships that use wind-assisted propulsion use less fuel.

4.6 Environmental Conservation

One renewable resource that does not contribute to GHG emissions or the depletion of natural resources is wind energy. It produces electricity without causing harmful waste or air pollution, in contrast to fossil fuels. The carbon footprint is greatly reduced with wind energy. By generating power without releasing carbon dioxide (CO_2), wind turbines contribute to reducing global warming and climate change [4, 14]. Generally speaking, wind farms have less of an adverse effect on nearby ecosystems than large-scale hydropower or the extraction of fossil fuels. Although there are worries about bird collisions, these problems are being resolved by improved technology and better siting.

4.7 Economic Development

Jobs in manufacturing, installation, operation, and maintenance are generated by the wind energy industry. This covers positions for engineers, technicians, turbine manufacturers, and construction workers. Jobs are also created by the need for supplies, transportation, and auxiliary services (such as catering and logistics). After a wind farm is up and running, continuous upkeep and technical assistance generate steady, long-term jobs. Local communities can benefit from the tax money and landowner lease payments that wind farms provide [20]. Electricity sales to the grid support regional and national energy networks by supplying a clean and reliable energy source that may draw in companies searching for reasonably priced and environmentally friendly electricity.

4.8 Educational and Research Applications

In the fields of research and education, wind energy holds great promise. It can be used to teach fundamental energy concepts or to develop innovative, sustainable technology. Numerous academic institutions are incorporating wind energy into their curricula to teach students about wind turbine physics, sustainable energy, and the advantages of wind power for the environment. Students gain real-world experience through projects like constructing miniature wind turbines or running wind energy simulations [61]. Students gain a better understanding of the scientific and engineering concepts underlying wind energy through these exercises. In order to raise knowledge and interest in renewable energy, educational initiatives like summer camps and seminars teach the public and students about the science and significance of wind energy [53]. In order to increase the reliability of wind energy, particularly in regions with varying wind conditions, researchers are looking into ways to integrate energy storage systems and increase wind turbine efficiency. To guarantee a steady supply of electricity, research is also being done on hybrid systems that combine wind energy with other renewable resources like solar or hydroelectric power.

4.9 Industrial Applications

With a wide range of industrial uses, wind energy provides a flexible and sustainable power source. Large-scale power production from wind turbines is a component of national grids. In certain situations, wind energy is utilized to supply direct mechanical energy for industrial operations like grinding and milling [10, 55].

5 Environmental, Economic, and Other Impacts of Wind Energy Technology

When compared to energy systems that rely on fossil fuels, wind energy technology generally has a favorable environmental impact. Nevertheless, there are several environmental issues related to its use and functioning. Both positive and negative impacts have been observed for wind energy technology [51, 59]. Figure 3 shows the positive and negative impacts of wind energy technology.

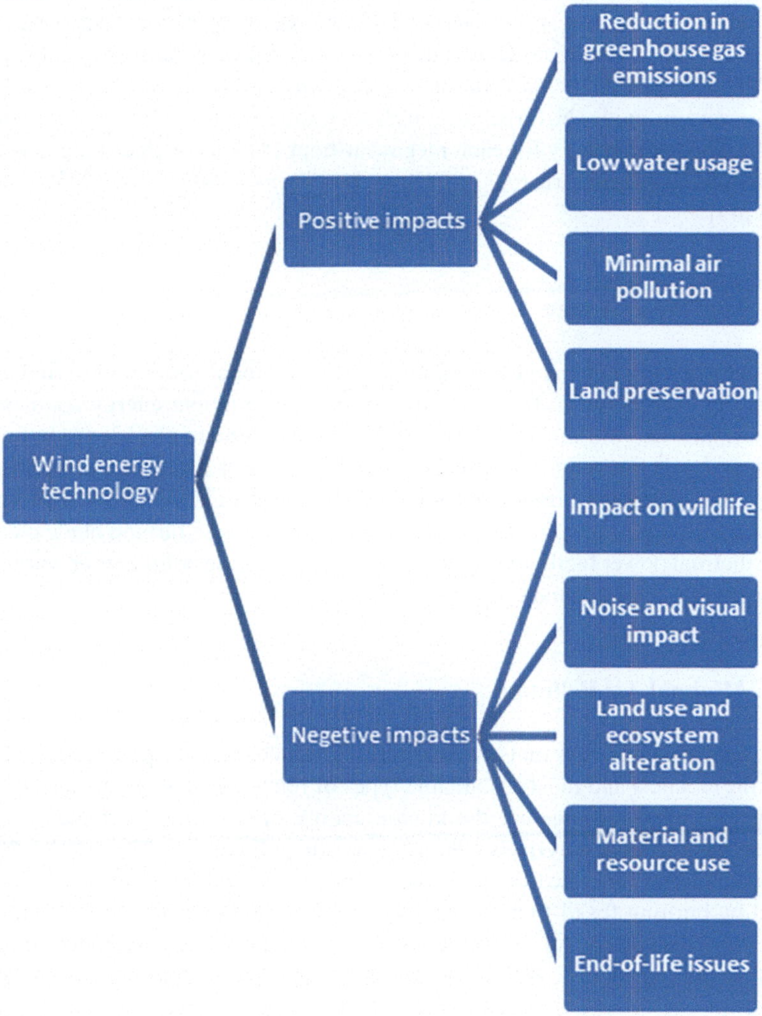

Fig. 3 Environmental impact of wind energy technology

5.1 Positive Impacts

Wind energy technology is an advanced technology and is growing day by day. Globally, it has a lot of impacts in various terms. The following are some positive impacts always observed by the experts.

5.1.1 Reduction in Greenhouse Gas Emissions

By taking the place of energy produced by fossil fuels like coal, oil, and natural gas, wind energy significantly lowers GHG emissions. Here are several ways that wind energy helps to reduce GHG emissions [12]. Since wind turbines don't burn fuel to produce power, they do not directly emit any carbon dioxide (CO_2), methane (CH_4), or nitrous oxide (N_2O). Emissions from traditional fossil fuel power plants are replaced with wind energy for each megawatt-hour (MWh) of electricity generated. For example, 0.4–0.9 metric tons of CO_2 are usually avoided for every MWh of wind energy [44].

5.1.2 Low Water Usage

When compared to nuclear power plants and conventional fossil fuels, wind energy is generally considered to be one of the most water-efficient energy sources [19]. Wind turbines use the kinetic energy of the wind to create electricity. In contrast to nuclear, coal, and natural gas plants, which use water for steam generation and cooling, this technology uses no water. The main uses of water in wind energy are for minor cooling in certain electrical systems and sporadic turbine blade cleaning. Unlike thermal power facilities, wind energy doesn't release pollutants or warm water discharges that could damage aquatic habitats.

5.1.3 Minimal Air Pollution

When compared to fossil fuels, wind energy produces less air pollution, making it one of the cleanest and most sustainable types of renewable energy. Instead of using combustion, wind turbines use the kinetic energy of the wind to create electricity [21]. Because of this, no GHGs (like CO_2) or air pollutants (such as NO_x, SO_2, or particulate matter) are released while the machine is operating. Energy that could be created by burning fossil fuels is replaced by wind energy for every kilowatt-hour of electricity produced. As a result, emissions related to electricity generation, such as carbon dioxide (CO_2), sulfur dioxide, NO_2, and particulate matter, are greatly reduced.

5.1.4 Land Preservation

In contrast to fossil fuel extraction techniques like mining or drilling, which can seriously affect the environment, wind farms require less land disturbance. Wind energy emits no GHGs and generates clean, renewable electricity. This lessens the effects of climate change on the ecosystem, including soil erosion, desertification, and deforestation brought on by changing climatic patterns [30]. When it comes to producing energy, wind turbines take up little room. Large tracts of land may be covered by wind farms, yet access roads and turbine bases only occupy a small percentage of the total area. The long-term preservation of land and ecosystems can be improved by using wind energy revenue to fund conservation initiatives like reforestation.

5.2 Negative Impacts

Every technology has both positive and negative impacts on human beings as well as the environment. Adoptability of the technology is a great concern in terms of the pros and cons of the technology. The following are the negative impacts that may be considered for wind energy technology.

5.2.1 Impact on Wildlife

Although wind energy is frequently regarded as a sustainable and clean energy source, its use may have unforeseen effects on animals. Rotating turbine blades can kill bats and birds when they collide with them [23]. Particularly at risk are migrating birds, raptors, and some bat species. Wind farm development has the potential to upset ecosystems by uprooting species or degrading their habitat [56]. Offshore wind turbine construction and operation have the potential to disturb marine life, especially for animals like dolphins that depend on echolocation. Turbine foundations have the potential to change local ecosystems by functioning as artificial reefs.

5.2.2 Noise and Visual Impact

Aerodynamic and mechanical noise are the two main forms of noise produced by wind turbines. While modern designs tend to reduce noise, larger turbines with higher capacities can produce more noise [22]. Distance dramatically reduces noise. In order to minimize disturbance, regulations frequently mandate that turbines be placed far enough away from residential areas.

5.2.3 Land Use and Ecosystem Alteration

Access roads, a foundation, and related infrastructure, including transmission lines and substations, are needed for every wind turbine. Although a wind farm's immediate physical footprint is quite tiny, supplementary infrastructure may have a considerable impact on the surrounding environment [17]. Huge wind farms have the potential to split habitats, affecting wildlife corridors and changing patterns of land use. Ecosystem connectedness may be diminished by roads and land cleared for wind turbine construction. Wind turbines may cohabit peacefully with pasture or agricultural land, which is where wind farms are frequently situated.

5.2.4 Material and Resource Use

Many components of wind energy technologies are made of fiberglass-reinforced polyester or epoxy, and more and more, carbon fiber for stronger, lighter blades. Mining for these resources may result in pollution and environmental damage [34].

5.2.5 End-of-Life Issues

Although wind energy technology is regarded as an environmentally beneficial and sustainable energy source, end-of-life (EOL) management is one of its concerns [3]. The materials and parts used in wind turbines, such as blades, towers, and other machinery, are the main cause of these problems. Composite materials like fiberglass and epoxy resins, which are strong but difficult to recycle, are the main components used to make wind turbine blades. Large numbers of wind farm blades reach the EOL as they get older, which makes waste disposal extremely difficult. Blades are frequently burned or dumped in landfills, both of which have negative effects on the environment [16]. Burning can generate toxic gases, and landfills take up area. Pollution of the soil and water can result from improper disposal of turbine parts.

6 Economic and Other Impacts of Wind Energy Technology

Wind energy technology is fast-growing technology globally. It provides energy everywhere in an eco-friendly way. Rather than environmental protection, it has a lot of economic impacts on the GDP of the country. The following are the major areas generating revenues and opportunities in the country [9].

6.1 Job Creation

With employment opportunities in a variety of industries, wind energy technology has emerged as a major global job creator. Skilled labor is needed to manufacture wind turbine components, which include towers, generators, gearboxes, and blades. It generates employment in sectors that manufacture the raw materials (steel, composites, etc.) needed to make turbines [58]. Turbine assembly and erection needed a high level of technical and engineering proficiency. Long-term positions for technicians who repair turbines and make sure the installed components operate efficiently, and that the components that produce wind energy function. Reducing the cost of the technology and improving the method's efficiency requires the development of more effective wind farm setups, materials, and turbine designs. Large-scale, intricate turbine component transportation can lead to the creation of specialist logistics professions. Project development, administrative and policy roles, supply chain, and many more are directly or indirectly related to the wind energy economic growth [54].

6.2 Cost Competitiveness

Advances in turbine technology, installation methods, resource availability, and supportive legislation are some of the elements that affect wind energy technology's cost competitiveness [25]. Onshore wind is now one of the least expensive sources of power due to a sharp decline in its Levelized Cost of Energy (LCOE) in recent years. Larger turbine blades and taller towers are examples of technological innovations that have boosted capacity and decreased prices. Larger and more effective turbines, better installation methods, economies of scale, and developments in floating wind platforms have all contributed to the recent decline in offshore wind costs, which were initially higher than those of onshore wind. Direct-drive systems and better blade aerodynamics are examples of innovations that lower maintenance costs and boost energy output. When combined with batteries or other energy storage devices, wind energy can provide reliable power and boost competitiveness in markets with fluctuating electricity demand [18]. The trend toward larger turbines makes it possible to access higher and more consistent wind speeds, which increases energy output. Cost reductions are facilitated by mass production and scaling up wind farms; supply chain efficiencies and global competition drive down prices for components like blades, towers, and nacelles. Wind farms have lower operational and maintenance (O&M) costs than fossil fuel plants because they do not require fuel purchases.

6.3 Local and National Economic Growth

Through a variety of channels, wind energy technology significantly contributes to the expansion of the local and national economies. Many jobs for engineers, technicians, and construction workers are created by the development of wind farms [40]. Contracts for services, logistics, and parts production boost local companies. Because wind farms broaden the tax base, more money may be allocated to community initiatives, infrastructure, and schools. For rural communities, wind energy offers a reliable source of income that lessens reliance on farming or the extraction of fossil fuels [28]. By lowering dependency on foreign fuels, wind energy keeps energy expenditures domestic. As wind technology improves over time, electricity costs decrease, which benefits both businesses and consumers. Wind energy investments encourage the development of a strong renewable energy industry, drawing both domestic and foreign investors.

The nation's trade balance experiences improved by exporting wind technology and experience. Innovation in engineering, materials science, and energy storage technologies is fueled by wind energy, which promotes industrial expansion. The growth of offshore wind farms encourages improvements in logistics and maritime technology. Pollution-related health care expenses can be reduced over time by lowering carbon emissions from fossil fuels. Access to green funding and subsidies is guaranteed, and penalties are avoided when international climate agreements are followed. The case studies that follow examine how wind energy technology can boost the nation's GDP and generate income [26]. In 2023, the wind energy industry helped sustain more than 120,000 jobs and brought about $20 billion annually for the US economy [36]. Denmark, a leader in wind technology worldwide, gets half of its electricity from wind, which boosts its GDP and energy independence. In India, wind energy plants have revitalized rural communities by generating jobs and extra income through leasing schemes [38].

In terms of energy security, community benefits, public perception and acceptance, and energy access, wind energy technology has an impact on society. Remote or underserved places can receive electricity via distributed wind energy systems. It is crucial for public acceptability and impression that there be some opposition because of noise, aesthetic impact, or land use issues. Nonetheless, increased local resilience through community-owned wind farm options and decentralized power generation, as well as encouraging local energy.

7 Challenges/Limitations and Future Prospects in Wind Energy

Despite the apparent simplicity, producing this kind of energy (wind energy) comes with a number of problems and limitations. As wind energy is a renewable energy source, it comes up with numerous challenges, including social, economic, technical, and environmental aspects, that may hinder its efficacy and implementation. The challenges can be:

i. The production of wind energy relies on the availability of wind, which is unpredictable and flexible. The days or times when there is no wind, then windmills or wind turbines are not able to generate electricity.
ii. The design of windmills or wind turbines is one of the most significant challenges. Slender and modern blades were designed by the researchers for the present wind turbines.
iii. The wind turbines, which are particularly located offshore, demand significantly expensive materials, building, and transportation expenses, and also require different substations to link the wind mills to the grid, which increases the overall expenses.
iv. Onshore wind farms and wind turbines demand a lot of land space, which might interfere with homes, preservation, and agricultural initiatives.
v. Turbines can change the environmental conditions and create noise, which leads the neighboring population to raise various issues. Thus, noise pollution can be one of the most critical challenges associated with wind energy.
vi. Ecosystem may be disturbed due to building and operating the wind farms.
vii. The battery technologies available currently in the market are not very effective which are widely used in conjunction with wind farms; thus, very large-scale energy production systems are required, which may be very expensive and technologically complex.
viii. Therefore, it is highly important to minimize the discussed negative impacts and key challenges on the ecosystem, human life, and environment so that wind energy can be used as a renewable energy source in the future.

8 Conclusion

Wind energy has enormous potential and promotes greater concentration, especially in large-scale systems. As an eco-friendly and sustainable substitute for fossil fuels and energy, wind energy has emerged as a key player in the worldwide shift to renewable energy. The wide adoption and acceptance have already produced a notable positive outcome in addition to prospects for more technological innovations. The present study examines the significance and applications of wind energy as well as the impacts and challenges in harnessing wind energy as a renewable energy source. Some research-based possibilities have been proposed to resolve such challenges, but such solutions and substitutes require more expenses and are less feasible. The technology of wind energy is crucial for the advancement of sustainability. It can significantly contribute to the fight against climate change and the development of sustainable and renewable energy sources around the world by resolving its drawbacks and maximizing its advantages. In conclusion, despite the environmental and technical challenges, wind energy remains a critical component in the path toward carbon neutrality. With continued innovation in turbine design, AI-driven efficiency, and energy storage, wind power can play a leading role in reducing global carbon emissions. As nations strive to meet climate targets, wind energy's contributions are expected to grow, helping to create a cleaner and more resilient energy future.

References

1. Akbari V, Naghashzadegan M, Kouhikamali R, Yaïci W (2023) Evaluation of regional elevation and blade density effects on the efficiency of a 1-kW wind turbine for operation in low-wind counties in Iran. Wind 3(3):320–342. https://doi.org/10.3390/WIND3030019
2. Al-Omari Z, Khlaifat N, Haddad M (2024) A feasibility study of combining solar/wind energy to power a water pumping system in Jordan's Desert/Al-Mudawwara Village. Environ Sustain Indicators. https://doi.org/10.1016/J.INDIC.2024.100555
3. Alavi Z, Khalilpour K, Florin N, Hadigheh A, Hoadley A (2025) End-of-life wind turbine blade management across energy transition: a life cycle analysis. Resour Conserv Recycl 213:108008. https://doi.org/10.1016/J.RESCONREC.2024.108008
4. Algburi S, Rendal E, Jaber ZK, Fakhruldeen HF, Amjad A, Sapaev IB, Farhan TT, Barakat M (2024) Evaluating the viability and potential of hybrid solar-wind renewable energy systems in relation to geographical and environmental factors. Results Eng. https://doi.org/10.1016/J.RINENG.2024.103473
5. Amfilochiou V, Debsharma T, De Baere I, Du Prez F, Van Paepegem W (2025) Interlaminar fracture toughness behaviour of a repairable glass-fibre-reinforced vitrimer for wind-energy applications. Compos B Eng 291:112023. https://doi.org/10.1016/J.COMPOSITESB.2024.112023
6. An J, Hong T (2024) Energy harvesting using rooftops in urban areas: Estimating the electricity generation potential of PV and wind turbines considering the surrounding environment. Energy Build 323:114807. https://doi.org/10.1016/J.ENBUILD.2024.114807
7. Assareh E, Saberipour M, Ahmadinejad M, Keykhah M, Agarwal N, Boudaghi F, Lee M (2024) Integrated wind farm solutions: harnessing clean energy for electricity, hydrogen, and freshwater production. Int J Hydrogen Energy. https://doi.org/10.1016/J.IJHYDENE.2024.10.055
8. Bilgili M, Yasar A, Simsek E (2011) Offshore wind power development in Europe and its comparison with onshore counterpart. Renew Sustain Energy Rev 15(2):905–915. https://doi.org/10.1016/J.RSER.2010.11.006
9. Can Şener ŞE, Anctil A, Sharp JL (2023) Economic and environmental factors of wind energy deployment in the United States. Renew Energy Focus 45:150–168. https://doi.org/10.1016/J.REF.2023.03.004
10. Castellani F, Natili F, Astolfi D, Vidal Y (2024) Wind turbine gearbox condition monitoring through the sequential analysis of industrial SCADA and vibration data. Energy Rep 12:750–761. https://doi.org/10.1016/J.EGYR.2024.06.041
11. Chaudhuri A, Datta R, Kumar MP, Davim JP, Pramanik S (2022) Energy conversion strategies for wind energy system: electrical, mechanical and material aspects. Materials. https://doi.org/10.3390/MA15031232
12. Chaurasiya PK, Patidar H, Shende V, Rajak U, Verma TN, Dwivedi G (2023) Evaluation of the reduction in greenhouse gas emissions attributable to wind energy: a retrospective evaluation of Indian Offshore and Coastal Site. Ocean Eng 281:114665. https://doi.org/10.1016/J.OCEANENG.2023.114665
13. Colvin RM, Witt GB, Lacey J (2016) How wind became a four-letter word: Lessons for community engagement from a wind energy conflict in King Island, Australia. Energy Policy 98:483–494. https://doi.org/10.1016/j.enpol.2016.09.022
14. Couce Montero L, Abramic A, Guerra Marrero A, Espino Ruano A, Jiménez Alvarado D, Castro Hernández JJ (2025) Addressing offshore wind farms compatibilities and conflicts with

marine conservation through the application of modelled benchmarking scenarios. Renew Sustain Energy Rev 207:114894. https://doi.org/10.1016/J.RSER.2024.114894
15. Dannemand Andersen P (n.d.) Review of historical and modern utilization of wind power. http://www.risoe.dk/rispubl/VEA/dannemand.htm. Accessed 4 Jan 2025
16. Desalegn B, Gebeyehu D, Tamrat B, Tadiwose T, Lata A (2023) Onshore versus offshore wind power trends and recent study practices in modeling of wind turbines' life-cycle impact assessments. Clean Eng Technol 17:100691. https://doi.org/10.1016/J.CLET.2023.100691
17. Dhar A, Naeth MA, Jennings PD, Gamal El-Din M (2020) Perspectives on environmental impacts and a land reclamation strategy for solar and wind energy systems. Sci Total Environ 718:134602. https://doi.org/10.1016/J.SCITOTENV.2019.134602
18. Ding JW, Fu YS, Lisa Hsieh IY (2024) The cost of green: analyzing the economic feasibility of hydrogen production from offshore wind power. Energy Convers Manage X 24:100770. https://doi.org/10.1016/J.ECMX.2024.100770
19. Elsakka MM, Refaat A, Alzahrani KM, Hee JL, Ma L, Elhenawy Y, Majozi T, Yosry AG, Amer A, Moustafa GH, Ahmed A (2024) Techno-economic assessment of a novel wind-powered RO system with a compressed air energy storage for water desalination. Energy 311:133296. https://doi.org/10.1016/J.ENERGY.2024.133296
20. Eze VHU, Mwenyi JS, Ukagwu KJ, Eze MC, Eze CE, Okafor WO (2024) Design analysis of a sustainable techno-economic hybrid renewable energy system: application of solar and wind in Sigulu Island, Uganda. Sci African 26:e02454. https://doi.org/10.1016/J.SCIAF.2024.E02454
21. Fallahpour M, Ghorbani Naeini H, Mirzaei PA (2024) Generic geometrical parametric study of wind-driven natural ventilation to improve indoor air quality and air exchange in offices. J Build Eng 84:108528. https://doi.org/10.1016/J.JOBE.2024.108528
22. Fatima B, Taqvi SAA (2024) Wind turbines noise pollution and challenges. In: Encyclopedia of renewable energy, sustainability and the environment, pp 597–606. https://doi.org/10.1016/B978-0-323-93940-9.00125-0
23. Garvin JC, Simonis JL, Taylor JL (2024) Does size matter? Investigation of the effect of wind turbine size on bird and bat mortality. Biol Cons 291:110474. https://doi.org/10.1016/J.BIOCON.2024.110474
24. Gipe P, Möllerström E (2022) An overview of the history of wind turbine development: Part I—the early wind turbines until the 1960s. Wind Eng 46(6):1973–2004. https://doi.org/10.1177/0309524X221117825
25. Hughes L, Longden T (2024) Offshore wind power in the Asia-Pacific: expert elicitation on costs and policies. Energy Policy 184:113842. https://doi.org/10.1016/J.ENPOL.2023.113842
26. Hussain B, Asif Ali Naqvi S, Anwar S, Usman M (2023) Effect of wind and solar energy production, and economic development on the environmental quality: Is this the solution to climate change? Gondwana Res 119:27–44. https://doi.org/10.1016/J.GR.2023.01.012
27. Johansen K (2021) Blowing in the wind: a brief history of wind energy and wind power technologies in Denmark. Energy Policy 152:112139. https://doi.org/10.1016/J.ENPOL.2021.112139
28. Juárez-Casildo V, Cervantes I, González-Huerta RG (2024) Strategic pathways for offshore wind in Mexico: geospatial insights and economic viability toward energy sustainability. Energy Sustain Dev 83:101565. https://doi.org/10.1016/J.ESD.2024.101565
29. Kaldellis JK, Zafirakis D (2011) The wind energy (r)evolution: a short review of a long history. Renew Energy 36(7):1887–1901. https://doi.org/10.1016/J.RENENE.2011.01.002
30. Kamal H, Mudassir MA, Kousar S, Makarem MA, Adnan Bodlah M, Murtaza S (2024) Wind energy investment analysis: design, parts, installation, and land costs. In: Encyclopedia of renewable energy, sustainability and the environment, pp 567–578. https://doi.org/10.1016/B978-0-323-93940-9.00011-6
31. Kamp LM (2004) Wind turbine development 1973–2000: a critique of the differences in policies between the Netherlands and Denmark. Wind Eng 28(4):341–354. https://doi.org/10.1260/0309524042886414
32. Kaygusuz K (2009) Wind power for a clean and sustainable energy future. Energy Sources Part B 4(1):122–133. https://doi.org/10.1080/15567240701620390

33. Kitzing L, Mitchell C, Morthorst PE (2022) Wind energy policy. Comprehen Renew Energy Second Edn 1–9(1–2):721–731. https://doi.org/10.1016/B978-0-12-819727-1.00160-6
34. Klyve ØS, Olkkonen V, Nygård MM, Lingfors D, Marstein ES, Lindberg O (2025) Retrofitting wind power plants into hybrid PV–wind power plants: Impact of resource related characteristics on techno-economic feasibility. Appl Energy 379:124895. https://doi.org/10.1016/J.APENERGY.2024.124895
35. Koholé YW, Wankouo Ngouleu CA, Vincelas Fohagui FC, Tchuen G (2024) Optimization and comparative analysis of hybrid renewable energy systems for sustainable and clean energy production in rural Cameroon considering the loss of power supply probability concept. Energy Convers Manage X. https://doi.org/10.1016/J.ECMX.2024.100829
36. Kotkin J (2011) The next hundred million: America in 2050, p 308
37. Kumar Y, Ringenberg J, Depuru SS, Devabhaktuni VK, Lee JW, Nikolaidis E, Andersen B, Afjeh A (2016) Wind energy: trends and enabling technologies. Renew Sustain Energy Rev 53:209–224. https://doi.org/10.1016/J.RSER.2015.07.200
38. Lacal-Arántegui R (2019) Globalization in the wind energy industry: contribution and economic impact of European companies. Renew Energy 134:612–628. https://doi.org/10.1016/J.RENENE.2018.10.087
39. Leung DYC, Yang Y (2012) Wind energy development and its environmental impact: a review. Renew Sustain Energy Rev 16(1):1031–1039. https://doi.org/10.1016/J.RSER.2011.09.024
40. Li J, Li G (2024) The utilization of renewable energy and the economic potential of offshore wind power supported by digital finance. Heliyon 10(16):e35175. https://doi.org/10.1016/J.HELIYON.2024.E35175
41. Li S, Chen Z, He X, Ye Y, Wan S, Dong L (2024) High performance hybrid omnidirectional wind energy harvester based on flutter for wireless sensing and hydrogen production applications. Nano Energy 132:110403. https://doi.org/10.1016/J.NANOEN.2024.110403
42. Lucas A (2006) Wind, water, work: ancient and medieval milling technology, p 439
43. Mahela OP, Shaik AG (2016) Comprehensive overview of grid interfaced wind energy generation systems. Renew Sustain Energy Rev 57:260–281. https://doi.org/10.1016/J.RSER.2015.12.048
44. Millstein D, O'Shaughnessy E, Wiser R (2024) Climate and air quality benefits of wind and solar generation in the United States from 2019 to 2022. Cell Reports Sustain 1(6):100105. https://doi.org/10.1016/J.CRSUS.2024.100105
45. Müller G, Jentsch MF, Stoddart E (2009) Vertical axis resistance type wind turbines for use in buildings. Renew Energy 34(5):1407–1412. https://doi.org/10.1016/J.RENENE.2008.10.008
46. Nguyen TTH, Phan GQ, Tran TK, Bui HM (2023) The role of renewable energy technologies in enhancing human development: empirical evidence from selected countries. Case Stud Chem Environ Eng 8:100496. https://doi.org/10.1016/J.CSCEE.2023.100496
47. Palmisano GO, De Boni A, Roma R, Acciani C (2021) Influence of wind turbines on farmlands' value: exploring the behaviour of a rural community through the decision tree. Sustainability 13(17):9630. https://doi.org/10.3390/SU13179630
48. Papi F, Bianchini A (2022) Technical challenges in floating offshore wind turbine upscaling: a critical analysis based on the NREL 5 MW and IEA 15 MW reference turbines. Renew Sustain Energy Rev 162:112489. https://doi.org/10.1016/J.RSER.2022.112489
49. Parker AM, Brown GTL (2001) Teacher evaluation of the second calibration of reading and writing assessments: Summary. asTTle Technical Report #19, pp 1–5. https://archive.org/details/windpowerrenewab0000gipe
50. Price T, Bunn J, Probert D, Hales R (1996) Wind-energy harnessing: Global, national and local considerations. Appl Energy 54(2):103–179. https://doi.org/10.1016/0306-2619(95)00073-9
51. Ramalho E, Lima F, López-Maciel M, Madaleno M, Villar J, Dias MF, Botelho A, Meireles M, Robaina M (2025) Understanding wind energy economic externalities impacts: a systematic literature review. Renew Sustain Energy Rev 209:115120. https://doi.org/10.1016/J.RSER.2024.115120
52. Reynolds TS (2023) Stronger than a hundred men. https://doi.org/10.56021/9780801825545

53. Sadi MA, Jamali A, Kamaruddin AMN bin A, Jun VYS (2024) Cascade model predictive control for enhancing UAV quadcopter stability and energy efficiency in wind turbulent mangrove forest environment. E-Prime Adv Electric Eng Electron Energy 10:100836. https://doi.org/10.1016/J.PRIME.2024.100836
54. Sooriyaarachchi TM, Tsai IT, El Khatib S, Farid AM, Mezher T (2015) Job creation potentials and skill requirements in, PV, CSP, wind, water-to-energy and energy efficiency value chains. Renew Sustain Energy Rev 52:653–668. https://doi.org/10.1016/J.RSER.2015.07.143
55. Steen M, Mäkitie T, Hanson J, Normann HE (2024) Developing the industrial capacity for energy transitions: resource formation for offshore wind in Europe. Environ Innov Soc Trans 53:100925. https://doi.org/10.1016/J.EIST.2024.100925
56. Stranddorf L, Colley T, Delefosse M, Svendsen JC, Irving Olsen S (2024) Marine biodiversity impact pathways for offshore wind farm decommissioning: implications for life cycle impact assessment development. Ecol Ind 167:112613. https://doi.org/10.1016/J.ECOLIND.2024.112613
57. Twidell J (2010) Book review: wind power. Proc Inst Civil Eng Energy 163(3):39–39. https://doi.org/10.1680/ENER.1000015
58. Wang X, Ali DA, Zhao W (2024) Urban energy transition and quality of life: a comprehensive statistical study on the nexus of wind clean energy adoption, job creation, and livability. Sustain Cities Soc 108:105436. https://doi.org/10.1016/J.SCS.2024.105436
59. Watson SCL, Somerfield PJ, Lemasson AJ, Knights AM, Edwards-Jones A, Nunes J, Pascoe C, McNeill CL, Schratzberger M, Thompson MSA, Couce E, Szostek CL, Baxter H, Beaumont NJ (2024) The global impact of offshore wind farms on ecosystem services. Ocean Coast Manag 249:107023. https://doi.org/10.1016/J.OCECOAMAN.2024.107023
60. Watson S, Moro A, Reis V, Baniotopoulos C, Barth S, Bartoli G, Bauer F, Boelman E, Bosse D, Cherubini A, Croce A, Fagiano L, Fontana M, Gambier A, Gkoumas K, Golightly C, Latour MI, Jamieson P, Kaldellis J et al (2019) Future emerging technologies in the wind power sector: a European perspective. Renew Sustain Energy Rev. https://doi.org/10.1016/J.RSER.2019.109270
61. Winsly BW, Ramalingam VK, Santhappan JS, Maria AJ, Chokkalingam MP, Rajendran V (2025) Optimal sizing and techno-economic-environmental evaluation of biomass-solar-wind-grid hybrid energy system: a case study of an institute in South India. Energy Convers Manage 325:119352. https://doi.org/10.1016/J.ENCONMAN.2024.119352
62. Yannopoulos SI, Lyberatos G, Theodossiou N, Li W, Valipour M, Tamburrino A, Angelakis AN (2015) Evolution of water lifting devices (Pumps) over the centuries worldwide. Water (Switzerland) 7(9):5031–5060. https://doi.org/10.3390/W7095031
63. Zhang C, Song P, Hou J, Xiao L, Wang X, Yang F, Wang X (2024) Technical and economic analysis of hydrogen production, storage and transportation by offshore wind power in different scenarios: a Guangdong case study. Int J Hydrogen Energy 94:829–837. https://doi.org/10.1016/J.IJHYDENE.2024.10.346
64. Zio E, Patil RB, Kara T, Şahin AD (2023) Implications of climate change on wind energy potential. Sustainability 15(20):14822. https://doi.org/10.3390/SU152014822

IoT-Based System for Hybrid Renewable Energy System and Its Application

Sashwat Kumar Singh and Debashis Kundu

Abstract The envision to effectively harness the potential of renewable energy is crucial due to the growing concerns of climate change and rising fossil fuel prices. Transitioning to renewables provides eco-conscious, cost-effective, and scalable solutions for global energy demands. Although promising, renewable energy systems face setbacks such as meager energy production, intermittency, and high initial costs. A Hybrid Renewable Energy System (HRES) is an integration of multiple energy sources like hydro, solar, and biomass, together with an energy storage system to provide a more reliable, efficient, and on-demand power supply. However, HRES includes multiple components that add layers of complexity in integration and management that limit its capacity to exclusively resolve challenges linked to renewable energy systems. This chapter provides an introductory idea on HRES and its composition, Internet of Things (IoT) fundamentals, and its application. It also aims to shed light on how an IoT-based approach to the HRES can help tackle these problems by enabling real-time monitoring and data analytics, improving energy management, and enhancing system efficiency. By leveraging IoT technology, HRES can adapt to changing conditions, ensuring a more resilient and effective energy solution in today's dynamic market.

Keywords Renewable energy · Hybrid renewable energy system · Internet of things

1 Introduction

The search for eco-friendly and sustainable energy resources is a paramount issue worldwide, especially in developing nations. The global energy demand has increased by 15% over the last decade, i.e., from 2013 to 2023, and has reached 640 exajoules. Even today, conventional energy sources (CES) such as coal, diesel, petrol, and

S. K. Singh · D. Kundu (✉)
Department of Chemical Engineering, Institute of Chemical Technology, Marathwada Campus, Jalna, Maharashtra, India
e-mail: d.kundu@marj.ictmumbai.edu.in

J. A. Siddique et al. (eds.), *Energy Sources*, Green Energy and Technology,
https://doi.org/10.1007/978-981-95-1209-6_10

natural gas (NG) continue to have a major share in energy production. However, the alarming reality is that the CES is no longer sufficient to meet the growing energy demand due to the exponentially growing population and industrialization surge [1, 2]. These compounding crises have led to the increased depletion of CES and rising prices, necessitating a shift toward renewable energy (RE). RE refers to energy derived from renewable sources that are naturally replenished on a human timescale, i.e., solar, wind, and tidal [3, 4]. Embracing renewable energy sources (RES) minimizes natural resource depletion and greenhouse gas emissions and reduces dependence on fossil fuels, creating a cleaner and more sustainable energy future [5]. Despite being a clean and eco-friendly energy source [6], challenges like variability, intermittency, and comparatively higher price of produced electricity with respect to grid electricity raise concerns about its overall reliability and effectiveness. However, the use of the complementary nature of different renewable energy systems (RES) and special design considerations together provide remarkable scope to mitigate these concerns. For instance, hybrid integration of solar/wind/hydro, etc. energy systems provides continuous output, with peak output during daytime from the sun, from wind at nighttime, and hydroelectric energy to maintain any supply variability with backup sources and storage systems like batteries [7, 8].

Hybrid renewable energy systems (HRES) are the integration of multiple energy sources to ensure a reliable and uninterrupted power supply [9, 10]. They are cost-effective solutions to the energy demand in off-grid and rural locations with limited access to conventional power grids [11]. The introduction of backup sources and storage systems to HRES greatly aided in maintaining a continuous power supply by minimizing the gap between power supply and requirements during peak hours [12, 13]. However, the presence of multiple RES and special designs adds significant layers of complexity and cost to the management and monitoring systems. To address this issue, a smart energy management system (EMS) is needed that uses the synergy between the Internet of Things (IoT) and recent technological advancements like artificial intelligence (AI) and cloud computing for real-time monitoring, prediction, and optimization of actions. The IoT is a network of physical objects—devices, instruments, etc.—capable of collecting and exchanging data to work in sync. This integration of smart technologies like HRES and IoT can manage the intermittent nature of integrated RES better. This increases the penetration of renewables in the total energy mix, leading to a more sustainable and resilient energy future [8, 14, 15].

2 Fundamentals of Hybrid Renewable Energy Systems

The term "hybrid" is used in multiple domains to refer to a system that combines and integrates multiple distinct constituents to leverage the unique characteristics of each component and the synergies that arise from their interactions [16].

HRES are coherent systems that combine at least two or more RES, like solar and wind, with or without a CES, together to provide a reliable and efficient power supply [10]. The most common RES available are wind, solar, hydro, tidal, etc. [17].

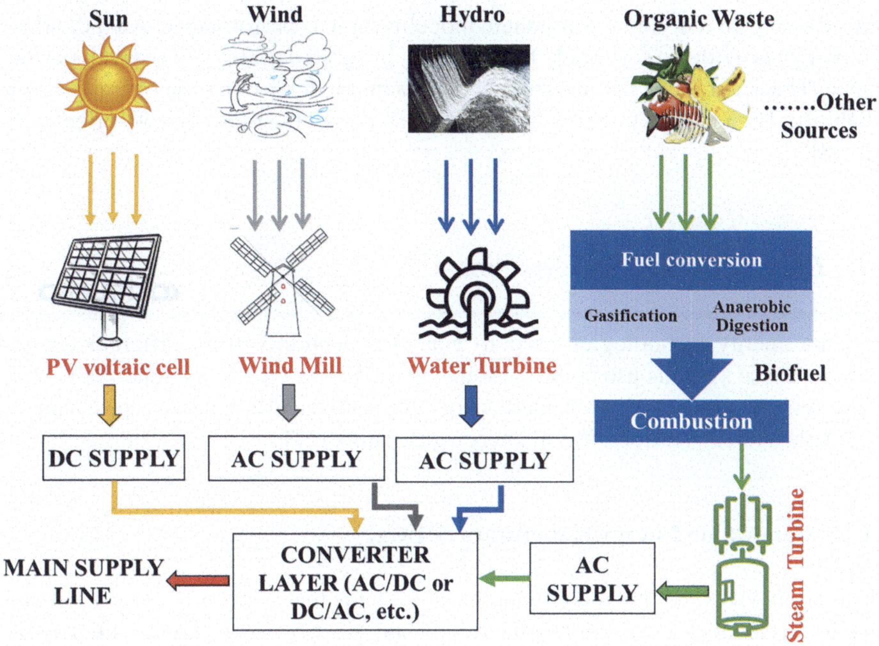

Fig. 1 General schematic of a hybrid renewable energy system

Some common examples of studied HRES are photovoltaic (PV)-wind diesel generator [18], PV/wind turbine (WT)/battery (BAT)/fuel cell [19], PV-wind turbine-fuel cell, and diesel generator (DG) [20]. HRES allows better resource utilization, low energy cost, and higher reliability in comparison to single-source energy systems, making it apt especially for remote and rural location energy demands [14, 21]. It has numerous applications in different fields such as transportation systems, i.e., electrification of railways [22], standalone power generation systems for electrification, power pumping, and telecommunication in rural areas [23], industrial utility systems for heating, cooling, and power supply [24], and electrification of commercial or residential buildings [25]. Figure 1 shows the general schematic of an HRES that indicates sources of energy and conversion technology used for electricity production. The nature of the output produced by different RES is different, which is then further transformed using a converter for the main supply.

3 Main Components of Hybrid Renewable Energy System

The main components of an HRES are energy generation systems (EGS), i.e., renewable or conventional, AC/DC/power electronic converters, energy storage systems (ESS), and load [12, 26, 27]. The design and accurate prediction of output for HRES depend solely on individual component modeling and its performance. An individual component is modeled separately first and then integrated to study the overall performance. The accuracy of the individual component model decides the overall system reliability. The overall system is finally optimized to deliver energy cost-effectively [12, 28].

3.1 Energy Generation Systems

EGS are energy technologies used to generate electricity from different energy sources. These systems use both renewable and CES for electricity generation [26]. These systems are sometimes named energy generators and are of two types, namely renewable and conventional/fossil energy generators [27].

3.1.1 Renewable Energy Generators (REGs)

REGs are devices that are a combination of systems that convert RE like wind and solar to mechanical energy and then, with the aid of a generator, convert it to a more useful form of energy, i.e., electricity. These sources behave as the primary source of energy for the HRES [27]. The most widely used REGs are the following.

(a) **Solar generators/generating systems**

Solar energy is one of the most abundant and freely available sources of energy, whose primary forms are heat and sunlight. The sun emits radiation at a rate of 3.8 $\times$ 10^{23} kW, from which Earth receives approximately 1.8 $\times$ 10^{14} kW. But only 60% of this energy, i.e., 1.08 $\times$ 10^{14} kW, reaches the earth's surface, and the rest either gets absorbed by the atmosphere or reflected back to space [29, 30]. The device mostly used to convert sunlight into electricity is the solar photovoltaic (PV) cell. It is the most used technology for capturing and converting sunlight into electricity due to its modular design that has no moving parts, zero emissions, and vibration- and noise-free working, along with efficient handling. It can be placed on a building rooftop or in open areas with ample sunlight for operation. The basic unit of PV is solar cells that are made up of semiconductor materials, mostly silicon-based, i.e., mono or polycrystalline silicon, to generate electricity. The sunlight falling on a PV excites the electron, causing the transition of electrons from a lower to a higher energy state. This transition of electrons, in turn, creates holes and free electrons in semiconductors, thus producing electricity [6, 26, 30]. The power output of a PV cell (P_s) is estimated using Eq. (1) [12, 31]

$$P_s = I_T \times Z \times A_{pv} \tag{1}$$

$$I_T = I_b R_b + I_d R_d + (I_b + I_d) I_r \tag{2}$$

where P_s is the power output (kW), Z is the efficiency of the PV cell (dimensionless), A_{PV} is the area of the PV cell or module (m^2), I_T is the total solar radiation incident on the PV surface (W/m^2), I_b, I_d is the direct normal solar radiation (sunlight directly from the sun) and diffuse solar radiation (sunlight scattered by clouds or atmospheric particles), respectively, R_b, R_d, R_r is the tilt factor for direct radiation, diffuse radiation, and reflected radiation, respectively.

(b) **Wind generators/generating systems**

Wind energy is a green, renewable source of energy, which is an indirect form of solar energy, as it is caused by the differential heating of the Earth's surface by the sun. The onshore wind energy potential to generate electricity is technically around 2×10^{13} to 5×10^{13} kWh per year [32, 33]. India itself has increased its wind energy generation capacity from 34,293 MW in July 2018 to 47,959.13 MW in November 2024 [34, 35]. The kinetic energy of moving air is converted to electricity using a turbine generator. The actual mechanical power (P_H) generated by the turbine is given by Eq. (3) [33]

$$P_H = \frac{1}{2}\rho A V_u^3 C_P \tag{3}$$

$$C_P = \frac{\left(1 + \frac{V_d}{V_u}\right)\left(1 - \left(\frac{V_d}{V_u}\right)^2\right)}{2} \tag{4}$$

where ρ is the air density, A is the area swept by the turbine blade, V_u and V_d denote upstream and downstream wind velocity, respectively, and Cp is the Betz limit (fraction of power captured by the rotor blade from upstream wind power.

(c) **Hydro generators/generating systems**

The term hydropower means "power of flowing water," which has the potential to cut through mountains, make canyons, and facilitate the transportation of goods. This power of flowing water can be harnessed by building dams over water bodies like rivers and lakes to generate electricity using water turbines. It includes mechanical parts like a rotor, shaft, and coupling, along with a generator, and hence is sometimes called a hydro generator, which converts the kinetic energy of falling water to mechanical energy and finally into electrical energy [36–38]. The equation is used to predict the actual power (P_W) produced is given by Eqs. (5) and (6) [39]

$$P_W = \eta \times \rho_{\text{Water}} \times g \times Q_t \times H \tag{5}$$

$$= (0.0027225 \times Q_t \times H) \times \eta \tag{6}$$

where η is overall efficiency, Q_t denotes total volumetric flow rate in time $t(Q_t = q_t \times t)$, and H is the head of the release of water. There are multiple other primary energy source generators that can be used in HRES, i.e., geothermal, tidal, and biomass.

3.1.2 Nonrenewable Energy Generators

These generators use conventional sources of energy like gas, oil, and coal to generate electricity and are also sometimes termed as fossil fuel-fired generators. These systems are used as a fail-safe component in HRES to maintain an uninterrupted power supply during intermittent conditions of RES, and also fulfill peak hour load demands. Some examples of these systems are diesel generators and NG generators. The nature and amount of load, pollutant emission, and cost of energy (COE) determine the selection of fossil fuel-fired generators [26, 40, 41]. Diesel generators are the most commonly used generators as a backup supply, and their power production in terms of fuel consumption is modeled using Eq. (7) [42]

$$P_0 = \frac{F_U - F_I G_r}{F_C} \tag{7}$$

where P_o, F_U, F_I, G_r, F_C represent output power generated (kW), amount of fuel used (l/kWh), intercept of fuel curve, generator ratings (kW), and slope of fuel curve.

3.2 AC/DC/Power Electronic Converters

The nature and magnitude of output from different REG are different like PV voltaic cell converts sunlight into direct current (DC) output, but wind turbines produce three-phase alternating current (AC) output. Table 1 shows the types of output based on generator type.

They behave as inverters to transform AC to DC or DC to AC, based on the load, since a grid supply requires AC, while a battery system requires DC to charge. It is also used to adjust the voltage of supply, i.e., to boost or buck as per demand. The two most common types of converters used are electromechanical power converters (EMCs) and electronic power converters (EPCs). Some examples of EMC devices are rotary converters and synchronous condensers, while EMC includes rectifiers and inverters. Therefore, it is a critical part of an HRES [27, 28].

Table 1 Types of output versus generator type

Generator type	Solar	Wind turbines	Hydro	Diesel/ gasoline	Biomass	Geothermal	Fuel cells	Inverter generators
Output type	DC	AC	AC	AC/DC	AC	AC	DC	Clean AC

3.3 Energy Storage Systems

The energy storage devices include batteries, compressed air storage, and supercapacitors, whose main aim is to manage the irregular power demands. It stores excess energy produced during nonpeak hours and then supplies it back as per the need of load, i.e., during peak hours, or during unavailability of renewable sources like solar and wind. It is simply a backup component, just like fossil fuel-fired generators, which help maintain an uninterrupted power supply. The most used ESS is a battery, especially lithium-based batteries, as it has long battery life, better efficiency, and high voltage and charge/discharge capacity rate. The battery is designed as per the load demand during intermittent working conditions of REG, with capacity based on Eq. (8) [17, 27, 42]

$$\text{Capacity} = \sum_{i=1}^{i=n} \frac{L_B S_{\text{Days}}}{(D_{\text{Depth}})_{\max} C_T} \tag{8}$$

where $L_B, S_{\text{Days}}, \left(D_{\text{Depth}}\right)_{\max}, C_T$ denote load on the battery, days of storage, maximum discharge depth, and temperature correction factor.

4 Architecture of Hybrid Renewable Energy System

The understanding of components only gives an idea about the nature and design of individual components of HRES. However, the architecture of HRES is studied to design a complete HRES, keeping the complexity of the framework and losses due to conversions minimum. It is achieved by optimized application of converters or by using bidirectional converters. The different types of architecture used are as follows [17].

4.1 DC-Coupled Isolated System

The design of this system is fairly simple with only one DC bus line that connects the RES, batteries (i.e., ESS), and load together. It includes DC output RES (i.e., solar), AC output RES integrated with AC/DC converters that provide DC, which is then connected directly to the DC bus, while the battery is connected through a bidirectional converter that can charge/discharge as per load requirements. The DC load is directly fulfilled by the DC bus supply, but to fulfill the AC load, a DC/AC converter is needed [17, 43].

4.2 AC-Coupled Isolated System

It has one AC bus line in the architecture, which is connected via an AC/AC converter to the AC output RES to supply constant-phase AC voltage to the bus, while for DC output RES, a DC/AC converter is needed to convert DC to AC. The connection of the battery is the same as in DC-coupled, but to fulfill the DC load, the AC must be converted. It is convenient for small-scale electricity production systems [43].

4.3 Hybrid-Coupled Isolated System

This system uses both AC and DC bus lines that use less number of converters, leading to fewer conversion losses but additionally increasing the complexity of the control system. It is an efficient but chaotic control system [17, 44].

4.4 Hybrid-Coupled Grid-Connected System

The main advantage of this system is storage capacity can be minimized, which directly reduces the initial capital cost of HRES. It is mainly used in the integration of the grid to reduce the load on the grid and to produce low-cost electricity by using RES in the most efficient way [17, 45].

Figure 2 shows the general architectures of HRES with different configurations of bus lines and converters using hybrid energy sources. The configuration in Fig. 2a and b shows lower conversion losses due to the presence of a single conversion process (i.e., AC to DC or DC to AC, respectively), making them an efficient choice. The presence of both AC and DC bus lines in Fig. 2c gives the added advantage of load flexibility and enhanced resilience to this hybrid configuration. Figure 2d has a similar configuration to Fig. 2c, except that it is grid-connected.

5 Application of Hybrid Renewable Energy System

Several national and international studies have been conducted to identify the best possible configurations of HRES that ensure a reliable and uninterrupted energy supply with the least cost and environmental impacts for different regions.

For a village named Damala in Haryana, India, HRES with three different configurations was studied using Hybrid Renewable Model for Electric Renewables (HOMER) software for electricity production at the lowest COE, net present cost (NPC). The study also aimed to minimize CO_2 emissions and get the maximum level of renewable fraction in electricity produced. Among all, the PV/WT/HYDRO/DG/

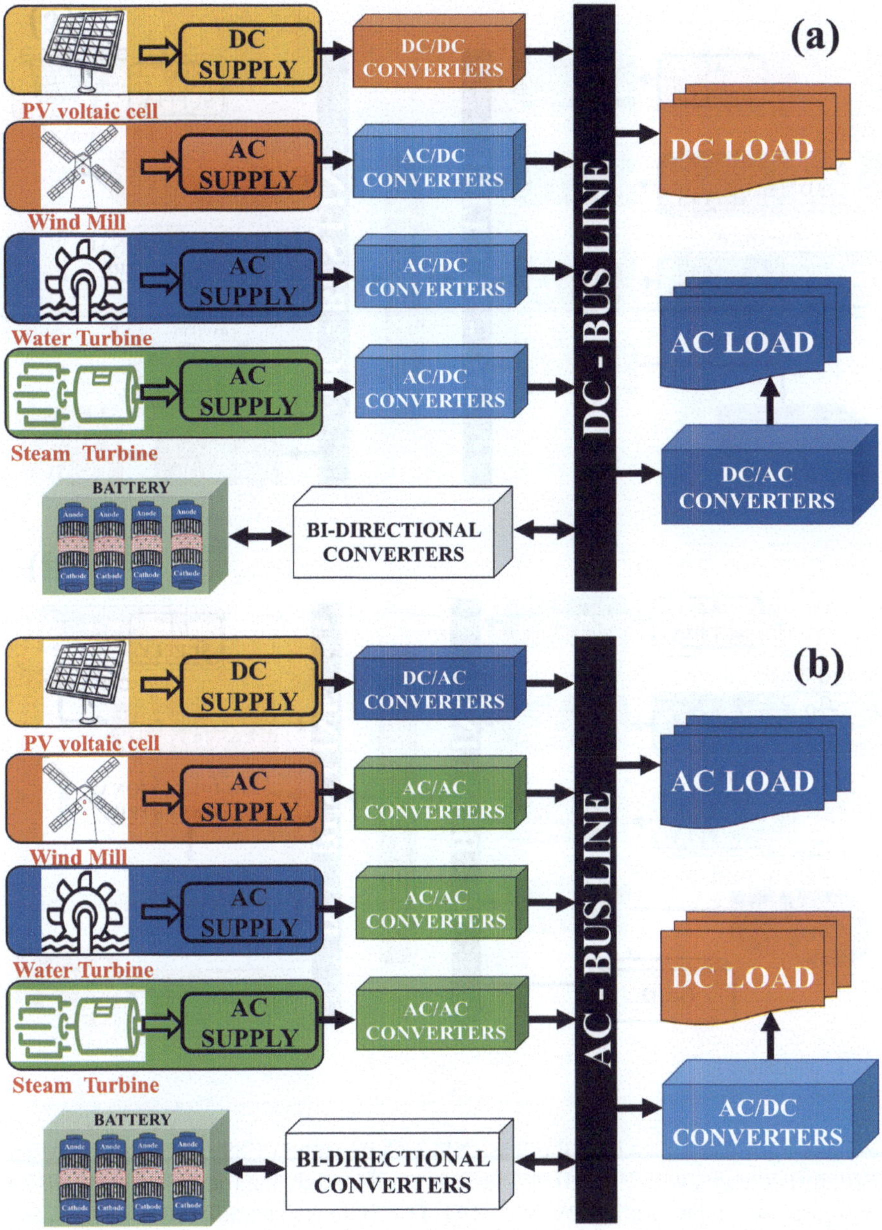

Fig. 2 Architecture of hybrid renewable energy system **a** DC-coupled isolated system, **b** AC-coupled isolated system, **c** hybrid-coupled isolated system, and **d** hybrid-coupled grid-connected system

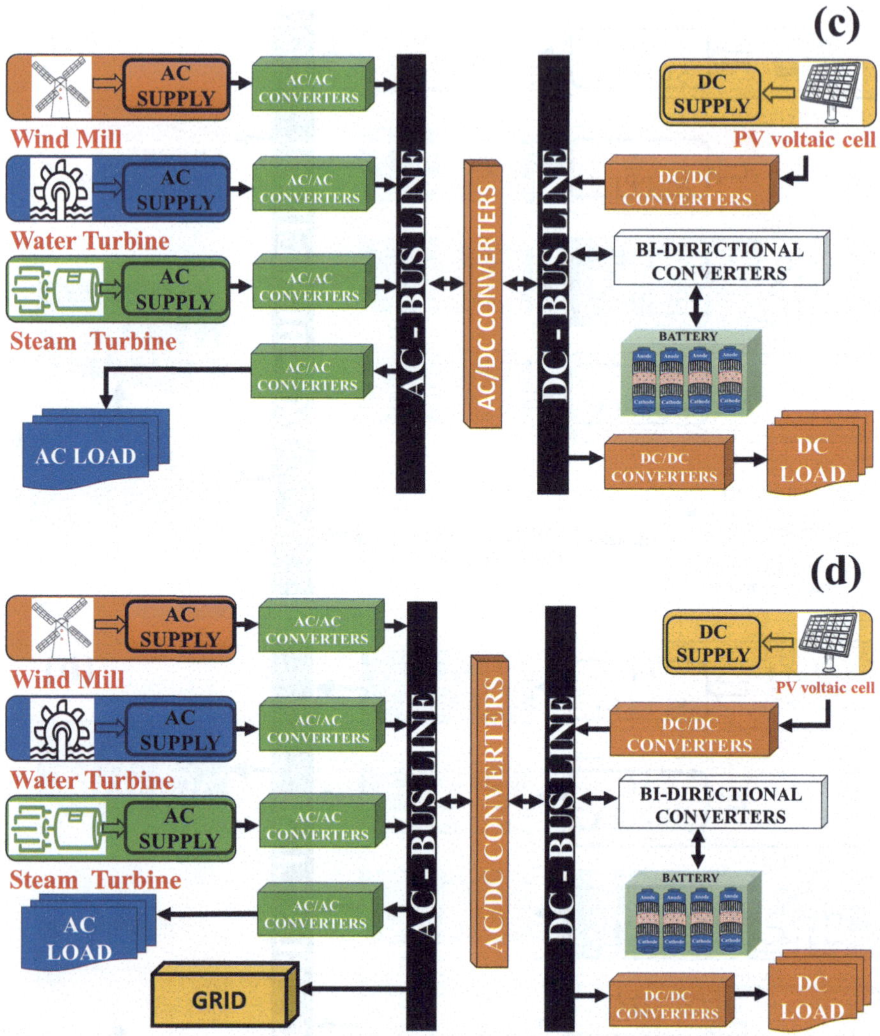

Fig. 2 (continued)

BAT configurations provided the least NPC, COE value of 899,186 $ and 0.046$/kWh, respectively, with RF of 0.997 [46]. A similar study having 11 different HRES configurations incorporating PV, WT, BAT, DG, biogas generator (BG), CONV, and electrolyser (ELEC) for a village Satjelia, in West Bengal, India, revealed that the PV/WT/BAT/converter (CONV)/DG configuration gives the lowest levelized cost of energy (LCOE) of 0.319$/kWh with NPC of 6.81M$ for the energy produced. However, considering both environmental impact and comparative economic performance (i.e., low LCOE and overall costs), the PV/WT/BAT/CONV/BG/ELEC configuration was deemed best among all 11 configurations [47].

Four different locations across Iran, Moaleman, Marvdasht, Nikouyeh, and Ghadamgah, were selected to study the feasibility of an off-grid energy system by varying components and their size, having components WT and PV systems for generation, with battery storage and fuel cells as backup. The hybrid system was found to be the most economical in all regions, with the least COE in the Moaleman region, equal to 0.54$/kWh [48]. A grid-integrated HRES was investigated for a hotel in Petra, Jordan, yielding a total NPC of USD 1.16 million and an energy cost of USD 0.0415 per kWh. The system provided an effective internal rate of return of 15.8% with a 77% reduction in carbon emissions relative to the baseline scenario [49]. These examples illustrate that HRES is a practically feasible solution for fulfilling energy needs at an affordable cost, even for remote locations.

6 Integration of HRES with IoT

HRES poses great potential to achieve reliable and low-cost energy, which is the primary aim of every country around the globe. However, HRES is facing challenges with its efficiency that can be cured using better architectures like a hybrid-coupled grid system. These HRES architectures introduce problems like complexity and stability because of the addition of multiple layers of devices [17, 43]. But with rapid technological upgrades like the introduction of IoT technologies and machine learning (ML) have paved a new path to deal with complex algorithms or systems seamlessly. IoT is characterized by its interconnecting ability with sensors, clouds, etc. to exchange data and enhance the analysis and computation capability of overall system performance. The integration of IoT with HRES architecture can make it a smart system that can provide cheap, reliable, and uninterrupted energy using real-time monitoring and algorithm-based decision-making on its own without any human intervention [50–52].

7 Concept of IoT

The name IoT simply means "internet built around things." IoT is a network of interconnected cyber-physical systems like sensors, actuators, clouds, automobiles, appliances, and Bluetooth to sense, collect, and manage data for a given system [51, 53]. It has applications in numerous fields like sustainable supply chain management [54], smart home monitoring systems [55], smart campus [43], food and agricultural sustainability [56], and hybrid EMS [57]. Integrating IoT with HRES ensures that the load supply remains uninterrupted under all conditions.

7.1 General Architecture of IoT

The architecture of an IoT is a structured framework of various layers that facilitates collaboration and interaction of devices, software, and embedded systems with each other to fulfill objectives like better handling and real-time monitoring. The basic architecture of an IoT integrated system is divided into three main layers. They are the device layer, communication layer, and application layer [31, 43, 58, 59].

7.1.1 Device Layer

This layer perceives data from the environment of the system and is hence also named as perception layer. It senses the variables or features required for study like the voltage and current, using components like sensors, actuators, and smart devices like thermostats and microcontrollers. The components in this layer are smart enough to sense, interact, and preprocess data using the computational capabilities of this layer, also called edge computing [58, 60].

7.1.2 Communication Layer

The communication layer is the aggregation of data via different communication networks like Message Queuing Telemetry Transport (MQTT), Wi-Fi, Ethernet, Bluetooth, ZigBee, or cellular network (e.g., 4G and 5G), called a connectivity protocol between device layer components, i.e., sensors and actuators, and the application layer. It simply acts as a gateway for the data [53].

7.1.3 Application Layer

This layer includes data analysis and processing elements like cloud computing, advanced algorithms like ML, and IoT platforms that compute and manage sensors, actuators, and other devices to attain the desired objective for which IoT is designed. This layer is responsible for decision-making based on certain algorithms and constraints provided by the user of the system [53, 58]. Figure 3 shows the three-layered architecture of the IoT system, detailing its devices, along with its application and protocol used within each layer. The IoT-based system also contains some additional layer that deals with the safety and security of the overall system from malware and other cyber threats.

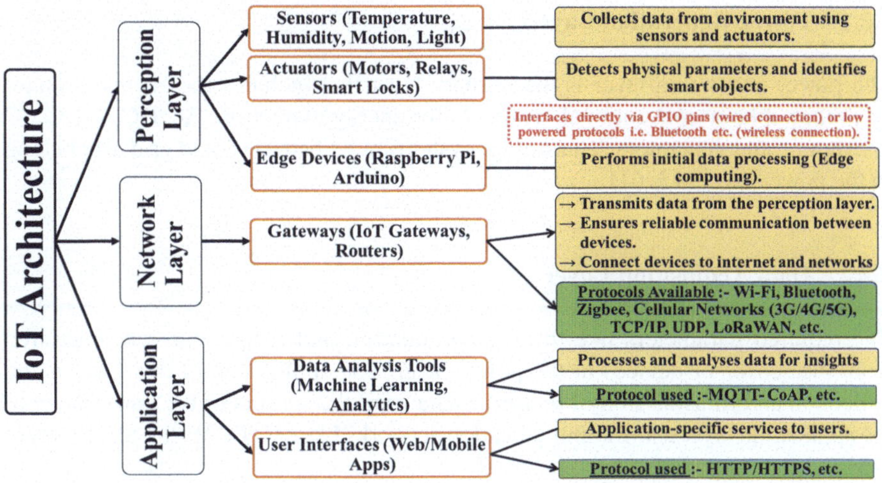

Fig. 3 Architecture of IoT

7.2 *Architecture of IoT-Integrated Hybrid Renewable Energy System*

The IoT architecture involves sensing, communication, and analysis tasks in the integrated framework, while the HRES adds the layer of a power generation system or generator system with different converters and ESS into it. This combination of IoT with HRES will result in a four-layer framework collectively called an IoT-integrated HRES. The system aims to provide a cheap, reliable, and continuous power supply by using the available RES in the best way possible [7, 43]. The different layers of IoT-integrated HRES are shown in Fig. 4.

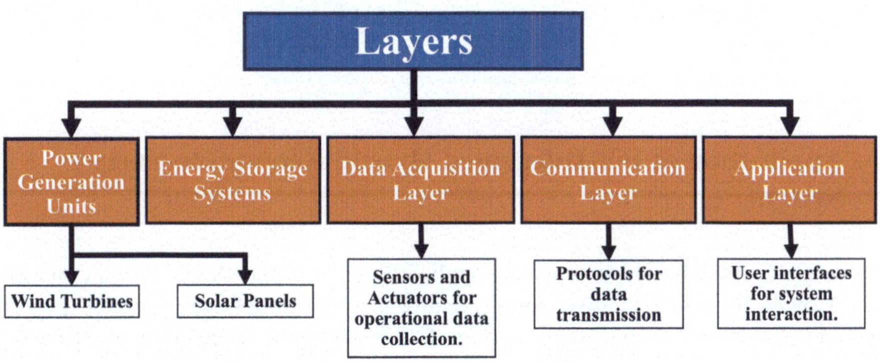

Fig. 4 Layers of IoT-integrated hybrid renewable energy systems

7.2.1 Power Generation Layer

The power generation layer is the primary layer that includes the energy sources, i.e., renewable or conventional, available for energy extraction, AC/DC or DC/AC converters, and generators. The storage systems (e.g., batteries) and grid also belong to the power layer [43, 61].

7.2.2 Data Acquisition Layer

The required parameters like the voltage, current, and temperature are measured from the sensor or devices in the power layer. The application of this layer is the collection and transmission of real-time data from the sensors of the power layer to the application layer using communication networks like MQTT, Wi-Fi, and Ethernet [43].

7.2.3 Communication Layer

The communication protocols or technologies used are either wired or wireless, which are managed by this layer. The different protocols depend on range, type, privacy and data security, and environmental conditions of the application [62].

7.2.4 Application Layer

The final layer performs applications like data analysis and processing, decision making, and prediction. The result of this analysis helps in determining the appropriate action to optimize the performance of HRES. Operations like energy consumption and demand response strategies are done by the application layer [63]. Figure 5 represents a detailed schematic of a four-layered IoT-integrated HRES system along with its devices.

8 Application of IoT-Integrated Hybrid Renewable Energy System

The merits of IoT have led to numerous applications in different domains of the energy sector, including efficient power supply, real-time monitoring systems, remote power generation, and smart home energy systems [43, 64–68]. The application of IoT-based devices, big data, and smart sensing systems to a residential complex with 20 houses with two different categories of households based on the number of family members living (i.e., 2 or 3) was studied to optimize resident power consumption without

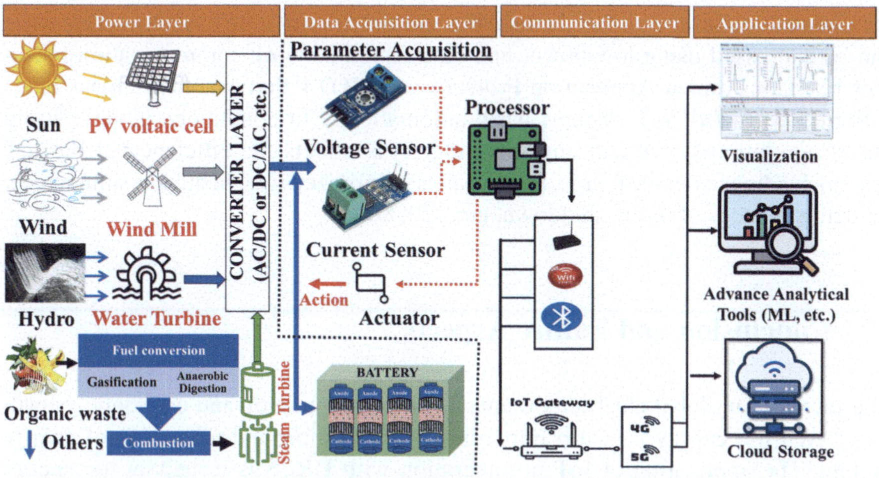

Fig. 5 Schematic diagram of IoT-integrated hybrid renewable energy system

affecting day-to-day life. The use of ML for forecasting the behavioral pattern of power usage gave insights to recommend suggestions to minimize power consumption and increase overall efficiency [64]. A smart highway power supply system was built using a PV and vertical axis wind turbine-based IoT system for lightning LED bulbs and emergency e-vehicle charging without the need for manpower. The system also featured services fault detection, preventive alerts, etc. [68]. IoT integration of HRES is studied for building smart infrastructure for electric vehicles (EVs) charging for local applications [65]. Development of IoT-driven smart microgrid integration system driven by Ethernet and Wi-Fi was done that enabling seamless integration of a grid at a university campus in Saudi Arabia. The system is powered using a hybrid system of PV, wind, and DG connected with a battery [43]. IoT-driven monitoring system was built for a remote NG pipeline station powered by a hybrid energy system with solar cells and supported by DG and battery system [66].

9 Challenges and Limitations

The architectural requirements of IoT are scalability, security, interoperability, and efficiency [69]. The IoT system faces scalability issues with the increase in connected devices, as it generates vast amounts of data that will require larger storage, automatic analysis, faster processors, and better management strategies, which adds to the cost of IoT [69–72]. The increased number of devices makes the system vulnerable to unauthorized access, causing data breaches, privacy violations, etc., leading to manipulation and information sniffing. It can be minimized using robust encryption, secure communication channels, etc. [73–75]. Different devices have varying dynamics and communication protocols that degrade the interoperability in IoT systems. This issue can be addressed using low-power and diminished network communication protocols like Constrained Application Protocol and MQTT that enable seamless interaction [59, 76–78]. IoT systems work automatically in continuous mode, getting on/off continuously without any servicing. This affects the efficiency of the IoT system, leading to device failure, which can be minimized by scheduled maintenance, reducing chances of direct shutdown, etc. [79, 80].

10 Conclusion and Future Aspects

The increase in global electricity demands due to population and economic growth has driven the energy scenario into constructing a more hybrid source of energy system. The application of IoT in integration with HRES is reshaping the energy landscape by addressing key challenges such as intermittency and inefficiency of RES. It also caters to capabilities for real-time monitoring, control, and optimization over the individual, as well as collective output from HRES. It also provides seamless coordination between multiple energy sources, making HRES a robust solution for sustainable power generation.

With the advancement of technologies like AI, ML, and smart grids, the prospects and potential of IoT in HRES will only grow with time. These advancements will help in increasing the predictive capabilities of IoT systems, which will help in effectively meeting the fluctuating demands and environmental needs. It also enhances resource allocation and adaptability of energy systems, which will indirectly increase the efficiency of hybrid RES. The optimization of HRES performance will increase renewable fraction in mixed energy, reducing dependency on conventional energy.

In conclusion, the application of multiple RES represents a crucial step toward achieving a sustainable and energy-efficient future. The integration of IoT with HRES can overcome the traditional limitations of RES, offering scalable and reliable solutions for diverse applications.

References

1. Kumar PH, Alluraiah NC, Gopi P, Bajaj M, Kumar S, Kalyan CHNS et al (2025) Techno-economic optimization and sensitivity analysis of off-grid hybrid renewable energy systems: a case study for sustainable energy solutions in rural India. Results Eng 25:103674. https://doi.org/10.1016/j.rineng.2024.103674
2. Peter Zeniewski AV, Alšauskas O, Arsalane Y, Buisson E, Chen O, Crow D, D'Ambrosio D, Dallard J, De Bienassis T, De Oliveira Bredariol T, Erdogan M, García V, Hays T-M, Hilaire J, Hopewell H, Kim T-Y, Kueppers M, Martinos A, Apostolos Petropoulos RP, Roge A, Saive G, Singh S, Spencer T, Ryota Taniguchi GT, Vautrin A, Zeniewski P (2024) World Energy Outlook 2024. IEA, pp 1–398

3. Dubey A, Tripathi A (2020) Review on grid connected solar wind hybrid power based IoT system. Int J Adv Res Electr Electron Instrum Eng 9(3):329–332. https://doi.org/10.15662/IJAREEIE.2020.0903007
4. Rupali Kindarle KG, Mayur Kodape,, Parimal Kanpartiwar PB, Sandip Thakre (2024) IoT based hybrid energy system for monitoring continuous power flow. Int Res J Modern Eng Technol Sci 06(03):1899–1908. https://doi.org/10.56726/IRJMETS50461
5. Nwagu CN, Ujah CO, Kallon DVV, Aigbodion VS (2025) Integrating solar and wind energy into the electricity grid for improved power accessibility. Unconvent Resour 5:100129. https://doi.org/10.1016/j.uncres.2024.100129
6. Panwar NL, Kaushik SC, Kothari S (2011) Role of renewable energy sources in environmental protection: a review. Renew Sustain Energy Rev 15(3):1513–1524. https://doi.org/10.1016/j.rser.2010.11.037
7. Mohd Farid MFN, Bohari AA (2024) Development of HRES with monitoring system. Prog Eng Appl Technol 5(2):254–265. https://doi.org/10.30880/peat.2024.05.02.026
8. Olatomiwa L, Mekhilef S, Ismail MS, Moghavvemi M (2016) Energy management strategies in hybrid renewable energy systems: a review. Renew Sustain Energy Rev 62:821–835. https://doi.org/10.1016/j.rser.2016.05.040
9. Dehghani F, Shafiyi MA (2023) Integration of hybrid renewable energy sources with the power system considering their economic complementarity. IET Renew Power Gener 17(15):3638–3650. https://doi.org/10.1049/rpg2.12871
10. Ajiboye AA, Popoola SI, Atayero AA (2018) Hybrid renewable energy systems: opportunities and challenges in sub-Saharan Africa. In: Proceedings of the international conference on industrial engineering and operations management, pp 1110–1116
11. Mohammed YS, Mustafa MW, Bashir N (2014) Hybrid renewable energy systems for off-grid electric power: review of substantial issues. Renew Sustain Energy Rev 35:527–539. https://doi.org/10.1016/j.rser.2014.04.022
12. Deshmukh MK, Deshmukh SS (2008) Modeling of hybrid renewable energy systems. Renew Sustain Energy Rev 12(1):235–249. https://doi.org/10.1016/j.rser.2006.07.011
13. Agajie EF, Agajie TF, Amoussou I, Fopah-Lele A, Nsanyuy WB, Khan B et al (2024) Optimization of off-grid hybrid renewable energy systems for cost-effective and reliable power supply in Gaita Selassie Ethiopia. Sci Rep 14(1):10929. https://doi.org/10.1038/s41598-024-61783-z
14. Sinha S, Chandel SS (2014) Review of software tools for hybrid renewable energy systems. Renew Sustain Energy Rev 32:192–205. https://doi.org/10.1016/j.rser.2014.01.035
15. Pandiyan P, Saravanan S, Usha K, Kannadasan R, Alsharif MH, Kim M-K (2023) Technological advancements toward smart energy management in smart cities. Energy Rep 10:648–677. https://doi.org/10.1016/j.egyr.2023.07.021
16. (DOE) USDoE (2021) Hybrid energy systems: opportunities for coordinated research. National Renewable Energy Laboratory, DOE/GO-102021-5447, CO
17. Singh R, Bansal RC (2018) Review of HRESs based on storage options, system architecture and optimisation criteria and methodologies. IET Renew Power Gener 12(7):747–760. https://doi.org/10.1049/iet-rpg.2017.0603
18. Ferrari L, Bianchini A, Galli G, Ferrara G, Carnevale EA (2018) Influence of actual component characteristics on the optimal energy mix of a photovoltaic-wind-diesel hybrid system for a remote off-grid application. J Clean Prod 178:206–219. https://doi.org/10.1016/j.jclepro.2018.01.032
19. Yahya W, Saied KM, Nassar A, Qader MR, Al-Nehari M, Zarabia J et al (2024) Optimization of a hybrid renewable energy system consisting of a of PV/wind turbine/battery/fuel cell integration and component design. Int J Hydrogen Energy 94:1406–1418. https://doi.org/10.1016/j.ijhydene.2024.11.187
20. Bourek Y, Messini EMB, Ammari C, Guenoune M, Chabira B, Saha BK (2024) A hybrid renewable energy system for Hassi Messaoud region of Algeria: modeling and optimal sizing. Energy Storag Saving. https://doi.org/10.1016/j.enss.2024.10.002

21. León Gómez JC, De León Aldaco SE, Aguayo AJ (2023) A review of hybrid renewable energy systems: architectures, battery systems, and optimization techniques. Eng. https://doi.org/10.3390/eng4020084
22. Ul Islam Z, Hossain Lipu MS, Karim TF, Fuad AM, Ali MMN, Shihavuddin ASM et al (2024) Optimizing hybrid renewable energy based automated railway level crossing in Bangladesh: techno-economic, emission and sensitivity analysis. Energy Convers Manag: X 24:100744. https://doi.org/10.1016/j.ecmx.2024.100744
23. Rekioua D, Bensmail S, Bettar N (2014) Development of hybrid photovoltaic-fuel cell system for stand-alone application. Int J Hydrogen Energy 39(3):1604–1611. https://doi.org/10.1016/j.ijhydene.2013.03.040
24. Walmsley TG, Philipp M, Picón-Núñez M, Meschede H, Taylor MT, Schlosser F et al (2023) Hybrid renewable energy utility systems for industrial sites: a review. Renew Sustain Energy Rev 188:113802. https://doi.org/10.1016/j.rser.2023.113802
25. Ataei A, Nedaei M, Rashidi R, Yoo C (2015) Optimum design of an off-grid hybrid renewable energy system for an office building. J Renew Sustain Energy 7(5). https://doi.org/10.1063/1.4934659
26. Kosmadakis G, Karellas S, Kakaras E (2013) Renewable and conventional electricity generation systems: technologies and diversity of energy systems. In: Michalena E, Hills JM (eds) Renewable energy governance: complexities and challenges. Springer, London, pp 9–30
27. Manwell J (2004) Hybrid energy systems. Encycl Energy 3:215–229
28. Sreeraj ES, Chatterjee K, Bandyopadhyay S (2010) Design of isolated renewable hybrid power systems. Sol Energy 84(7):1124–1136. https://doi.org/10.1016/j.solener.2010.03.017
29. Thirugnanasambandam M, Iniyan S, Goic R (2010) A review of solar thermal technologies. Renew Sustain Energy Rev 14(1):312–322. https://doi.org/10.1016/j.rser.2009.07.014
30. Kannan N, Vakeesan D (2016) Solar energy for future world—a review. Renew Sustain Energy Rev 62:1092–1105. https://doi.org/10.1016/j.rser.2016.05.022
31. He B, Ismail N, Leng KKK, Chen G (2023) Techno-economic analysis of an HRES with fuel cells, solar panels, and wind turbines using an improved Al-Biruni algorithm. Heliyon 9(12). https://doi.org/10.1016/j.heliyon.2023.e22828
32. Joselin Herbert GM, Iniyan S, Sreevalsan E, Rajapandian S (2007) A review of wind energy technologies. Renew Sustain Energy Rev 11(6):1117–1145. https://doi.org/10.1016/j.rser.2005.08.004
33. Manyonge AW, Ochieng R, Onyango F, Shichikha J (2012) Mathematical modelling of wind turbine in a wind energy conversion system: power coefficient analysis. Appl Math Sci 6(91):4527–4536
34. India Go (2018) Executive Summary on Power Sector (July-2018). Executive Summary on Power Sector, pp 1–54
35. India Go (2024) Executive Summary on Power Sector (November-2024). Executive Summary on Power Sector, pp 1–65
36. Gatte MT, Kadhim RA (2012) Hydro power. Energy Conserv 9(51000):95–124. https://doi.org/10.5772/52269
37. Chakraborty S, Ahmad MI, Guin A, Mukherjee S, Goswami RP, Roy R (2015) Hydropower: its amazing potential—a theoretical perspective. J Civ Environ Eng 2(12):56–60
38. Chen J, Yang HX, Liu CP, Lau CH, Lo M (2013) A novel vertical axis water turbine for power generation from water pipelines. Energy 54:184–193. https://doi.org/10.1016/j.energy.2013.01.064
39. Loucks DP, van Beek E (2017) River Basin Modeling. In: Loucks DP, van Beek E (eds) Water resource systems planning and management: An introduction to methods, models, and applications. Springer, Cham, pp 469–526
40. Ramesh M, Saini RP (2024) Effect of different batteries and diesel generator on the performance of a stand-alone hybrid renewable energy system. Energy Sour Part A: Recov Utiliz Environ Effects 46(1):7983–8005. https://doi.org/10.1080/15567036.2020.1763520
41. Basnet S, Deschinkel K, Le Moyne L, Cécile PM (2023) A review on recent standalone and grid integrated hybrid renewable energy systems: system optimization and energy management strategies. Renew Energy Focus 46:103–125. https://doi.org/10.1016/j.ref.2023.06.001

42. Akinyele D (2017) Techno-economic design and performance analysis of nanogrid systems for households in energy-poor villages. Sustain Cities Soc 34:335–357. https://doi.org/10.1016/j.scs.2017.07.004
43. Eltamaly AM, Alotaibi MA, Alolah AI, Ahmed MA (2021) IoT-based hybrid renewable energy system for smart campus. Sustainability. https://doi.org/10.3390/su13158555
44. Arul PG, Ramachandaramurthy VK, Rajkumar RK (2015) Control strategies for a hybrid renewable energy system: a review. Renew Sustain Energy Rev 42:597–608. https://doi.org/10.1016/j.rser.2014.10.062
45. Nehrir MH, Wang C, Strunz K, Aki H, Ramakumar R, Bing J et al (2011) A review of hybrid renewable/alternative energy systems for electric power generation: configurations, control, and applications. IEEE Trans Sustain Energy 2(4):392–403. https://doi.org/10.1109/TSTE.2011.2157540
46. Krishan O, Sathans (2018) Design and techno-economic analysis of a HRES in a rural village. Procedia Comput Sci 125:321–328. https://doi.org/10.1016/j.procs.2017.12.043
47. Roy D, Wang R, Roy S, Smallbone A, Roskilly AP (2024) Hybrid renewable energy systems for sustainable power supply in remote location: techno-economic and environmental assessment. Energy Convers Manag: X 24:100793. https://doi.org/10.1016/j.ecmx.2024.100793
48. Hosseinalizadeh R, Shakouri GH, Amalnick Mohsen S, Taghipour P (2016) Economic sizing of a hybrid (PV–WT–FC) renewable energy system (HRES) for stand-alone usages by an optimization-simulation model: case study of Iran. Renew Sustain Energy Rev 54:139–150. https://doi.org/10.1016/j.rser.2015.09.046
49. Al-Rawashdeh H, Al-Khashman OA, Al Bdour JT, Gomaa MR, Rezk H, Marashli A et al (2023) Performance analysis of a hybrid renewable-energy system for green buildings to improve efficiency and reduce GHG emissions with multiple scenarios. Sustainability. https://doi.org/10.3390/su15097529
50. Hamed Al Hajri N, Naji Al Harthi R, Pasam GK, Natarajan R (2024) IoT and machine learning based green energy generation using hybrid renewable energy sources of solar, wind and hydrogen fuel cells. E3S Web Conf 472. https://doi.org/10.1051/e3sconf/202447201008
51. Munirathinam S (2020) Chapter Six—Industry 4.0: industrial internet of things (IIOT). In: Raj P, Evangeline P (eds) Advances in computers. Elsevier, pp 129–164
52. Mohammed AM, Sadeeq SZ (2021) Energy management for internet of things via distributed systems. J Appl Sci Technol Trends 2(02):80–92. https://doi.org/10.38094/jastt20285
53. Domínguez-Bolaño T, Campos O, Barral V, Escudero CJ, García-Naya JA (2022) An overview of IoT architectures, technologies, and existing open-source projects. Internet Things 20:100626. https://doi.org/10.1016/j.iot.2022.100626
54. Khan Y, Su'ud MB, Alam MM, Ahmad SF, Ahmad AYAB, Khan N (2023) Application of Internet of Things (IoT) in sustainable supply chain management. Sustainability. https://doi.org/10.3390/su15010694
55. Al-Ali AR, Zualkernan IA, Rashid M, Gupta R, Alikarar M (2017) A smart home energy management system using IoT and big data analytics approach. IEEE Trans Consum Electron 63(4):426–434. https://doi.org/10.1109/TCE.2017.015014
56. Morchid A, El Alami R, Raezah AA, Sabbar Y (2024) Applications of internet of things (IoT) and sensors technology to increase food security and agricultural sustainability: benefits and challenges. Ain Shams Eng J 15(3):102509. https://doi.org/10.1016/j.asej.2023.102509
57. Eltamaly AM, Ahmed MA (2023) Performance evaluation of communication infrastructure for peer-to-peer energy trading in community microgrids. Energies. https://doi.org/10.3390/en16135116
58. Lombardi M, Pascale F, Santaniello D (2021) Internet of things: a general overview between architectures, protocols and applications. Information. https://doi.org/10.3390/info12020087
59. Hou L, Zhao S, Xiong X, Zheng K, Chatzimisios P, Hossain MS et al (2016) Internet of things cloud: architecture and implementation. IEEE Commun Mag 54(12):32–39. https://doi.org/10.1109/MCOM.2016.1600398CM
60. Laghari AA, Wu K, Laghari RA, Ali M, Khan AA (2022) RETRACTED ARTICLE: A review and state of art of internet of things (IoT). Arch Comput Methods Eng 29(3):1395–1413. https://doi.org/10.1007/s11831-021-09622-6

61. Stefanov A, Liu CC (2012) ICT modeling for integrated simulation of cyber-physical power systems. In: 2012 3rd IEEE PES innovative smart grid technologies Europe (ISGT Europe), pp 1–8
62. Shakerighadi B, Anvari-Moghaddam A, Vasquez JC, Guerrero JM (2018) Internet of things for modern energy systems: state-of-the-art, challenges, and open issues. Energies. https://doi.org/10.3390/en11051252
63. Abir SMAA, Anwar A, Choi J, Kayes ASM (2021) IoT-enabled smart energy grid: applications and challenges. IEEE Access 9:50961–50981. https://doi.org/10.1109/ACCESS.2021.3067331
64. Paredes-Valverde MA, Alor-Hernández G, García-Alcaráz JL, Salas-Zárate MdP, Colombo-Mendoza LO, Sánchez-Cervantes JL (2020) IntelliHome: an internet of things-based system for electrical energy saving in smart home environment. Comput Intell 36(1):203–224. https://doi.org/10.1111/coin.12252
65. Banik A, Sengupta A, Das P (2010) Smart hybrid next generation charging infrastructure for EVs. IJSART 7(3):123–130
66. Waqas M, Jamil M (2024) Smart IoT SCADA system for hybrid power monitoring in remote natural gas pipeline control stations. Electronics. https://doi.org/10.3390/electronics13163235
67. Rao CK, Sahoo SK, Yanine FF (2024) A literature review on an IoT-based intelligent smart energy management systems for PV power generation. Hybrid Adv 5:100136. https://doi.org/10.1016/j.hybadv.2023.100136
68. Hans MR, Tamhane MA (2020) IoT based hybrid green energy driven street lighting system. In: 2020 fourth international conference on I-SMAC (IoT in social, mobile, analytics and cloud) (I-SMAC), pp 35–41
69. Samizadeh Nikoui T, Rahmani AM, Balador A, Haj Seyyed Javadi H (2021) Internet of Things architecture challenges: a systematic review. Int J Commun Syst 34(4):e4678. https://doi.org/10.1002/dac.4678
70. Tekeste T, Saleh H, Mohammad B, Ismail M (2019) Ultra-low power QRS detection and ECG compression architecture for IoT healthcare devices. IEEE Trans Circuits Syst I Regul Pap 66(2):669–679. https://doi.org/10.1109/TCSI.2018.2867746
71. Soltesz S, Pötzl H, Fiuczynski ME, Bavier A, Peterson L (2007) Container-based operating system virtualization: a scalable, high-performance alternative to hypervisors. In: Proceedings of the 2nd ACM SIGOPS/EuroSys European conference on computer systems 2007. Association for Computing Machinery, Lisbon, Portugal, pp 275–287
72. Vinueza Naranjo PG, Baccarelli E, Scarpiniti M (2018) Design and energy-efficient resource management of virtualized networked Fog architectures for the real-time support of IoT applications. J Supercomput 74(6):2470–2507. https://doi.org/10.1007/s11227-018-2274-0
73. Usamentiaga R, Fernandez MA, Villan AF, Carus JL (2018) Temperature monitoring for electrical substations using infrared thermography: architecture for industrial internet of things. IEEE Trans Ind Inf 14(12):5667–5677. https://doi.org/10.1109/TII.2018.2868452
74. Pape S, Rannenberg K (2019) Applying privacy patterns to the internet of things' (IoT) architecture. Mob Netw Appl 24(3):925–933. https://doi.org/10.1007/s11036-018-1148-2
75. Vučinić M, Tourancheau B, Rousseau F, Duda A, Damon L, Guizzetti R (2015) OSCAR: object security architecture for the Internet of Things. Ad Hoc Netw 32:3–16. https://doi.org/10.1016/j.adhoc.2014.12.005
76. Sarkar C, Akshay Uttama Nambi SN, Prasad RV, Rahim A, Neisse R, Baldini G (2015) DIAT: a scalable distributed architecture for IoT. IEEE Internet Things J 2(3):230–239. https://doi.org/10.1109/JIOT.2014.2387155
77. Raza U, Kulkarni P, Sooriyabandara M (2017) Low power wide area networks: an overview. IEEE Commun Surv Tutor 19(2):855–873. https://doi.org/10.1109/COMST.2017.2652320
78. Jones CE, Sivalingam KM, Agrawal P, Chen JC (2001) A survey of energy efficient network protocols for wireless networks. Wireless Netw 7(4):343–358. https://doi.org/10.1023/A:1016627727877

79. Catarinucci L, Donno Dd, Mainetti L, Palano L, Patrono L, Stefanizzi ML et al (2015) An IoT-aware architecture for smart healthcare systems. IEEE Internet Things J 2(6):515–526. https://doi.org/10.1109/JIOT.2015.2417684
80. Nikoui TS, Rahmani AM, Tabarsaied H (2019) Data management in fog computing. Fog and edge computing: principles and paradigms, pp 171–190

Introduction and Classification of Fuel Energy

Priti Gupta, Shilpa Sharma, Shagufta Jabin, and Ritu Bir

Abstract The significance of energy cannot be overstated. It may be sufficient to illustrate this argument by imagining where humanity would be without it. Similar to the past, energy will be crucial in determining the future of cities, countries, and the entire world. This thought experiment underscores just how fundamental energy is to every facet of modern life and demonstrates why its significance is so profound. This chapter delves into the definition of energy and its various forms in order to have a better understanding of the role that energy plays in the world, both now and in the future, with global impact. There will likely be a significant change in the relative positions of the main engines driving the global economy in the near future. Fuel energy is one of the most diverse and significant kinds of energy. The introduction and classification of fuel energy have been discussed in this chapter exclusively.

Keywords Sustainable energy · Fossil fuels · Hydrogen energy · Wind energy · Solar energy · Tidal energy

P. Gupta · S. Sharma
Department of Chemistry, Manav Rachna University, Faridabad, Haryana, India
e-mail: priti@mru.edu.in

S. Sharma
e-mail: shilpa@mru.edu.in

S. Jabin (✉)
Department of Applied Sciences, Manav Rachna International Institute of Research and Studies, Faridabad, Haryana, India
e-mail: shagufta.set@mriu.edu.in

R. Bir
Department of Applied Sciences, Greater Noida Institute of Technology, Greater Noida, India

J. A. Siddique et al. (eds.), *Energy Sources*, Green Energy and Technology,
https://doi.org/10.1007/978-981-95-1209-6_11

1 Introduction

One of the significant drivers of growth and prosperity of any country is its energy resources and distribution among all stakeholders. Therefore, Energy is one of the most important and critical aspects of society and is closely linked to mankind. In the broadest meaning of the word, energy fuels our daily lives, and the issues of sustainable energy have permeated everybody's lives. This advances to the fact that the use of fuel energy is becoming increasingly popular, and its preservation is the need of the hour.

Fuel energy is the major source of energy in the world. Consciousness about the need for alternative sources of energy has improved multiple-fold globally. Better alternative energy resources are being explored to reduce carbon footprint and carbon emission [1]. Studies are being done on various factors influencing the consumption of fuel energy using demand modelling modules. This has been done due to an understanding of how fuel energy contributes to the growth of a sustainable energy system. The fuel energy can be stored and can be released whenever required. Energy can be derived from fuel and can be broadly classified as fossil fuel, biodiesel, renewable energy, hydrogen fuel, nuclear fuel, etc. [2]. Fossil fuels include coal (38%), oil (14%), and natural gas (28%). The utilization of fossil fuels lowers the capital expenditure of those countries that depend massively on fossil fuels as compared to other energy sources [3]. The demand for energy is increasing day by day, and it is expected that it will rise to 1.3% every year till 2040 due to an increase in population, global economic growth, shifting of the larger population from rural areas to urban areas, especially in developing nations [4]. Presently, 35 billion tons of CO_2 are released in the environment due to very large energy demand [5]. Out of 35 billion tons of CO_2, 90% is generated from fossil fuels only. The remaining 10% is contributed by all other kinds of energy resources. The leading countries generating high amounts of CO_2 into the environment are Canada and Iran, followed by many other countries. The major CO_2 contributors are illustrated in Fig. 1 [5].

Carbon emissions and energy consumption are directly correlated with one another. So, the "International Renewable Energy Agency" (IRENA) is setting the target to reduce the carbon footprint up to 30% by 2030 [6]. IRENA is more focused on greener and cleaner fuel energy as compared to fossil fuels. The World Energy Outlook in 2020 has targeted net zero emissions by the year 2050 for a sustainable and safe environment [7]. So, cleaner fuel energy like wind energy, hydrogen energy, solar energy, and tidal energy is the most vital type of green energy resource. So, multiple studies are going on to accelerate the world toward greener and cleaner energy to minimize carbon footprint and to make the environment sustainable [1].

Fuel energy based on fossil fuels is a source of carbon footprint. So, replacing it with other clean fuel energy is a challenging task. So that CO_2 emissions and other atmospheric pollutants, viz., SO_2 and Nox, present in the climate can be minimized. There are different policies that are aimed at reducing pollution so that the health of human beings can be improved. However, fuel energy development is being done globally with respect to installation, generation, and storage capacity. As per latest

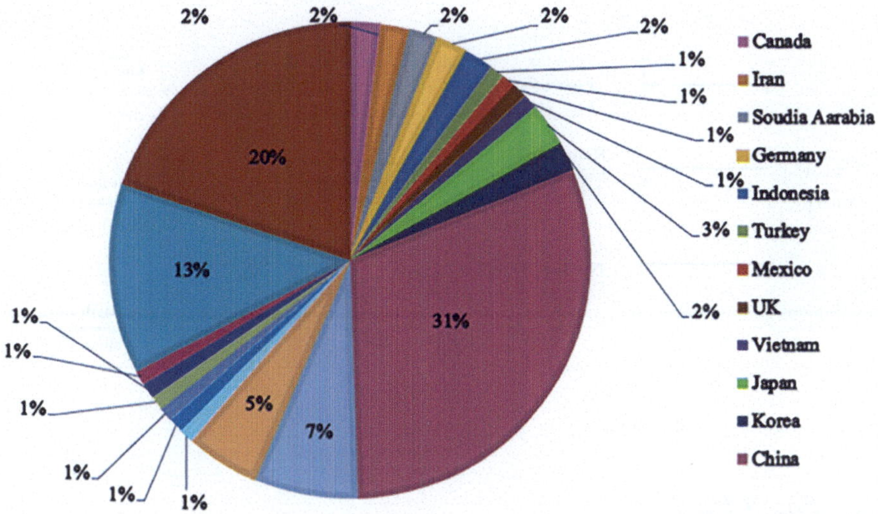

Fig. 1 Major countries with major CO_2 emissions globally [5]

studies, the growth of fuel energy is going on at a remarkable speed. Solar energy is more feasible in South and North Africa as well as in the Middle East [4]. The complete cycle from procuring raw materials to the generation of fuel energy has a direct correlation with human health. Depending upon human health, the fuel energy is divided into three categories. The first category is involved in the generation of fossil fuels such as natural gas, wood, and coal, whose usage causes air pollution and global warming. The second category involves renewable energy, which includes biofuel energy, solar energy, hydrogen energy, wind energy, and hydro energy. The second category results in low emission of toxic gases. The last category involves nuclear energy. Nuclear energy is mainly based on nuclear fission techniques. It produces minimum waste compared to the first and second categories.

2 Classifications of Fuel Energy

Over the last decade, there has been a significant shift in the requirement of fuel energy due to overpopulation and huge energy requirements. The large amount of requirement for fuel energy affects water resources, land resources, and the climate. Fuel energy includes fossil fuels, renewable energy, and nuclear energy. Renewable energy is further categorized into various kinds. In current study, biofuel energy, solar energy, hydrogen energy, wind energy, and hydro power energy have been discussed under renewable energy. The classification of fuel energy has been shown in Fig. 2.

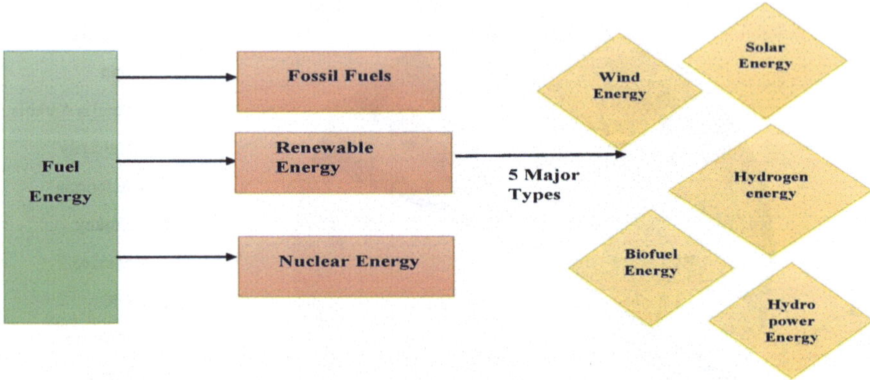

Fig. 2 Various types of fuel energy

2.1 *Fossil Fuel*

Fossil fuels are formed by animals and plant preservatives buried underground. Their backbone is mainly made up of carbon and hydrogen. Fossil fuels can produce massive energy and are easily accessible [8]. Petroleum is made up of blue-green algae, zooplankton, and phytoplankton. However, coal tar, coal gas, and coke are byproducts of coal. Coal generates around 40% of global electricity and supplies 30% of primary energy. They produce fuel energy, but at the same time, they have limitations. They generate $CO_{2,}$ which is primary source of greenhouse gases. In addition to it, fossil fuels also generate SO_2 gases, which are hazardous to the environment. The acidification of the environment is also caused due to fossil fuels. Further, they cause lots of diseases in miners. Coal, natural gases, and petroleum are the main sources of fossil fuels [9].

In power plants, fossil fuels generate huge amounts of heat, which are massively available in nature, and their extraction process is simple to perform [10]. The fossil fuels available for the generation of steam are blast furnace gases, natural gases, liquefied petroleum gases, coke oven gases, high and light diesel, furnace oil, fuel oil, liquefied natural gases, anthracene, peat, bituminous coal, biomass, and many others. They have high amounts of carbon, oxygen, and hydrogen, along with sulfur and nitrogen. They contain hydrogen in high amounts (HHV of 142 $MJkg^{-1}$) and carbon with HHV of 32.80 $MJKg^{-1}$. However, the presence of sulfur may cause corrosion in the fuel. The HHV of sulfur is found to be 9.30 $MJKg^{-1}$. Although the content of sulfur is very low in fuel generated from fossil fuels. However, coal is utilized in the production of steel up to 70% globally. In South Africa and Mongolia, almost all fossil fuels are the main energy sources. The countries mostly dependent on fossil fuels are Germany, Indonesia, the United Kingdom, the United States, and China. Fossil fuels are mostly available in all countries, but they are unevenly distributed. The highest production of coal is in China, followed by the United States, Australia, Indonesia, Germany, and South Africa.

Petroleum includes a huge quantity of liquid hydrocarbons. The significant types are natural gas, tar, and oil. Petroleum is mostly present in porous rock present on the upper layer of the earth's crust [11]. Petroleum also exists in oil sands and tar sands. However, oil consumption from fossil fuels has increased from 32 to 53% across different countries in the world [12]. The major users of oil from fossil fuels are South America, North America, and Africa.

Natural gas is the main source of electricity through the use of steam turbines and gas turbines. They are a source of cleaner fuel energy as compared to other fossil fuels [13]. Natural gas generates around 30% less CO_2 as compared to petroleum and 45% less as compared to coal. They are available commercially in natural gas fields. Around 65% of gas reserves are present in Russia and the Middle East [11]. The largest gas field is available in Iran, followed by Qatar and Russia. As per the International Energy Agency, global production of natural gas, more than 90 countries in the world utilize fuel energy in the form of natural gas [14]. The United States and Russia are the largest users of natural gas.

2.2 *Renewable Energy*

Rapid increase in the global population leads to immense pressure on energy production. This ongoing energy issue faced by many countries across the globe is due to limited infrastructure, underdeveloped power grids, and a shortage of dependable and economical access to modern energy services. The energy sector has a direct relationship with CO_2 emissions, causing a greenhouse effect, due to which there has been serious concern about opting for sustainable solutions globally. Dependency on fossil fuels for the production of electricity has contributed to environmental conditions, human health, and ecosystems. The perfect answer to this is the shift toward the sustainable, cleaner, and renewable form of energies like biofuel energy, solar energy, wind energy, hydrogen energy, and hydropower energy, which are environmentally friendly, do not affect the natural environment, mitigate climate change, and create economic opportunities [15].

2.2.1 Biofuels

Biofuel is an inexpensive, secure, and eco-friendly fuel derived from renewable resources. It is a sustainable substitute and a cure-all for the problems caused by the consumption of fossil fuels. The use of biomass resources for the production of biofuel has drawn more attention globally. Due to rising prices of oil and changes in climatic conditions, researchers have been interested in renewable technology for biofuel production [15]. Furthermore, biofuels have unique properties like non-toxicity, biodegradability, cost-effectiveness, and availability [16].

Biofuels were divided into first-generation and second-generation categories by researchers [17, 18]. They claim that biodiesel, bioethanol, and biogas are instances of first-generation biofuels (1GB), whereas animal fats, vegetable oils, etc. are examples of second-generation biofuels (2GB). However, researchers have classified biofuels into three generations: 1GB, 2GB, and 3GB. These biofuels are produced from food crops, which are edible feedstocks, aquatic materials as cultivated feedstocks, and nonfood crops as inedible feedstocks, respectively [19].

However, biofuels were divided into primary and secondary categories as studied by researchers [20, 21]. They define primary biofuels as biomasses that are used unmodified like grass, wood, and wood chips and are typically burned to produce heat for the purpose of cooking and warming. On the other hand, secondary biofuels are divided into 1GB (made from grains), 2GB (lignocellulosic biomass), and 3GB (algae) categories. [22, 23], and [24] categorized biofuels into the following categories: natural biofuels, 1GB, 2GB, 3GB, and 4GB.

Consequently, primary biofuels are produced directly by burning woody plants and dry animal waste, while secondary biofuels are further separated into first-, second-, third-, and fourth-generation biofuels that are discussed below.

Primary Biofuels

Primary biofuels, sometimes also known as natural biofuels [23] or zero-generation biofuels. They are unique and utilized precisely and directly as they are, without any pretreatment, processing, or modification. Firewood, pellets, wood chips, animal waste, crop forest leftovers, and gases produced from landfills are a few types of primary biofuels. Cooking, roasting, home heating, drying, and electricity production are notable applications of primary biofuels. Nevertheless, their usage is impolite, degrades the quality of the air, and may be harmful to the user's health [25].

Secondary Biofuels

The first generation generates ethanol from starch-rich food crops like barley, wheat, corn, sugarcane, potato, or biodiesel from soybean, sunflower, and animal fat. However, the second generation produces biodiesel and bioethanol from a variety of plant species, including cassava, jatropha, miscanthus, straw, grass, and wood. The third generation produces biodiesel from microalgae and microbes [26].

First-Generation Biofuels (1 GB)

Corn, sugarcane, soybean oil, and palm oil are the main sources of first-generation biofuels, often known as conventional biofuels. Bioethanol and biodiesel are the two primary forms of first-generation biofuels; they are produced by fermenting sugars and transesterifying oils, respectively. When edible plant parts are utilized to produce biofuels, the first generation's production of biofuels has a detrimental effect on food security [27]. First-generation biofuels (1GB) were developed in response to the demand for a viable and sustainable substitute for finite energy sources. Biogas, bioalcohols, biodiesel, biomethanol, and bioethanol are some prominent examples. Edible (food) crops such as corn, soybeans, wheat, edible vegetable oil and palm oil [28], *Moringa oleifera, Jatropha curcas*, rapeseed, Karanja [29], sugar cane, grains, wood, straw, household waste, and dried manure [30] are important feedstocks for the manufacturing of 1GB. These types of biofuels are already commercially available and represent the most advanced forms of biofuel production. According to estimates since 2000, over 90% of all investments in biofuel feedstock in tropical nations have gone toward oil palm and sugarcane [31]. First-generation biofuels have the capability to reduce CO_2 emissions and increase domestic energy security. However, there are some issues with feedstock sources, such as potential effects on biodiversity, land use, and competition with food crops. The ability to be mixed with petroleum-based fuels, burn in internal combustion engines, and be distributed through existing infrastructure are characteristics of "first generation" biofuels, such as biodiesel (bio-esters), bioethanol, and biogas. It can also be utilized in existing alternative vehicle technology, such as FFVs ("Flexible Fuel Vehicle") or natural gas vehicles [18].

However, scientists appear to be rather skeptical of first-generation biofuels. The growing output of first-generation biofuels is constrained by worries about carbon balances and environmental effects. The primary drawback of first-generation biofuels is the debate between food and fuel; the rise in food costs is partly caused by the increased production of these fuels [32].

Second-Generation Biofuels (2 GB)

Second-generation biofuels, also known as advanced biofuels, were created in response to the problems of the first generation using lignocellulosic biomass from agricultural and woodland residues, as well as other waste streams (such as from the food sector like wheat bran, animal fats, or wastes of cooking and frying oil). There may be other possible sources of second-generation biofuels like drought-resistant tree or shrub Jatropha curcas, which can also be generated in wastelands [33]. In addition to this conversion of biomass to liquid biodiesel, biohydrogen, bio-oil, bio-dimethylfuran, bioalcohols, bio-Fischer–Tropsch, bio-ethanol, and biosyngas [34]. Therefore, second-generation biofuels avoid the need to alter agricultural land use and therefore do not pose a threat to food supplies. But second-generation waste streams are frequently more complicated feedstocks than sugarcane or palm oil, and they may contain substances like lignin that can lower fermentation efficiency. As a result, using extra pretreatment processes is typical, which lengthens the process and raises expenses [35].

Third-Generation Biofuels (3 GB)

The difficulties with first-generation (1GB) and second-generation (2GB) led to the origination of third-generation (3GB), especially in the area of feedstock selection. Third-generation biofuels are produced from aquatic feedstocks like algae, which are resilient enough to withstand harsh environments. They are a prime example of 3GB and are separated into microalgae and macroalgae. Figure 3 demonstrates

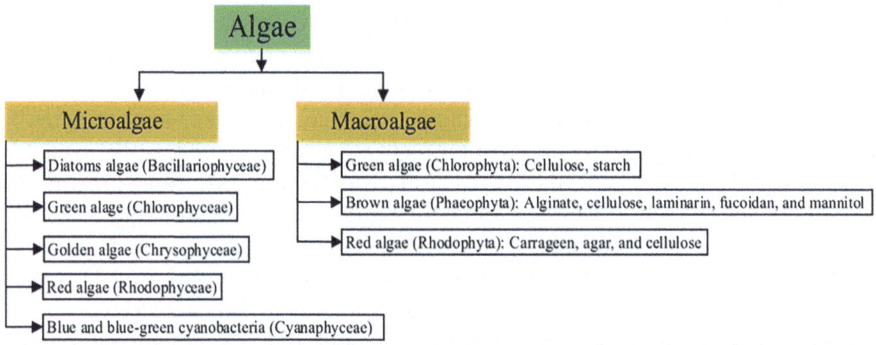

Fig. 3 Characteristics of microalgae and macroalgae [36]

the characteristics of both macroalgae and microalgae [36]. Heterotrophic aerobic fermenters, open pond cultivation, and photobioreactors (phototrophic reactors) are methods for producing artificial microalgae and macroalgae biomass. Examples of 3GB include biodiesel, bioethanol, biomethanol, vegetable oil, and jet fuels; Other feedstocks for 3GB consist of yeast, fungus, and cyanobacteria. More investment has been made in 3GB in recent years, especially in technology for algae culture and conversion [37].

Fourth-Generation Biofuels (4 GB)

Genetically engineered or optimized feedstocks, which are mainly intended for absorbing massive amounts of carbon utilizing genomically modified bacteria that can effectively create biofuels, are paired with fourth-generation technologies [32]. The manufacture of this generation of biofuels, in contrast to 2GB and 3GB, guarantees sustainable production and captures CO_2 emissions from the combustion of oxygenated fuel at each phase of the production process [38]. Production is now economically competitive due to the significant cost reductions brought about by production technologies. Hydrogenated renewable diesel, green aviation fuel, biodiesel, vegetable oil, and bio-gasoline are some of the main types of 4 GB. The complete classification of biofuels is shown in Fig. 4.

2.2.2 Solar Energy

Solar energy is a renewable form of energy. It is abundant and will remain till the sun shines. Today, research is being conducted globally to maximize the use of this greener alternative. Solar energy is now becoming an alternative to other nonrenewable forms of energy as it generates minimal greenhouse gas emissions, which contribute to air pollution, posing respiratory issues. Moving toward this greener energy eliminates the release of particulate matter and oxides, leading to improved air quality and helping combat climate change. Solar energy systems use very little

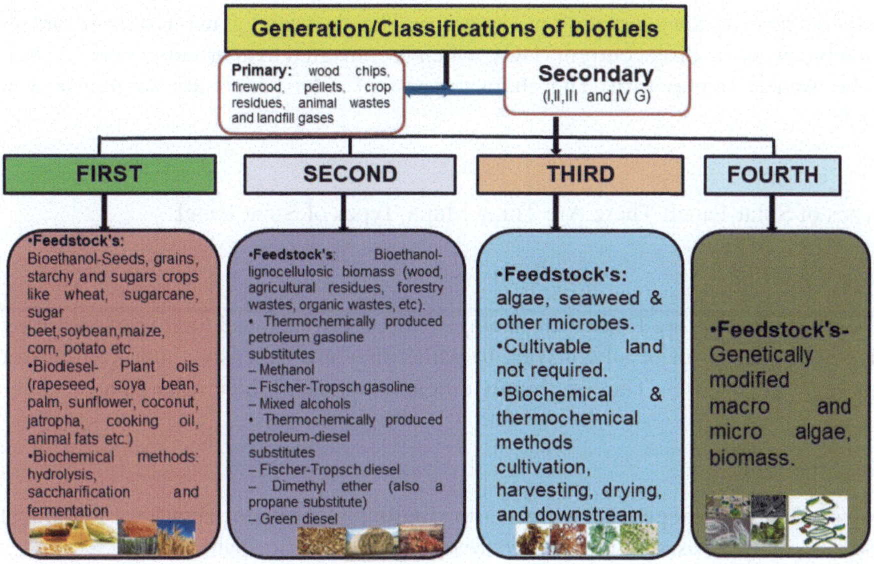

Fig. 4 Classification of biofuels [39]

water compared to fossil fuel power plants, which use enormous quantities of water for cooling and other activities. Solar energy conserves valuable water resources and reduces the burden on aquatic ecosystems, which is especially significant in water-scarce countries. The only disadvantage of using solar energy is that the initial cost may be high, but solar energy-based appliances can significantly reduce the daily cost over time with low maintenance expenses. The installation of solar panels on rooftops reduces land disruption. Fossil fuel extraction, on the other hand, results in habitat deterioration, drop in biodiversity, and land degradation due to mining. Globally, many different countries, like Australia, Germany, Japan, the United States, and the Netherlands, have been exploring and utilizing solar energy for different applications. Solar energy has diverse applications in different fields like agriculture, electricity generation, transportation, water distillation, air drying, and robotics.

Various techniques are used to capture the sun's abundant energy through different methods. The main technologies are: (1) photovoltaic (PV) systems and (2) concentrated solar power. In the former method, the sunlight is captured directly using solar panels and converted sunlight to electricity, whereas in the latter, lenses or mirrors are used to concentrate the large area of sunlight onto a small area, which produces heat followed by steam generation to drive turbines for electricity generation. Apart from electricity generation, PV systems have applications like powering calculators, spacecraft, providing energy in rural areas, and offering off-grid solutions. This usually happens due to absorption of light in which photons from sunlight hit the semiconductor, typically made from silicon, in the photovoltaic cells, resulting in electron excitation. In electron excitation, the energy from photons gives energy to electrons, allowing them to flow freely. The free flow of electrons creates an electric field that leads to the generation of electricity. The electricity generated from various solar panels is the direct current (DC), which is converted to alternating current (AC) to be used in homes [40]. The characteristics of solar panels are summarized in Table 1.

Types of Solar Panel: There Are Three Major Types of Solar Panel

Monocrystalline

These solar panels are formed from single silicon crystals, and their manufacturing process wastes a lot of silicon, due to which they are expensive and are used for commercial purposes. They are highly efficient among all three solar panels.

Polycrystalline

Polycrystalline solar panels are made from melting raw silicon. They are economical and can easily be used in residential areas with shorter life spans.

Thin Film

These thin-film solar panels are best suited for large-scale industries. They are made by depositing a thin layer of photovoltaic substances, like amorphous silicon, indium gallium selenide, and cadmium telluride, on the solid glass surface.

Technologies Used to Improve the Efficiency of Solar Panels

Tunnel Oxide Passivated Contact (TOPCon)

Passivation, minimization of electron recombination, is improved by incorporating a tunnel oxide buffer layer between heavily doped polysilicon and base wafer, which is due to an indirect metal contact with crystalline silicon wafer and 1-D current flow. There is an effective electron-selective contact due to asymmetric tunnel barriers for

Table 1 Various kinds of solar panels

Characteristics	Monocrystalline	Polycrystalline	Thin Film
Types of crystals	Single silicon crystals	Multiple silicon crystals	By depositing thin layer of photovoltaic substance onto the surface like glass
Efficiency	17–22%	Around 15%	7–13%
Life span	25 years		10– 20 years
Cost	Expensive	Budget friendly	Economical
Aesthetic	Solid Black	Blue	All black

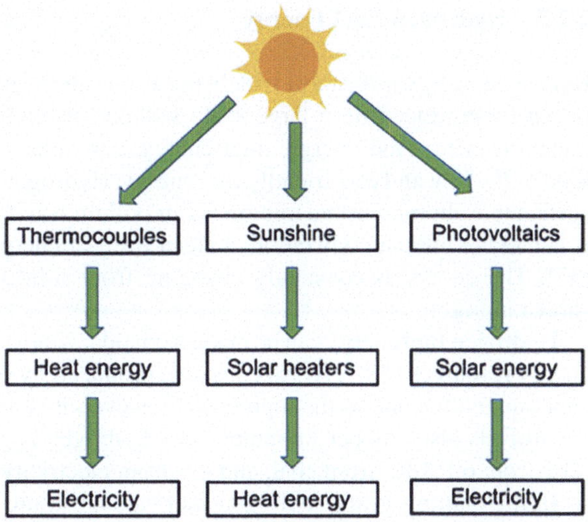

Fig. 5 Different ways and purposes of harnessing solar energy [44]

electrons and holes. This creation of asymmetric tunnel barriers for electrons and holes is due to the different band offsets between Si and SiO_2 (4.7 eV for valence band and 3.2 eV for the conduction band) [41]. TOPCon solar cells offer great efficiency, leading to 24% in pilot line and volume production.

Phase Change Materials (PCMs)

PCMs are materials that can regulate the temperature by either absorbing or releasing heat, which can increase the life of solar panels. PCMs serve as a remarkable material in protecting PV panels under extreme weather conditions. Phase change materials are used on PV panels to alter the temperature, which is done by integrating the PCMs with solar PVs, which can improve panel efficiency. Adding PCMs leads to a decrease in surface temperature of PV panels with 6% higher efficiency and 16% increase in electrical output. The important aspect in PCMs is the selection of material, which affects panel temperature. As the temperature increases by each degree, the efficacy of the PV cells is affected by 0.45% [42], but when PV is used in combination with PCM, the temperature drops to 12 °C, which gives 1.6 times more energy than the traditional. Different PCMs have been explored by researchers in order to find the PCM that can regulate the temperature and passive cooling of PV panels. PCMs like RT21, RT35, and RT 44 with melting points 20–23 °C, 34–36 °C, and 43–44 °C, respectively, are chosen for their efficiency and mainly used for commercial purposes [43]. Inorganic and organic PCM materials are also used in synthesizing solar panels. Different ways and purposes of harnessing solar energy are shown in Fig. 5.

2.2.3 Hydrogen Fuel Energy

Hydrogen energy is the most widely used and significant type of clean energy source. It can be produced on a large scale without depending upon the environment in comparison to wind energy, tidal energy, and solar energy [45, 46]. It is another kind of flexible and eco-friendly fuel energy. Hydrogen energy is an environmentally sustainable alternative that minimizes air pollution and global warming. As the world is achieving human civilization, a clean form of energy is available for humankind [47]. The energy is constantly changing from solid to liquid and vapor phase as shown in Fig. 6.

Hydrogen fuel energy can be made from tide, wind, and solar, which are renewable energy resources [48]. Hydrogen is also produced by biomass and sewage sludge in small quantities due to the high cost. However, it can be generated easily by water electrolysis also. As per literature, 48% hydrogen is generated from natural gases, 30% from oil, 18% from coal, and 4% from electricity [49, 50].

In the coming years, hydrogen fuel can substitute hydrocarbons because of its adaptability and huge benefits. Recently, a water-splitting method has been explored to produce green and clean energy. During this process, no carbon dioxide and carbon monoxide are liberated. Hydrogen as a fuel has tremendous advantages, including huge storage capacity, purity, low transportation cost, low emission energy, versatility, and renewability [47]. As per literature, hydrogen fuel energy will play a big role in various industrial applications, including electronic technology and air technology in the coming years [52]. As the world is progressing toward green energy for the benefit of mankind and to launch energy projects for the development of nations

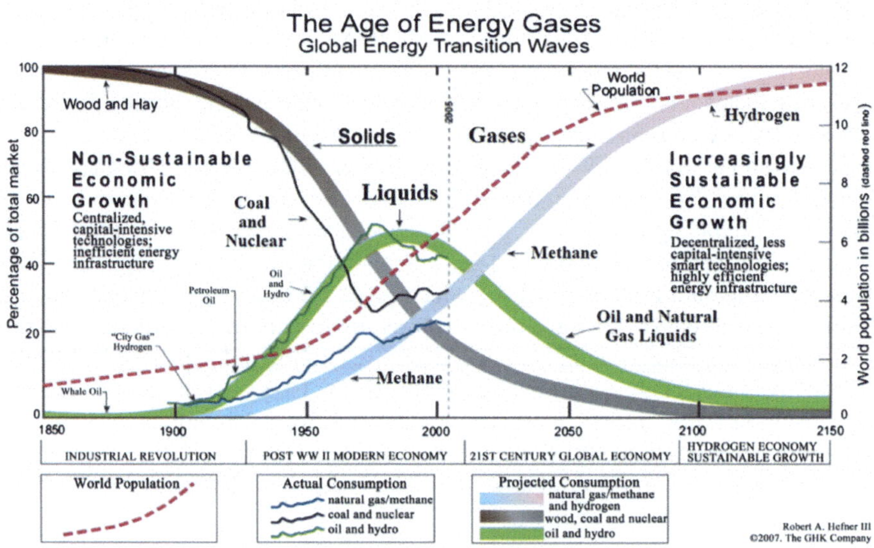

Fig. 6 Transition of energy from solid to liquid and vapor [51]

and to cater to a huge population, the countries are engaging themselves for extensive production of hydrogen fuel. The hydrogen fuel is utilized mainly for heavy industries, viz. concrete, steel production, and the transportation industry. Requirement for green hydrogen is increasing day by day at the global level, especially in Asia and Europe, which is enhancing the growth of renewable energy to minimize dependency on fossil energy supplies. The strong investment is also being made by various industries, which can support the production and storage of hydrogen fuel [53].

Hydrogen fuels are classified into different categories based on origin and color. The green type of hydrogen fuel is generated from renewable energy, and the blue type is produced from fossil fuel with CO_2 and CO capture [54, 55]. Furthermore, grey type of fossil fuel is generated by fossil sources without any gas capture. However, the red type is generated by nuclear energy. The most important types of hydrogen fuel are blue and grey. Hydrogen fuel is produced by two main techniques such as reforming and pyrolysis.

Reforming

Most hydrogen fuel is generated by reforming techniques. They are mostly produced by the steam reforming of hydrocarbons. In this technique, hydrocarbon and steam combine at high temperature to form hydrogen and CO_2 [56]. Natural gases such as naphtha and liquefied petroleum gases are extensively utilized in the reforming process to achieve hydrogen fuel. It involves the following reactions:

$$CH_4 + H_2O \rightarrow CO + 3H_2$$

$$CO + H_2O \rightarrow CO_2 + H_2$$

$$CH_4 + H_2O \rightarrow 2CO + H_2O$$

$$C_mH_n + mH_2O(g) \rightarrow mCO + (m + 0.5n) + H_2$$

Pyrolysis

It is a thermal decomposition process in anaerobic conditions that converts hydrocarbons to hydrogen fuel and other carbon-based materials [57]. The breaking down of hydrocarbons into different carbon and hydrogen-based materials takes place at high temperature. Methane breaks down at a temperature of about 1400 °C. The temperature can be lower in the presence of transition metal catalysts such as Fe, Ni, and Co. The following reactions take place during the process of pyrolysis [51]:

$$C_mH_n \rightarrow mC + 0.5nH_2$$

$$C_mH_n + (2m - 0.5n)H_2 \rightarrow mCH_4$$

$$CH_4 \rightarrow C + 2H_2$$

Hydrogen fuel is produced by two techniques, including hydrogasification and methane cracking gasification. This process is found to be efficient as it enables the formation of useful byproducts such as carbon nanofibers, carbon nanotubes, and carbon spheres.

2.2.4 Wind Energy

Wind power is a sustainable and inexhaustible form of energy with a smaller environmental impact. Wind power has a historical descent initially by sail, followed by windmills and wind pumps. In the ninth century few countries used the power of wind to grind grain and pump water. Ancient sailors sailed long distances with the help of wind energy. In the current era, wind energy has been harnessed to generate electricity, in which wind turbines are connected in order to generate electricity. Wind turbine utilizes the force of air, which is also called aerodynamic force, to produce electricity. This aerodynamic force is generated from rotor blades like airplane wings or helicopter blades. When wind blows through the blade, air pressure on one side of the blade decreases, whereas it increases on the other side. This change or difference in the pressure of wind creates lift (stronger force) and drag, resulting in the spinning of the rotor. The rotor is further connected to the generator via a shaft and gears. These gears in the gearbox speed up the rotation and enable a compact generator. Hence, the electricity is generated by conversion of aerodynamic force into electricity production [58].

Factors Affecting Wind Power

The various factors responsible for wind power are as follows.

Emplacement Specific Factors

Flow of wind Wind is explained as the flow of air from a high-pressure region to a low-pressure region, as shown in Fig. 7. It differs from solar power on the basis of availability. The wind is available throughout the day and night, whereas solar power is available only during daytime. Wind speed is an important aspect in electricity generation. Higher or stable wind speed leads to increases in the electric power as stronger winds rotate the blades faster, resulting in more mechanical and electrical

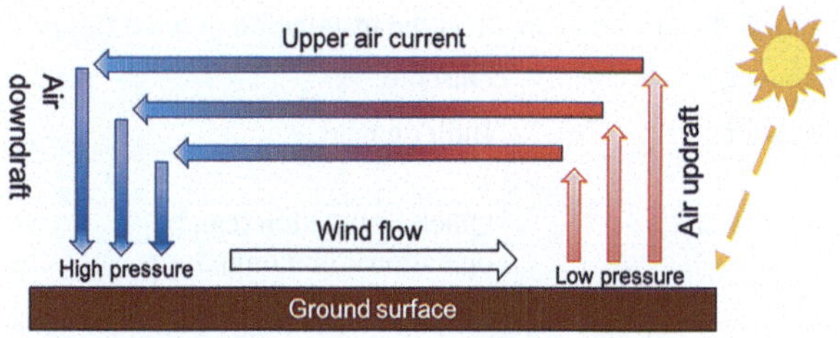

Fig. 7 Wind flow diagram [44]

power. If wind speed doubles, the power output increases eightfold. An anemometer is used to measure wind speed and direction at various heights.

Air Density and Temperature Air density affects wind power because the denser the air higher the pressure on the rotors. Air density is affected by the altitude, temperature, and pressure. Air temperature also plays an important role in a wind turbine's efficiency. The thermal expansion and contraction of materials are influenced by the air temperature, which directly affects the wind power.

Altitude The altitude of wind turbine towers is another parameter that affects the efficiency [59] of power generation. Wind speed often rises with altitude, so when the hub height or altitude of the wind turbine tower is raised, there are fewer barriers from the ground. These fewer obstructions permit the wind to flow freely, and due to this, the taller wind turbine towers are able to capture more energy.

Geography and Topography Places like hills, valleys, coastal areas, and elevated wind terrains, where high winds are widespread, are the best choice geographically for wind power generation.

Turbine Design Factors

Blade Radius The blade radius is engineered to be as large as possible in order to collect a greater percentage of kinetic energy of the wind. Small turbines with wide or large blades help them to harness more wind, resulting in higher-power generation. Therefore, the circular width swept by turbine's rotating blades significantly influences the power captured by the blades [60].

Pitch and Yaw Control There is a correlation between control techniques and efficiency, which includes the methods that are utilized to regulate the blade pitch, rotor speed, generator torque, and yaw angle of the turbine. The capability of wind turbines to generate power across a wide range of wind speeds is attainable by the optimization of pitch and yaw control [61].

Table 2 Parameters that affect the installation of wind farms

Parameters	Onshore	Offshore
Location	Built on land	Constructed on water bodies like the ocean and large lakes
Installation	Quick installation (can be operational in months)	Complex installation (need platforms, underwater cables, specialized safety features)
Energy production	Lower energy production	Higher energy production
Environmental impact	Impacts local wildlife	Impact marine ecosystem can enhance marine life by serving as artificial reefs
Energy generation	Less effective in energy generation due to variable winds	More effective in energy generation due to robust and strong winds
Turbine size	Smaller turbines due to height restriction	Large turbines
Maintenance	Less expensive	More expensive

Cut-In and Cut-Out Speed Cut-in and cut-out speed are important in generating electricity. Cut-in point is the point at which the turbine starts generating power and shuts down at cut-out to avoid any damage.

Advantages of Wind Power

Cleaner Energy Wind energy is a renewable form of energy and causes no pollution or harm to the environment. It is an alternative to an inexhaustible form of energy.

Economical It is one of the low-cost electricity sources in per unit of energy produced. It is definitely cheaper than energy produced from fossil fuels.

Wind Farms Offshore and onshore wind farms are usually used to harness wind energy. Onshore wind farms are often cheaper as they are built on land with easy installation, whereas offshore wind farms are generally more expensive due to complex installation in water bodies like oceans and large lakes. The former wind farm has a lower capacity factor because of variable wind conditions, resulting in the generation of around 1 MW of power, whereas the latter benefits from consistent and robust wind speeds that move in one direction, leading to higher energy production. In addition to this, local wildlife may be impacted by onshore wind farms and may trigger problems about visual and noise disruption. Offshore wind farms are located far from population areas, resulting in less visual and noise disruptions on human populations. They can also act as artificial reefs, which may enhance marine life and create less harmful impact on marine animals [62]. The different parameters affecting the installation of wind energy are tabulated in Table 2.

2.2.5 Hydropower Energy

Hydropower is a renewable form of energy that generates electricity. Hydropower is also known as hydroelectric power, which is the oldest and largest form of energy. Electricity is an important aspect in the societal and economic development of a country. The conventional use of fossil fuels for electricity generation has contributed to increasing global warming. So, hydropower generation serves as an alternative to it and helps in alleviating the environmental crisis. In 2023, around 4200 TWh of electricity was supplied by hydropower globally, making it the highest power-generating form of renewable energy. The hydropower works on the principle of converting the kinetic energy of flowing water, usually from dams, into electricity using turbines and generators. There are different types of hydropower facilities, namely conventional hydropower, run-of-river hydropower, pumped storage hydropower, and small-scale hydropower. Conventional hydropower dams can store water to meet electricity needs [63]. In run-of-river hydropower, a portion of the stream, which later rejoins the main river, is channelized through a power house, whereas water can be stored and pumped for both pumping and generating electricity in pumped storage hydropower. Small-scale hydropower, as the name suggests, is operational at small scale to carry out micro-hydropower projects.

3 Nuclear Energy

Nuclear energy is the safest kind of energy and has the lowest carbon footprint compared to other types of fuel energy. Nuclear energy can be harnessed to produce electricity via nuclear reactions, mainly nuclear fission and fusion. Nuclear energy is not actually renewable energy in the real sense [64]. It is extracted from uranium mining. Uranium is exploited for the production of fuel, and it breaks down into two/more nuclei [65]. The energy produced during the splitting brings heat. Nuclear reactors mostly depend upon uranium to initiate a nuclear reaction. Nuclear energy is the safest and cleanest energy. But at the same time, it generates radioactive wastes. Researchers are taking an interest in nuclear energy due to its very low carbon emissions. Because high carbon emission poses a big threat to the well-being of human beings and the safety of biodiversity. It is a great possibility that nuclear energy may be regarded as green energy in the future due to low CO_2 emission and can be a great replacement for renewable energy [66]. However, the limitation of high cost is a big challenge for it. Hilliard et al. [67] explored the advantages and limitations of nuclear energy in broader aspects [67].

Nuclear energy is one of the leading kinds of energy in most of the developed nations, especially the United States, which is generating around 770 billion KW of electricity [68]. As per latest research, due to rigorous utilization of nuclear energy, the United States is able to inhibit the release of around 470 million metric tons of CO_2 into the climate. Due to CO_2-free energy, nuclear energy improves air quality. The Department of Renewable Energy in the United States emphasizes majorly on safeguarding the nuclear reactors, evolving cutting-edge innovations, improving the nuclear fuel for viable energy sources, and supporting the economy [68]. But at the same time, nuclear energy has a few limitations.

The disposal and management of nuclear waste is a big challenge. Because radioactive waste poses a big threat to the environment and human health. Nuclear power plants are designed with lots of safety measures. Otherwise, accidents can lead to a big loss of life. The nuclear power plant can cause big environmental and health hazards in the vicinity of the human population [69]. So its installation faces big opposition from the public in most of the countries. This is the main cause and challenge in expansion of nuclear energy usage. Huge quantity of water is required to run a nuclear power plant. So it cannot be run in water-scarce areas.

4 Conclusion

The future of fuels is undeniably intertwined with the global pursuit of sustainability. At this crucial juncture, the pressing need to combat climate change, provide energy security, and safeguard the environment is posing a threat to our reliance on conventional fossil fuels. As a result, there is now a strong drive for sustainable and renewable energy sources, opening the door to a cleaner and safer energy future.

In summary, despite its complexity, the move away from fossil fuels is necessary for a sustainable future. With their own benefits and drawbacks, hydrogen, wind, solar, biofuels, and advanced nuclear technologies all present intriguing avenues for greener energy. However, continual technological advancement, encouraging governmental regulations, and significant funding are necessary for the effective adoption of these alternatives. By embracing these alternatives and fostering international cooperation, we can mitigate climate change, enhance energy security, create economic opportunities, and ensure a more prosperous and equitable future for all. While the journey is long, decisive action today is paramount to securing a cleaner and healthier planet for generations to come.

References

1. Bazzanella A, Ausfelder F (2017) Low carbon energy and feedstock for the European chemical industry: technology study. DECHEMA, Gesellschaft für Chemische Technik und Biotechnologie eV, A study by The European chemical industry council
2. Hao N, Alper K, Tekin K, Karagoz S, Ragauskas AJ (2019) One-pot transformation of lignocellulosic biomass into crude bio-oil with metal chlorides via hydrothermal and supercritical ethanol processing. Bioresour Technol 288:121500
3. Guliyev F (2022) The new global energy order: shifting players, policies, and power dynamics. In: Public responses to fossil fuel export. Elsevier, pp 25–44
4. Ahmad T, Zhang D (2020) A critical review of comparative global historical energy consumption and future demand: the story told so far. Energy Rep 6:1973–1991. https://doi.org/10.1016/j.egyr.2020.07.020

5. Adu D, Jianguo D, Asomani SN, Abbey A (2024) Energy generation and carbondioxide emission—the role of renewable energy for green development. Energy Rep 12:1420–1430. https://doi.org/10.1016/j.egyr.2024.07.013
6. IRENA, RES (2020) International renewable energy agency. Abu Dhabi, 2020.
7. Ritchie H, Roser M, Rosado P (2020) Renewable energy published online at OurWorldinData.org. Retrieved from: https://ourworldindata.org/renewable-energy [Online resource]
8. Hou H, Lu W, Liu B, Hassanein Z, Mahmood H, Khalid S (2023) Exploring the role of fossil fuels and renewable energy in determining environmental sustainability: evidence from OECD countries. Sustainability 15(3):2048. https://doi.org/10.3390/su15032048
9. Paramati SR, Shahzad U, Doğan B (2022) The role of environmental technology for energy demand and energy efficiency: evidence from OECD countries. Renew Sust Energ 153:111735
10. Apergis N, Payne JE (2014) Renewable energy, output, CO_2 emissions, and fossil fuelprices in Central America: evidence from a nonlinear panel smooth transition vector error correction model. Energy Econ 42:226–232
11. Sharma B, Shrestha A (2023) Petroleum dependence in developing countries with an emphasis on Nepal and potential keys. Energ Strategy Rev 45:101053
12. Yoshino N, Alekhina V (2019) Empirical analysis of global oil price determinants at the disaggregated level over the last two decades, Asian development report July 2019
13. Gunnella V, Jarvis V, Morris R, Tóth M (2022) Natural gas dependence and risks to activity in the euro area. Econ Bull Boxes 1
14. Di Bella G, Flanagan MJ, Foda K, Maslova S, Pienkowski A, Stuermer M, Toscani FG (2022) Natural gas in Europe: the potential impact of disruptions to supply. IMF
15. Aransiola EF, Betiku E, Ikhuomoregbe DIO, Ojumu TV (2012) Production of biodiesel from crude neem oil feedstock and its emissions from internal combustion engines. Afr J Biotechnol 11(22):6178–6186. https://doi.org/10.5897/AJB11.2301
16. Pugh S, McKenna R, Moolick R, Nielsen DR (2011) Advances and opportunities at the interface between microbial bioenergy and nanotechnology. Canadian J Chem Eng 89(1):2–12. https://doi.org/10.1002/cjce.20434
17. Akhlaghi S, Gedde UW, Hedenqvist MS, Braña MTC, Bellander M (2015) Deterioration of automotive rubbers in liquid biofuels: a review. Renew Sust Energ Rev 43:1238–1248. https://doi.org/10.1016/j.rser.2014.11.096
18. Naik SN, Goud VV, Rout PK, Dalai AK (2010) Production of first and second generation biofuels: a comprehensive review. Renew Sust Energ Rev 14(2):578–597. https://doi.org/10.1016/j.rser.2009.10.003
19. Jamwal VL, Kapoor N, Gandhi SG (2020) Biotechnology of biofuels: historical overview, business outlook and future perspectives. Biotechnol Bus-Concept Deliv 109–127. https://doi.org/10.1007/978-3-030-36130-3_6
20. Algayyim SJM, Wandel AP, Yusaf T, Hamawand I (2018) Production and applicationof ABE as a biofuel. Renew Sustain Energy Rev 82:1195–1214. https://doi.org/10.1016/j.rser.2017.09.082
21. Rodionova MV, Poudyal RS, Tiwari I, Voloshin RA, Zharmukhamedov SK, Nam HG, Allakhverdiev SI (2017) Biofuel production: challenges and opportunities. Int J Hydrogen Ener 42(12):8450–8461. https://doi.org/10.1016/j.ijhydene.2016.11.125
22. Mahjoub N, Sahebi H, MazdehM TA (2020) Optimal design of the second and third generation biofuel supply network by a multi-objective model. J Clean Poduc 256:120355. https://doi.org/10.1016/j.jclepro.2020.120355
23. Noraini MY, Ong HC, Badrul MJ, Chong WT (2014) A review on potential enzymatic reactions for biofuel production from algae. Renew Sust Energ Re 39:24–34. https://doi.org/10.1016/j.rser.2014.07.089
24. Chowdhury H, Loganathan B (2019) Third-generation biofuels from microalgae: a review. Curr Opin Green Sustain Chem 20:39–44. https://doi.org/10.1016/j.cogsc.2019.09.003
25. Knapczyk A, Francik S, Wójcik A, Ślipek Z (2020) Application of methods for scheduling tasks in the production of biofuels. In: Renewable energy sources: engineering, technology, innovation: ICORES 2018. Springer International Publishing, pp 863–873. https://doi.org/10.1007/978-3-030-13888-2_83

26. Dragone G, Fernandes B, Vicente AA, Teixeira JA (2010) Third generation biofuels from microalgae. Appl Microbiol Biotechnol 2:1355–1366. https://www.researchgate.net/publication/236006059
27. Nadeem H (2019) Microbial production of ethanol. Materials research foundations, 46. In: Book: microbial fuel cells: materials and applications. Publisher: Materials Research Forum LLC, USA. https://doi.org/10.21741/9781644900116-12
28. Rajak U, Verma TN (2018) Effect of emission from ethylic biodiesel of edible and non-edible vegetable oil, animal fats, waste oil and alcohol in CI engine. Energ Conver Manag 166:704–718. https://doi.org/10.1016/j.enconman.2018.04.070
29. Ahamed MS, Dash, SK, Kumar, A, Lingfa P (2020) A critical review on the production of biodiesel from Jatropha, Karanja and Castor feedstocks. Bioresour Utilization Bioprocess 107–115. https://link.springer.com/chapter/10.1007/978-981-15-1607-8_12
30. Hadin A, Eriksson O (2016) Horse manure as feedstock for anaerobic digestion. Waste Mgt 56:506–518. https://doi.org/10.1016/j.wasman.2016.06.023
31. Wang X, Nordlander E, Thorin E, Yan J (2013) Microalgal biomethane production integrated with an existing biogas plant: a case study in Sweden. Applied Ener 112:478–484. https://doi.org/10.1016/j.apenergy.2013.04.087
32. Chisti Y (2008) Biodiesel from microalgae beats bioethanol. Trends Biotechnol 26(3):126–131. https://doi.org/10.1016/j.tibtech.2007.12.002
33. Kumar A, Sharma S (2008) An evaluation of multipurpose oil seed crop for industrial uses (*Jatropha curcas* L.): a review. Ind Crop Pod 28(1):1–10. https://doi.org/10.1016/j.indcrop.2008.01.001
34. Du C, Zhao X, Liu D, Lin CSK, Wilson K, Luque R, Clark J (2016) Introduction: an overview of biofuels and production technologies. Handb Biofuels Prod 3–12. https://doi.org/10.1016/B978-0-08-100455-5.00001-1
35. Malode SJ, Prabhu KK, Mascarenhas RJ, Shetti NP, Aminabhavi TM (2021) Recent advances and viability in biofuel production. Energy Convers Manag X 10:100070. https://doi.org/10.1016/j.ecmx.2020.100070
36. Awogbemi O, Kallon DVV, Onuh EI, Aigbodion VS (2021) An overview of theclassification, production and utilization of biofuels for internal combustion engine applications. Energies 14(18):5687. https://doi.org/10.3390/en14185687
37. Veeramuthu A, Ngamcharussrivichai C (2020) Potential of microalgal biodiesel: challenges and applications. Renew Energy-Technol Appl. https://doi.org/10.5772/intechopen.91651
38. Chew BC, Shen XB, Ansell J, Hamid SR, Oh YP (2019) Review a decade of BP's technology roadmap on the next generation biofuels development. IOP Conf Ser: Earth Environ Sci 268(1):012009. https://doi.org/10.1088/1755-1315/268/1/012009. IOP Publishing
39. Malik K, Capareda SC, Kamboj BR, Malik S, Singh K, Arya S, Bishnoi DK (2024) Biofuels production: a review on sustainable alternatives to traditional fuels and energy sources. Fuels 5(2):157–175. https://doi.org/10.3390/fuels5020010
40. Kannan N, Vakeesan D (2016) Solar energy for the future world: a review. Renew Sust Energ Rev 62:1092–1105. https://doi.org/10.1016/j.rser.2016.05.022
41. Kim YJ, Kweon IS, Min KH, Lee SH, Choi S, Jeong KT, Kim KH (2022) Thermal annealing effects on tunnel oxide passivated hole contacts for high-efficiency crystalline silicon solar cells. Sci Rep 12(1):15024. https://doi.org/10.1038/s41598-022-18910-5
42. Stritih U (2016) Increasing the efficiency of PV panels with the use of PCM. Renew Ener 97:671–679. https://doi.org/10.1016/j.renene.2016.06.011
43. Durez A, Ali M, Waqas A, Nazir K, Kumarasamy S (2023) Modelling and optimization of phase change materials (PCM)-based passive cooling of solar PV panels in multi-climate conditions. Front Energy Res 11:1121138. https://doi.org/10.3389/fenrg.2023.1121138
44. Majeed Y, Khan MU, Waseem M, Zahid U, Mahmood F, Majeed F, Raza A (2023) Renewable energy as an alternative source for energy management in agriculture. Energy Rep 10:344–359. https://doi.org/10.1016/j.egyr.2023.06.032
45. Li C, Liao G, Fang B (2022) Magnetic-field-promoted photocatalytic overall water-splitting systems. Chem Catal 2(2):238–241. https://doi.org/10.1016/j.checat.2022.01.007

46. Megía PJ, Vizcaíno AJ, Calles JA, Carrero A (2021) Hydrogen production technologies: from fossil fuels toward renewable sources. A mini review. Energ Fuels 35(20):16403–16415. https://doi.org/10.1021/acs.energyfuels.1c02501
47. Liao G, Fang B, Li C (2023) A high-pressure-tolerant vapor-fed photocatalytic system for efficient water splitting. Chem Catalysis 3(6):100671. https://doi.org/10.1016/j.checat.2023.100671
48. Owusu PA, Asumadu-Sarkodie S (2016) A review of renewable energy sources, sustainability issues and climate change mitigation. Cogent Eng 3(1):1167990. https://doi.org/10.1080/23311916.2016.1167990
49. Axelsson M, Marchiori CF, Huang P, Araujo CM, Tian H (2021) Small organic molecule based on benzothiadiazole for electrocatalytic hydrogen production. J Am Chem Soc 143(50):21229–21233. https://doi.org/10.1021/jacs.1c10600
50. Uddin MN, Daud WW (2014) Technological diversity and economics: coupling effects on hydrogen production from biomass. Energy Fuels 28(7):4300–4320. https://doi.org/10.1021/ef5007808
51. Le PA, Trung VD, Nguyen PL, Phung TVB, Natsuki J, Natsuki T (2023) The current status of hydrogen energy: an overview. RSC Adv 13(40):28262–28287
52. Le PA, Le VQ, Tran TL, Nguyen NT, Phung TVB (2022) Computation and investigation of two-dimensional $WO_3 \cdot H_2O$ nanoflowers for electrochemical studies of energy conversion and storage applications. ACS Omega 7(12):10115–10126. https://doi.org/10.1021/acsomega.1c06150
53. Noh YS, Lee KY, Moon DJ (2019) Hydrogen production by steam reforming of methane over nickel based structured catalysts supported on calcium aluminate modified SiC. Int J Hydrogen Energy 44(38):21010–21019. https://doi.org/10.1016/j.ijhydene.2019.04.287
54. Bauer C, Treyer K, Antonini C, Bergerson J, Gazzani M, Gencer E, Gibbins J, Mazzotti M, McCoy ST, McKenna R, Pietzcker R, Ravikumar AP, Romano MC, Ueckerdt F, Ventek J, Spek M (2022) On the climate impacts of blue hydrogen production. Sustain Energ Fuels 6:66–75. https://doi.org/10.1039/D1SE01508G
55. Howarth RW, Jacobson MZ (2021) How green is blue hydrogen? Energy Sci Eng 9(10):1676–1687. https://doi.org/10.1002/ese3.956
56. Palo R, Dagle RA, Holladay JD (2007) Methanol steam reforming for hydrogen production. Chem Rev 107(10):3992–4021. https://doi.org/10.1021/cr050198b
57. Wang S, Dai G, Yang H, Luo Z (2017) Lignocellulosic biomass pyrolysis mechanism: a state-of-the-art review. Prog Energy Combust Sci 62:33–86. https://doi.org/10.1016/j.pecs.2017.05.004
58. Acakpovi A, Adjei P, Nwulu N, Asabere NY (2020) Optimal hybrid renewable energy system: a comparative study of wind/hydrogen/fuel-cell and wind/battery storage. J Elec Comp Eng (1):1756503. https://doi.org/10.1155/2020/1756503
59. Li J, Yang S, Hao J, Gao G, Wang F, Bai H, Xue X (2024) Advances and applications of wind engineering in exceptional terrain. J Trans Eng. https://doi.org/10.1016/j.jtte.2024.09.002
60. Miller LM, Brunsell NA, Mechem DB, Gans F, Monaghan AJ, Vautard R, Kleidon A (2015) Two methods for estimating limits to large-scale wind power generation. Proc Natl Acad Sci 112(36):11169–11174. https://doi.org/10.1073/pnas.1408251112
61. Zhang F, Gao S, Gao G, Dai J, Yang S, Wang W (2024) Evaluation of aging characteristics in wind turbine performance based on yaw power loss. Sustain Energy Techn 72:104094. https://doi.org/10.1016/j.seta.2024.104094
62. Mohanty S, Ramasam A, Mohanty A, Mohanty PP, Soudagar MEM, Khan TY, Cuce E (2024) Optical power monitoring systems for offshore wind farms: a literature review. Sustain Energy Technol Assess 72:104029. https://doi.org/10.1016/j.seta.2024.104029
63. Karre RK, Srinivas K, Mannan K, Prashanth B, Prasad CR (2022) A review on hydro power plants and turbines. In: AIP conference proceedings, vol 2418. AIP Publishing, p 1. https://doi.org/10.1063/5.0081709
64. Hossain R (2022) A short review of the nuclear energy. J Elec Eng 7(2)

65. Dungan K (2017) Uranium from seawater: infinite resource or improbable aspiration? Prog Nucl Energy 100:235–243
66. Ramana MV (2009) Nuclear power: economic, safety, health, and environmental issues of near-term technologies. Annu Rev Env Res 34:127–152. https://doi.org/10.1146/annurev.environ.033108.092057
67. Hilliard GH, Hasnain FU, Irfan M (2014) Phase change material for the cooling of solar panels—an experimental study. Eng Proc 45(1):43. https://doi.org/10.3390/engproc2023045043
68. Bhaskar S, Rajput SR (2018) Nuclear power: a comprehensive analysis of its benefits and drawbacks. Int J Creative Res Thought 6(1):83–88. ISSN: 2320-2882
69. Marques JG (2014) Environmental characteristics of the current generation III nuclear power plants. WIRE Ener Environ 3(2):195–212. https://doi.org/10.1002/wene.81

Artificial Intelligence Techniques in Wind Energy Prediction for Sustainable Development

Sadiqa Abbas, Shagufta Jabin, Sonal Saluja, Anupama Chadha, and Jyoti Chawla

Abstract The modern era is witnessing an evolution in technology that is reshaping how we approach problems. The Sustainable Development Goals (SDGs) serve as a global framework encouraging both developed and developing nations to shift toward renewable technologies to address the rising energy demands. Wind energy has a rich history dating back to ancient times. Over the centuries, the utilization and development of wind energy have evolved, leading to its application in various capacities by different countries worldwide. The integration of Big Data and Artificial Intelligence (AI) has indeed sparked a revolution in the wind power industry. Trillions of data points given by wind turbines can be utilized in various aspects after proper analysis, including forecasting, operational management, condition monitoring, and facilitating modifications in wind energy turbine components' operation. AI technologies also have the capability to forecast wind power generation even before actual energy production. This predictive ability enables operators to plan the incorporation of wind energy into the power grid well in advance. In this chapter, the authors will explore the use of various AI/Big Data techniques in harnessing the power of wind energy. Leveraging these technologies facilitates many ways, including optimization of system performance and anticipation of potential downtime in components, consequently enhancing plant productivity by curbing costs related to operation and component replacements. Application of AI for wind power forecasting helps to utilize wind energy in the best possible way and supports achieving SDGs. Efficient

S. Abbas · S. Saluja
Department of Civil Engineering, School of Engineering & Technology, Manav Rachna International Institute of Research and Studies, Sector-43, Aravalli Hills Campus, Faridabad, Haryana, India

S. Jabin (✉) · J. Chawla
Department of Applied Sciences (Chemistry), School of Engineering & Technology, Manav Rachna International Institute of Research and Studies, Sector-43, Aravalli Hills Campus, Faridabad, Haryana, India
e-mail: shagufta.set@mriu.edu.in

A. Chadha
Department of Computer Applications, School of Computer Applications, Manav Rachna International Institute of Research and Studies, Sector-43, Aravalli Hills Campus, Faridabad, Haryana, India

J. A. Siddique et al. (eds.), *Energy Sources*, Green Energy and Technology,
https://doi.org/10.1007/978-981-95-1209-6_12

generation of wind energy will contribute to clean and healthy environment with a lot of infrastructure development and employment opportunities.

Keywords Renewable energy · Wind energy · Sustainable development goals · Artificial intelligence · Machine learning · Big data · Deep learning

1 Introduction

The global population's expansion has led to an increased demand from consumers, necessitating a corresponding rise in supply to meet diverse consumer needs [79]. Wind energy stands out as a renewable source that holds promise in addressing the challenges of energy security and environmental degradation. The realization of this potential is contingent on countries with substantial wind energy resources incorporating it significantly into their energy portfolios. The electricity sector emerges as a promising domain where wind energy can be harnessed to meet substantial energy needs. However, the current reality reveals a predominant reliance on fossil fuels in the energy profiles of most global economies, even in advanced nations, with wind renewable energy contributing only a small fraction to the overall energy mix. Notably, Germany is among the few countries showcasing promising opportunities in fully leveraging wind energy [90]. Renewable energy sources are highly favored due to their crucial role in mitigating greenhouse gas (GHG) emissions and ensuring national energy security, all while avoiding the limitations associated with nuclear power [78].

The rise of global temperatures and the widespread impact of severe climate change have heightened environmental apprehensions. Presently, over 90% of the world's electricity is derived from fossil fuels [97]. Energy production plays a crucial role in contributing to global warming, any alterations in this sector can exert a substantial influence on the environment [15]. The escalating issue of global warming, largely attributed to conventional energy sources, has prompted governments and private sector to institute policies aimed at minimizing environmental impact. Consequently, renewable energy has emerged as a key focus of development and deployment. Among the array of renewable energy sources, wind energy stands out as the fastest-growing and one of the cleanest resources globally. Its rapid expansion is a testament to the growing recognition of the need for sustainable and environmentally friendly energy alternatives [79].

Wind and solar energy are currently the focal points of attention in the realm of renewable energy, and they are anticipated to play pivotal roles in the future. A key advantage of renewable energy sources lies in their freedom from GHG emissions. The reduction of these gases is crucial to prevent the escalation of temperatures and the associated adverse impacts on the global environment. Harnessing wind and solar power as renewable energy sources holds promise in mitigating these environmental impacts and contributing to a more sustainable and environmentally friendly energy landscape [79].

The global adoption of wind energy has surged in recent years as part of efforts to decrease GHG emissions. While wind power itself is cost-free, the installation and upkeep of wind turbines incur substantial expenses. The scale of a wind turbine installation is influenced not just by the wind patterns at a specific site but also by the costs associated with turbine infrastructure and maintenance. The period required for the turbine to recoup its costs is contingent upon the expenses related to turbine energy [15].

Despite its numerous advantages, wind energy is contingent upon the highly variable nature of wind, which manifests both geographically and temporally [7]. Therefore, the substantial uncertainty associated with the highly variable nature of wind makes the integration of wind power into the power grid a complex task [52, 75]. Utilizing reserve power plants is a practical yet costly solution to offset the fluctuations in wind power [25].

Figure 1 depicts the global wind energy installation sourced from the Global Wind Report 2023 [46]. The wind energy generation across the globe in the years 2020, 2021, and 2022, along with the installed wind turbine capacity sourced from the Energy Institute [49] Statistical Review of World Energy, is summarized in Table 1.

An analysis of data from last three years reveals that the Asia–Pacific region maintains the highest percentage of wind energy generation, which is around 42.8%, followed by Europe at 26.34% and North America at 23.61% for the year 2022. Furthermore, countries such as Germany and India are demonstrating a robust interest in wind energy generation [37].

Compared to fossil fuels, renewable energy sources have a significantly lower environmental impact. These sources, such as solar panels and wind turbine generators, produce no direct carbon dioxide (CO_2) or nitrogen oxide (NO_x) emissions. A diverse array of devices is available to convert ambient energy into a more useful form, such as electricity [14]. Among these devices, wind turbines stand out as some

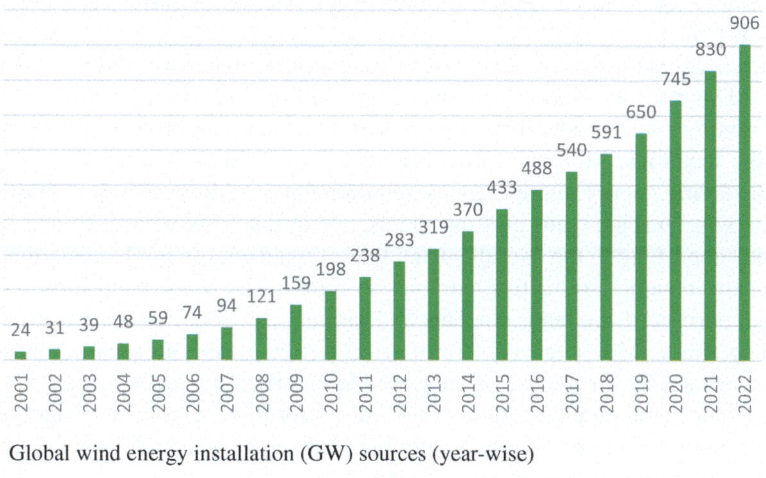

Fig. 1 Global wind energy installation (GW) sources (year-wise)

Table 1 Wind energy generation and installed turbine capacity

Terawatt-hours	Wind energy generation			Installed wind turbine capacity gigawatts		Share 2022 (%)
	2020	2021	2022	2021	2022	
North America	396.7	437.8	497	154.7	163.5	18.20
South & Central America	85.5	106.2	118.4	31.8	35.5	4.00
Europe	512.7	500.4	554.5	232.1	251.8	28.00
CIS	2.5	4.7	7.1	2.2	2.5	0.30
Middle East	2.6	2.9	3	1.03	1.05	0.10
Africa	21.3	22.8	23.9	7.3	7.7	0.90
Asia Pacific	575.1	779.2	900.9	394.86	436.77	48.60
Total world	1596.4	1861.9	2104.8	824.9	898.82	100

Source Statistical Review of World Energy, 2023, 72nd Edition

of the most popular and accessible means of converting ambient energy into electricity [101]. Nevertheless, wind energy, like many other renewable energy sources, has historically incurred high capital costs. However, over the past decade, there has been a remarkable shift in this trend. Statistics indicate a significant reduction in the cost of wind production in recent years, decreasing from two million dollars per megawatt (MW) to one million dollars in the last decade [66]. This accomplishment has led to a growing prevalence of wind power plants in both developed and developing countries [86].

Wind energy is considered a green source of power since wind turbines do not directly produce emissions. This characteristic aids nations in meeting their emission reduction objectives and addressing climate change. Studies have demonstrated that the utilization of wind energy contributes to a reduction in carbon dioxide (CO_2) emissions [57].

Wind energy stations typically demand less space compared to conventional power stations. This shift to wind energy opens up additional space and land that can be repurposed for other productive projects. Wind turbines can be strategically placed in remote locations. Moreover, the capacity of wind farms is adaptable, aligning with fluctuating energy and electricity requirements. Additionally, it is feasible to integrate wind turbines into existing ranches or farms. This versatility and spatial efficiency contribute to the overall advantages of wind energy implementation. This significantly benefits productive activities in rural areas. Since wind turbines only occupy a portion of the land, ranchers and farmers can continue their work on the remaining space. Moreover, owners of wind power plants often pay rents to the ranchers or farmers for the use of the leased land, providing landowners with additional income. This arrangement not only allows for the coexistence of agricultural activities and wind energy infrastructure but also serves as an additional source of revenue for those in rural communities [90].

Consequently, predicting wind power generation has become a crucial challenge. The erratic variability of wind energy poses difficulties in accurately forecasting wind power, and assessing the precision of predictive models is no straightforward task. Inaccuracies in wind power predictions can result in complications in the transmission and distribution of electricity to various consumers. To address this, precise forecasting methods are employed, often leveraging a combination of artificial intelligence (AI) algorithms and statistical approaches. These advanced techniques play a pivotal role in enhancing the accuracy of wind power forecasts, ensuring a more reliable and efficient electricity supply to meet the growing demands of consumers [79].

Effectively harnessing wind energy requires understanding its patterns of power generation. Instances where power generation exceeds demand can lead to power grid failures, resulting in disconnections for dependent parties and potentially costly incidents. To mitigate such occurrences, it is crucial to implement power balancing and frequency management. Having prior knowledge of the expected energy output facilitates effective power balancing, allowing power engineers to redirect surplus energy to storage systems, thus averting accidents [21]. The significance of wind power prediction is escalating as the wind penetration rate continues to rise. In every power system, there exists a reasonable capacity to adjust to changes in demand, given that demand estimation is inherently imprecise. In instances where the penetration rate is relatively low, power systems may not necessitate a heightened focus on the fluctuations in wind power supply [82]. Commonly employed statistical methods for various applications include the artificial neural network (ANN) [19], convolutional neural network (CNN) [59], support vector regression (SVR) [16], and back propagation neural network (BPNN) [1].

The expanding wind energy market is driving the need for both logistical and economic enhancements. Technologically, researchers are focused on maximizing the efficiency of wind turbines through various optimization strategies, including aerodynamic optimization [102], optimization of blade shapes [13], and positioning optimization of wind turbines within a wind farm [39].

In general, wind turbines employ a diverse array of sensors to measure ambient variables such as humidity and temperature, as well as turbine parameters like vibration and speed. These measurements are captured by a supervisory control and data acquisition (SCADA) system, which plays a crucial role in monitoring the condition of the overall system [31, 65].

Wind power forecasting plays a crucial role in enabling power grid dispatching departments to proactively acquire information about the expected wind power generation. This allows for precise adjustments to dispatching plans, leading to the reduction of reserve capacity in the electrical power system. Additionally, accurate wind power forecasting contributes to eliminating unnecessary operating costs for the power grid. Furthermore, it helps mitigate the adverse impacts on power grid security caused by the integration of wind power. Ultimately, this improvement in forecasting can enhance the installed capacity of wind power within the system [35].

Wind farm design is a complex process involving numerous and diverse parameters. Ekonomou et al. developed an ANN model to calculate the optimal number of wind turbines in a wind farm. This model considers various parameters, including terrain morphology, wind speed and direction, turbine type, and costs, to determine the most effective number of turbines for the given conditions [28].

Peiris et al. conducted research centered around the "Pawan Danawi" wind farm situated in Sri Lanka. Their study involved the application of various numerical, statistical, and machine learning (ML) methods to predict wind power generation. Through their investigation, the researchers identified that ANN emerged as highly favored and efficient techniques in the realm of wind power prediction. Remarkably, the model developed by the researchers demonstrated optimal accuracy when subjected to a validation percentage of 15% [76]. It is not feasible to eliminate the variability of wind, but the application of ML techniques offers the potential to significantly enhance the predictability and value of wind power. Utilizing ML can introduce greater data precision to wind farm operations, enabling operators to swiftly and intelligently assess how their power output aligns with electricity demand. This approach facilitates faster, data-driven decision-making, contributing to more efficient and responsive management of wind farm operations [30].

Leveraging AI and ML technologies holds the potential to optimize the performance of wind turbines and minimize downtime. These advanced technologies can be employed for predicting maintenance requirements, fine-tuning turbine settings, and enhancing overall efficiency. By utilizing AI and ML, wind turbine operators can proactively address maintenance issues, adjust turbine parameters for optimal operation, and ultimately improve the overall performance and reliability of wind energy systems (WES) [78].

Eliminating the inherent variability of wind is impractical; however, the application of ML techniques can significantly enhance the predictability and value of wind power. This approach also contributes to strengthening the data rigor in wind farm operations. ML enables wind farm operators to make faster, data-driven, and more intelligent assessments of how their power output aligns with electricity demand, thereby improving overall operational efficiency [30].

Deep learning models, particularly the multilayer perceptron (MLP) for weight optimization and CNN for mining spatial feature structures, have demonstrated a notable enhancement in wind power forecasting accuracy when compared to existing ML algorithms [24]. Hybrid methods aim to enhance predictions by combining diverse forecasting techniques through the application of weighting, preprocessing or decomposition, feature selection or optimization, and post-processing techniques [40, 99].

The unpredictable and intermittent nature of wind power underscores the importance of accurate wind power forecasting, especially in the context of reliable and efficient power grid operation with significant wind power integration. In this study, six-month-ahead wind power forecasting models were developed using tree-based learning algorithms. The investigation focused on understanding the impact of input data on forecasting accuracy, resulting in the creation of three distinct models. Experiments conducted in the study revealed a noteworthy decline in forecasting accuracy when utilizing longer time intervals and height extrapolation in the models. The findings emphasize the significance of selecting appropriate input data parameters for maintaining accurate wind power forecasts. To assess the generalization ability of the developed models, they were tested against wind data collected at different heights and locations than those used during the model training phase. The simulation results confirmed the efficacy of tree-based learning algorithms not only for long-term wind power forecasting but also for predicting wind power at various heights and geographical locations. This suggests the potential versatility and applicability of these algorithms in addressing the challenges posed by the variable nature of wind power [4].

The United Nations introduced the Sustainable Development Goals (SDGs) in 2015, also known as the Global Goals, as a collective effort to eliminate poverty, safeguard the environment, and foster global peace and prosperity by 2030 [45]. The four pillars of sustainable development encompass society, environment, culture, and economics [32].

All 17 SDGs are interconnected, reflecting the reality that actions in one domain impact outcomes in others, necessitating adjustments for social, financial, and ecological sustainability. Figure 2 exhibits the compliance of the London Array wind power project with different SDGs. Nations have committed to collaborative efforts, particularly prioritizing advancements for those facing the most significant challenges. As part of these goals, there is a collective aim to increase the proportion of renewable energy in the global electricity mix to at least 70% by 2030. Within the framework of SDGs, the development of renewable energy serves to enhance energy security while contributing to community development and environmental preservation on a global scale [71].

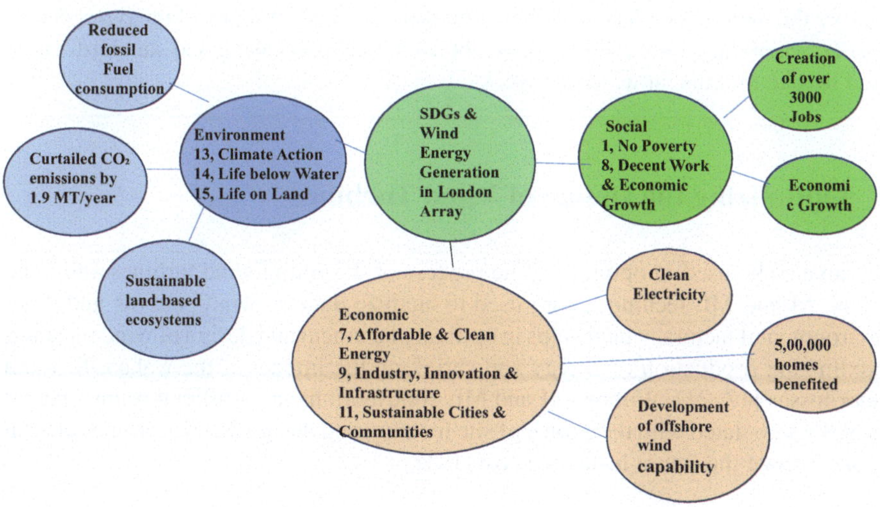

Fig. 2 SDGs and wind energy generation in London Array

2 AI in Wind Energy Industry

Gone are the days when terms like data warehouse, Big Data, AI, and ML were buzzwords. Businesses used to believe that these are the technologies to solve all the data-related problems without having enough knowledge of their benefits. Broad and expensive solutions were designed or implemented in hopes that they would add value to businesses. Nowadays, these technologies are evolving, and the usage of these technologies in various areas is becoming more sophisticated. AI and ML, especially, are being used in businesses to great success. In this chapter, the authors have tried to explore the usage of AI and ML in wind energy industry. AI provides solutions for analyzing data, generating patterns and solving performance issues, but it has its own challenges [64].

3 Monitoring of Performance of Wind Turbines

Wind various external sources. This data is required to be effectively analyzed to understand and improve turbine performance, AI and ML play an important role to process and analyze all this data (Marugánturbines produce trillions of data points which include performance data, environment and climate data, and data from [5, 39, 63]). ML algorithms accurately label and categorize data streams to provide deeper insights. This enables efficient analytics and visualizations for monitoring the performance of turbines. The wind industry is now following standard formats to store the data, which has made benchmarking against similar turbine types easier. Earlier, when these technologies were not used, data processing and standardization used to take months, now, within weeks.

4 Optimizing the Design of Wind Turbines

AI can also be used in optimizing the process of designing wind turbines and wind farms. AI and ML techniques are used to analyze data on wind patterns and other environmental factors, which helps in optimal placement and design of wind turbines, maximizing production of energy and shrinking the impact of the wake effect and other losses [13, 21]. Without AI and ML contributions, it is difficult to understand the risks associated with the choice of site location and design. Site location is crucial to drive down the future insurance costs [84].

5 AI in Wind Energy and Associated Challenges

5.1 *Utilization of New Technologies*

Businesses need to understand that AI is a tool not a solution to the problems. Simply implementing them is not enough to solve data challenges. On the other hand, without a proper planned strategy, businesses can suffer huge losses implementing these AI-based techniques. Businesses also need to understand the problems or issues they want solutions for. Without a detailed insight and defined strategy, it can be impossible to get solutions.

5.2 *Collaboration with Stakeholders in the Wind Energy Industry*

In the wind energy sector, apart from operators and owners, there are many parties involved in managing wind assets. In the wind energy industry overall, there are very few Original Equipment Manufacturers (OEMs)—each with hundreds of gigawatts of installed energy. This means that these OEMs have massive data that could be used to train AI and ML algorithms. Unfortunately, OEMs are unwilling to share data, and this results in a lack of insights into industry-wide issues.

5.3 *Unavailability of Well-Labeled Data*

The AI and ML techniques require clean and well-labeled data to make accurate predictions. Data in the wind industry varies significantly across OEMs. Lack of industry standardization, and codes vary between OEMs. This results in a huge investment of time in cleaning the data and putting it in the standard and labeled format.

5.4 *Data Privacy and Ethical Concerns*

Data Privacy is one of the major challenges that businesses are facing nowadays. It is crucial to protect the AI systems from cyber-attacks and unauthorized access for the security and integrity of energy systems. Encryption, frequent security audits, and highly secured data storage systems are a few of the measures that should be taken to prevent the AI-based systems from vulnerabilities [3, 17, 23].

Usage of AI systems requires that the following ethical concerns should be taken into consideration: potential bias in decision-making and allocation of work to human resources. Businesses should develop ethical guidelines for using AI models in energy systems [3, 17, 23].

6 AI and ML Algorithms in the Field of Wind Energy

Like all renewable sources, wind energy also poses the challenge of unpredictability. To make the best use of wind energy, it becomes necessary to learning the trends in which it generates power. The power grids fail if the amount of power generated is more than the power demand. This can lead to expensive mishaps. This problem can be solved by power balancing and handling frequency. The information about the amount of energy generated enables power balancing. This information will help power engineers to divert excess energy to storage mechanisms to prevent such accidents. Various ML algorithms can be used to forecast the amount of energy generated [21]. The researchers have also explored that, compared to individual systems, the hybrid AI system gives more accurate findings [104].

In this section, the authors have discussed the various ML techniques that can be used in the wind energy sector. As can be seen in Fig. 3, there is a constant growth in power generation using wind energy.

The AI techniques, especially ML and deep learning methods, are gaining momentum in the field of wind energy, as shown in Fig. 4. ML methods use thousands of data points to train the machine while deep learning methods use millions of data points. ML algorithms are designed by using explicit programming, while deep learning algorithms are designed based on the layers of neural networks, making the

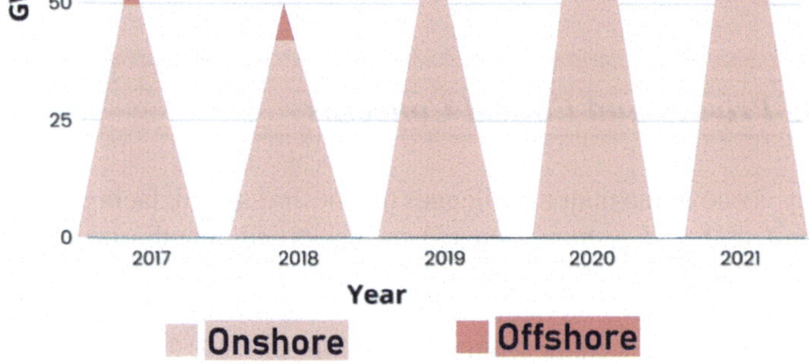

Fig. 3 Growth of wind power capacity over the last decade by region

Fig. 4 Various AI techniques in wind energy

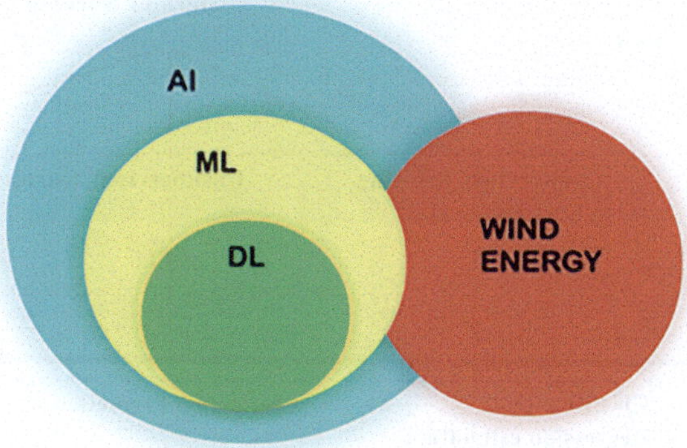

machine think like human brain. ML methods give results in less time as compared to deep learning methods. With the advancements in the domain of image processing, ML, and deep learning models are providing solutions in image recognition and making predictions.

Many statistical models have been used to predict wind power. Statistical modes deal with finding a relationship between variables to predict an outcome. These methods include mean variance estimation (MVE) technique, principal component analysis, and fuzzy logic. ML models have been used to predict the wind speed and energy. A statistical model is the use of statistics to build a representation of the data and then conduct analysis to infer any relationships between variables or discover insights ML is the use of mathematical and or statistical models to obtain a general understanding of the data to make predictions.

Figure 5 exhibits the categories of ML algorithms. These algorithms are categorized depending on the type of mathematical or statistical method used. All the statistical and ML algorithms have their own advantages and disadvantages. A brief description of different models, along with advantages and disadvantages, is illustrated in Fig. 6.

6.1 Statistical Models

There are several statistical models designed for specific research issues, data formats, or applications. Some of the commonly used statistical models and their applications in the field of wind energy are described below.

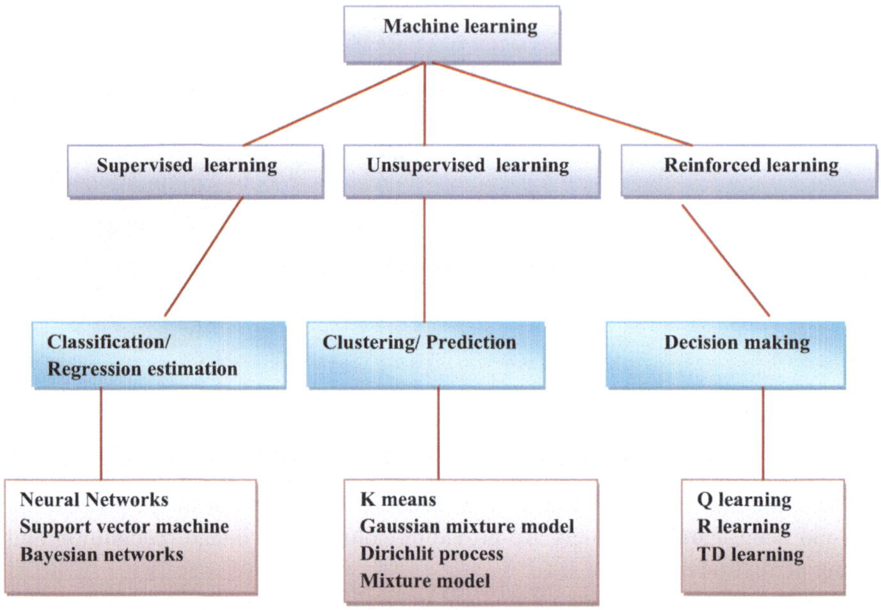

Fig. 5 Categories of machine learning algorithms

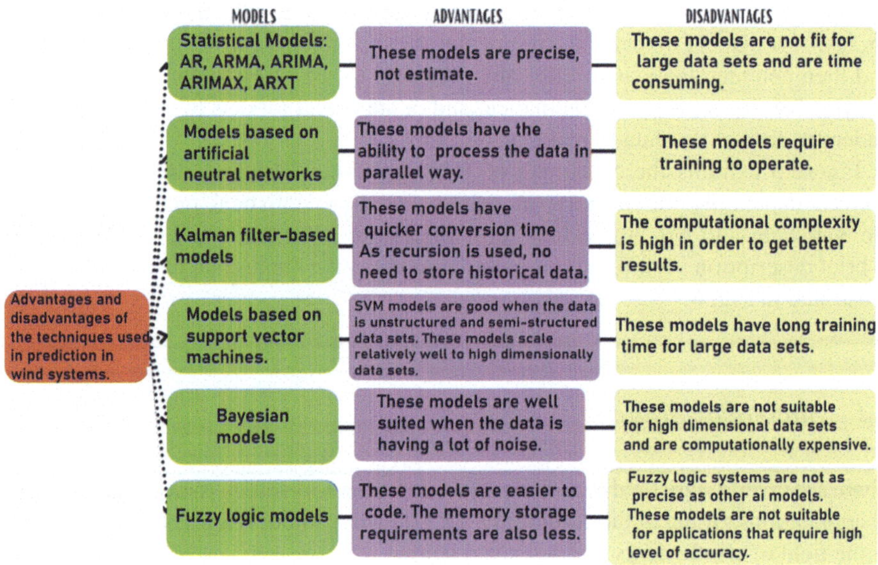

Fig. 6 Classification of models used in wind forecasting, along with advantages and disadvantages

6.1.1 Linear Regression Models

These models represent the connection or relation between a set of variables. There are two types of variables: continuous variables and one or more predictor variables. For example, in the dataset of the personal details of the employees—height, age, and gender, a linear regression model may be used to estimate the weight of the employees.

In the field of wind energy, this model can be used to determine the amount of energy generated from the average wind speed. This can help in forecasting the production volume and balance the energy system.

6.1.2 Logistic Regression Models

Logistic regression models like regression models are used to represent the relationship between the variables. The difference is that in this model, the outcome variable is binary (for example, yes/no). For example, depending on the personal data of the patients, age, blood pressure, and cholesterol levels, a logistic regression model can predict whether the patient would have a heart attack.

In the wind energy sector, a logistic regression model can help in predicting the suitable sites for wind energy development. The end users of the model can predict the suitable site depending on their region, project size, and other predictors of interest.

6.1.3 Time Series Models

Time series models work on data that changes over time. Examples of such kind of data are stock prices and weather trends. These types of models can help in discovering trends, seasonal patterns, and other forms of temporal correlations.

In the wind energy sector, time series models can help in wind power forecasting. Wind power forecasting is a challenging task because of the uncertainties and fluctuations in wind speeds. Wind power forecasting can improve the efficiency of the power plant by planning activities like maintenance scheduling appropriately.

6.1.4 Clustering Models

Clustering models group together comparable observations or records based on their similarities in terms of their attributes. Clustering models can be used to reveal patterns in datasets that would be difficult to detect using other approaches.

Clustering models can group the wind turbines in a wind farm by identifying similar parameters based on wind speed and output power.

6.1.5 Artificial Neural Network Models

An ANN is a computational model whose working replicates the working of human brain's neural structure. This model consists of interconnected nodes organized into layers, just like human brain, where neurons are arranged in layers. Information flows through this network of nodes during the training phase of the model, enabling it to recognize patterns and make predictions.

ANN models are used to find faults in wind turbines. The gearboxes and bearings, which are the essential parts of the wind turbines, have a high failure rate. The ANNs can predict complicated nonlinear relationships through pattern recognition, which, in turn, helps in fault detection in these parts of the wind turbines.

6.1.6 Kalman Filter-Based Models

Kalman filter-based model provides an estimate of x k at time, given the initial estimate of x 0. This model is called a "filter" because it filters out measurement noise. Kalman filter has many applications in areas like robotics, GPS sensors, and radars. Kalman filter was also used in Apollo navigation computer that took humans to the Moon and back.

In the wind energy sector, this algorithm can be used to predict the characteristics of systems subject to noise, such as wind turbines.

6.1.7 Support Vector Machine Models

SVM model is used for linear or nonlinear classification, regression, and even outlier detection tasks. SVM has a variety of applications like image classification, spam detection, and handwriting identification. SVMs are efficient in a variety of applications because they can manage high-dimensional data.

SVMs predict the wind power of the wind turbine generators indirectly. This model first predicts the wind speed. The wind speed prediction helps to predict the wind power by using the power-wind speed characteristics.

6.1.8 Bayesian Models

Bayesian ML models are statistical models based on Bayes' theorem as given in Eq. 1.

$$p(\theta|x) = p(x|\theta)p(\theta)p(x) \tag{1}$$

where

$p(\theta|x)$ is the posterior probability of event θ given event x.

$p(x|\theta)$ is the likelihood of event x given event θ.

$p(\theta)$ is the prior probability of event θ.
$p(x)$ is the total probability of event x.
Bayesian models can be used for system reliability modeling and decision-making under uncertain conditions in the field of wind energy.

6.1.9 Fuzzy Logic Models

The term fuzzy refers to things that are not clear or are vague. In the real world, many times we encounter a situation when we cannot determine whether the state is true or false; fuzzy logic provides very valuable flexibility for reasoning. In this way, we can consider the inaccuracies and uncertainties of any situation.

The fundamental concept of fuzzy logic is the membership function, which defines the degree of membership of an input value to a certain set or category. The membership function is a mapping from an input value to a membership degree between 0 and 1, where 0 represents nonmembership and 1 represents full membership.

In the Boolean system, truth value 1.0 represents the absolute truth value, and 0.0 represents the absolute false value. But in the fuzzy system, there is no logic for the absolute truth and absolute false value. But in fuzzy logic, there is an intermediate value too present, which is partially true and partially false.

Fuzzy logic controllers can be used for tracking the maximum power of wind generator. This, in turn, can be used to balance the power demand using a fuel cell without the use of sensors.

The researchers have concluded that the ANNs are efficient in modeling linear and nonlinear functions with arbitrary shapes. The prediction done by the ANNs has more accuracy as compared to other models. They also explored that by using ensemble techniques (ensemble modeling combines several very similar models in order to improve the results) and hybridization of models (hybrid modeling combines different modeling technologies to improve the results) [9, 30, 42, 58, 67, 93, 98].

7 Applications of Artificial Intelligence in Wind Power Sector

7.1 1Wind Resource Assessment

Site identification is the most important part for the economic feasibility of the project. AI techniques can be used to analyze wind data and can predict the location of placement of wind turbines for maximum energy production from a particular wind farm.

7.1.1 Maintenance of Turbines

AI techniques can analyze data received from sensors and other sources to identify the problems that may occur and can suggest the maintenance needs, thus reducing downtime and improving efficiency [27].

7.1.2 Smart Component Integration

AI models can suggest the components to be integrated to design a turbine with good efficiency. The information received from various electronic, mechanical, and electrical components used in a wind turbine can be analyzed to select the components to extract the best energy output from a wind turbine at the lowest cost.

7.1.3 Employee Training

The companies invest a lot of effort and money in creating and maintaining user manuals for wind turbine operations and maintenance. Large investments are made in staff training to make them well versed with the turbine operations.

7.1.4 Protecting the Environment

AI models can monitor the movement of birds using various sensors and can reduce the harmful impact of wind farms on local wildlife.

7.1.5 Energy Storage Optimization and Energy Trading

AI models can predict the usage of energy and also the need for backup power. AI models are now a days are also for designing trading strategies in a competitive energy market.

Some of the real-life examples where companies are using AI to improve efficiency, performance, and reliability of renewable energy production and management systems [50].

- Google AI is developing AI algorithms to forecast wind energy generation and solar energy generation.
- Vestas is using AI to develop predictive maintenance solutions for wind turbines.
- Siemens Gamesa Renewable Energy is using AI to develop wind turbines that are more efficient and have a longer lifespan.

Xcel Energy, a US-based regulated electric utility and natural gas delivery company, is using AI techniques to forecast the power output of a given wind farm 3 days ahead with quite good accuracy [56].

Utility Engie, a leading European player in on-site utilities, has signed a partnership with Google Cloud for the development of an AI-based business solution to optimize the wind portfolio of Engie (https://www.saurenergy.com/solar-energy-news/engie-google-cloud-to-seek-more-efficient-wind-energy-through-ai-based-solutions, retrieved on 29.04.2024).

8 Wind Energy and Conventional Energy: Pros and Cons

The consistent rise in the utilization of petroleum, coal, and natural gas to meet energy demands has been steadily elevating the release of carbon dioxide (CO_2) and other harmful gases, contributing to the phenomenon of global warming. Global CO_2 emissions linked to energy are projected to increase from 31 billion metric tons in 2010 to 36 billion metric tons in 2020, and further escalate to 45 billion metric tons in the 2040s [38], indicating a 46% surge over the next four decades. One-fifth of the world's GHG emissions are caused by the usage of coal. One of the few short-term solutions to drastically cut these emissions is to replace coal with biomass fuels [103]. Unlike many other renewable production choices, biomass burning is usable in places without major wind, solar, or hydroelectric resources and does not have the problem of being intermittent. It has been demonstrated that biomass cofiring lowers net GHG emissions and reduces emissions of SO_x and NO_x [83]. In order to decarbonize the energy sector, wind energy is crucial to scaling up renewable electricity sources produced by wind turbines is a sustainable and clean energy source. The wind power sector is growing quickly due to advancements in technology and innovative business models, which are leading to an increase in installed capacity [96]. Power generation for a wind farm with a big installed capacity swings drastically whenever the wind speed changes by 1 m/s. The nonlinear connection between wind speed and wind power generation is the cause of this volatility. As a result, accurate and timely wind energy forecasting is essential to ensuring sustainable energy development and optimizing wind farm site selection [10, 26, 70, 100]. However, a variety of elements, including wind direction and atmospheric pressure, influence wind speed since the origins of wind in nature are complicated and unclear [69, 77]. Additionally, wind energy prediction is challenging because of several elements that contribute to the unpredictable, volatile, and intermittent character of wind power output [60, 62]. The increasing capacity of WES suggests the need for innovative methods and approaches to mitigate the unpredictability of wind energy production and enhance its dependability, upkeep, and financial returns [43]. Wind turbines are not constantly powered at their rated capacity, as wind energy is a fluctuating renewable energy source. Advanced technologies like automation, data analysis, robotics, AI, and ML have been sought after by wind turbine manufacturers and producers to optimize wind power generating systems and reduce manufacturing and O&M costs [91]. AI forecasts wind energy production and consumption, allowing smart grids to effectively store and transfer energy [34, 51]. Furthermore, the advancement of certain generators and the use of power electronic devices have made gearless turbine designs possible [44]. Horizontal-axis wind turbines (HAWTs) and vertical-axis wind turbines (VAWTs) are the two categories into which modern wind turbines fall. The bulk of the wind business is dominated by HAWTs because of their higher energy output and efficiency when compared to VAWTs [54, 55]. Due to their installation near the ground and thus reduced wind exposure, which reduces power production, VAWTs are intrinsically troublesome. VAWTs are much more expensive since they require more material and a larger size in order to provide the same output as HAWTs. VAWTs do, however, still have certain benefits, such as reduced speed cuts that allow them to generate some energy even under low wind conditions. They can catch the wind coming from every direction and also make very little noise. VAWTs have demonstrated efficacy in small-scale and rooftop applications. The cost of wind energy has dropped dramatically since 1980 because of advancements in turbine technology and better efficiency. These days, wind energy makes clean, inexpensive, and economical electricity available to people everywhere. In addition to massive power systems, wind energy is also used in the following fields.

8.1 Water Pumping

Large-scale wind water pump installation has the potential to strengthen the economies of agricultural nations like India. Agricultural regions are typically far from grids; therefore, wind pumps save electricity in addition to reducing the expense of installing gearbox systems. For instance, in India, farmers make up 50% of the population, yet only 20% of them have electric water pumps [68].

8.2 Remote and Domestic Electricity Supply

Small wind turbines can provide electricity in remote regions that are inhabited, including hills and rural areas. Small wind turbines that can power a family and are readily erected in the garden or on top of a roof are offered by several manufacturers [92].

8.3 Street Lighting

In many regions of the world, streetlights driven by wind are operating satisfactorily. There is the use of an LED panel and a little wind turbine. A more modern design, the fire winder is a hanging LED light that runs on tiny air currents [54, 55].

8.4 Wind–Solar-Powered EV Charging Stations

A quick plug-in electric vehicle (PEV) charging station that is powered by solar and wind is a completely novel idea. A trend toward electric vehicles might make these kinds of eco-charging stations increasingly popular and energy-efficient. These simple and convenient EV charging stations may be put in parking lots or along highways [48].

Along with these advantages, there are several financial, technological, social, and environmental difficulties with wind energy, even though it is a clean and promising energy source. Although these consequences may not seem significant, they have a lasting impact on nearby communities; thus, caution must be exercised. WES often require significant funding from the public or government due to their high startup costs [89]. The integration of large-scale wind generation plants with the electrical grid presents significant technical obstacles. Efficiency, power safety and security, power imbalance, power quality and dependability, and load control are some of these. Problems with grid stability and dependability are typically brought on by the wind's erratic, uncontrolled nature. These variations may result in voltage spikes and subsequent power losses in transmission lines if the grid is less stable. Moreover, harmonics cause voltage distortion [47, 89]. The primary socio-environmental barriers to wind turbines include visual effects on the landscape, noise pollution, radar, and telecommunications interference (if within line-of-sight), and threats to animals [87]. One of the main ecological concerns of WES is bird and bat mortality. However, in comparison to other more frequent threats like predators and building accidents, the incidence of bird deaths linked to wind turbines is incredibly low [105]. Wind turbines also pose a risk to human safety [53].

9 Wind Energy and Sustainable Development Goals

The well-being of human life majorly depends upon environmental sustainability and ecological factors. However, ecology and environment depend upon the quality of air, water, and energy available for different purposes in life. Human beings are harnessing wind energy for significant purposes such as using windmills, sailing ships, and generating electricity [2, 72]. According to recent research, more than 55% energy efficiency remains unavailable for mankind, and only 25–45% energy is utilized in a constructive way [41].

The United Nations set SDGs to achieve a sustainable future for mankind [8, 95]. These goals came into existence in 2015, and implementation was scheduled to be completed by 2030 [54, 55]. There are 17 SDGs set by the United Nations to end hunger and poverty and to attain gender equality, make sure to have healthy lives, empower women of all sectors, and promote well-being for all age groups. SDGs were established earlier, also by the United Nations, as global goals in 1987 [33, 71]. The Brundtland Commission issued the statement in 1987 for global goals, which defines sustainable development as "development that fulfills existing requirements without harming future generations" [32, 36]. The four important elements of sustainable development are social, cultural, environmental, and economic [32]. Developed nations can offer resources, food, and clean air to their citizens to maintain a healthy and hearty environment.

All 17 SDGs are correlated with each other, and the truth of sustainability actions of one SDG affects the outcomes in another SDG, requiring adjustments to social, financial, and ecological sustainability [85]. Renewable energy helps us to achieve security in the field of environmental protection, mechanical work, construction, Internet of Things (IOT), and economy [88]. In recent years, clean energy has been considered as an alternative solution to energy problems being faced by different nations. Renewable energy includes geothermal energy, wind energy, solar energy, hydropower energy, tidal energy, and other forms of energy [81].

Compared to other forms of alternative energy, wind energy has more benefits in terms of better technology and being economical [2]. All 17 SDGs have been illustrated in Fig. 7. Many developed nations are dependent on wind energy for better sustainability. Wind energy plays an important part in meeting the target in terms of SDGs.

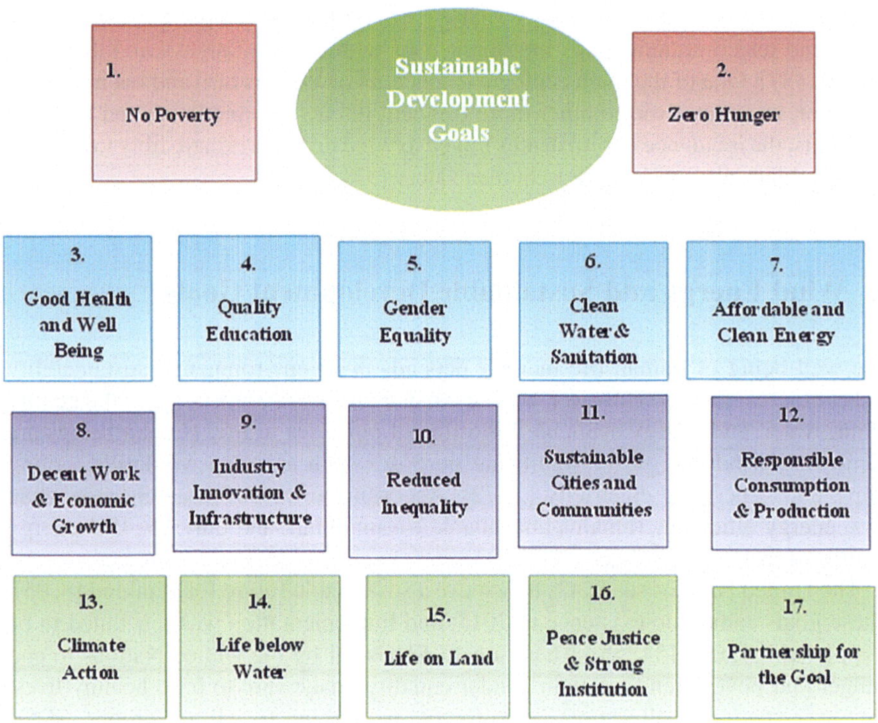

Fig. 7 Seventeen SDGs set by the United Nations

Wind energy is a significant element of renewable energy, and it has advantages in terms of cost, sustainability, and clean environment [18, 22]. Efficiency of turbines has been improved in recent periods and is helping to generate more energy. Further, wind turbines are safe to operate and have a much lower impact on the environment. Wind energy is being used for different applications such as pumping water, milling grains, sawing wood, and generating electricity [74, 80].

SDGs offer a strong foundation for collaboration in attaining a sustainable future for mankind. Without clean energy, it is not possible to achieve the goals of SDGs. Hence, good health of human beings (SDG-1), clean water and sanitation (SDG-3), and climate action (SDG-13) cannot be achieved without clean and affordable energy. These SDGs are directly correlated with clean energy. However, there are a few other SDGs that are indirectly connected with clean energy [29]. Implementation of clean energy ensures that the entire SDGs is achieved. Among all kinds of renewable energies, wind energy is one of the most vital and economical [18]. It is much easier to install wind energy plants because of their flexibility, modularity, and easy installation process [12]. So, wind energy can minimize the inequality in terms of the distribution of energy among urban and rural populations [11]. Wind energy can also provide energy to developing nations and energy can be accessible to all the citizens of the countries. Table 2 summarizes the correlation of wind energy with different SDGs.

The most interesting component is that wind energy can be helpful not only for life on land but also for life under the water. Wind energy cannot only maintain the marine ecosystem but also reduce ocean acidification [6]. As we are aware that military confrontations and geopolitical tension are interrelated to access fossil fuel resources. Nations powered by wind energy are more peaceful than compared to those without

Table 2 Wind energy correlation with different SDGs

SDGs	Correlations
SDG-1	Help to access clean energy and create more employment for local people
SDG-3	Help to improve public health by reducing the quantity of air pollution and emissions of greenhouse gases. They are directly related to adverse health effects
SDG-6	Consumption of water is not required
SDG-7	Renewable and pure form of energy, and it does not generate greenhouse gas emissions. It helps to achieve clean energy sources
SDG-8	Help to generate new jobs which help to improve economic growth
SDG-9	Help to get energy security and decrease carbon footprint of the industries
SDG-11	Reduce the requirement of traditional power plants, which results in reducing pollution
SDG-12	Help to reduce water and pollution by providing safer energy
SDG-13	Responsible for reducing greenhouse gas emissions by supplying a low-emission energy source
SDG-14	Reduces adverse effects on conventional resources
SDG-15	Help to utilize unused regions for clean energy production

Table 3 Advantages and limitations of wind energy in terms of SDGs

Advantages	Limitations
• Reducing air and water pollution – SDG 3: good health and well-being – SDG 14: life below water – SDG 15: life on land • Increasing access to clean energy – SDG 7: affordable and clean energy • Minimizing the impact of climate change on mankind and environment – SDG 13: climate action • Improving the output of wind energy – SDG 7: affordable and clean energy • Improving the diversification of grid – SDG 11: sustainable cities and communities – SDG 12: responsible consumption	• Creating noise pollution in adjoining areas – SDG 15: life on land – SDG 3: good health and well-being • Causing a threat to wildlife animals, especially small animals and birds – SDG 15: life on land • Need to fix location for installation; terrain is unsuitable for installation – SDG 7: affordable and clean energy • Completely dependent on speed of wind and direction – SDG 7: affordable and clean energy

access to wind energy. Wind energy is considered one of the most environmentally sustainable and purest energy resources, which helps in attaining SDGs [45]. Wind energy does not cause any kind of pollution in the environment. It has a good number of benefits such as creating jobs, being accessible to developing nations, and reducing carbon emissions. Installation of wind energy plants also provides employment to local people. However, the usage of wind energy also has some disadvantages. Table 3 tabulates the advantages and disadvantages of wind energy in terms of SDGs.

9.1 Contribution of Wind Energy in Sustainable Development

SDGs can be achieved by exploring the significance and contribution of wind energy in various fields. They have been highlighted in the following points.

9.2 Contribution Toward Environmental Sustainability

As we know, wind energy plays a significant role in the generation of renewable energy. Wind energy changes kinetic energy to electrical energy and helps in decreasing carbon footprint [73]. But at the same time, it also has some negative impacts [8]. The installation of wind energy plants also contributes to the destruction of habitat and causes noise pollution. Birds are also killed by wind turbines. GHGs are also emitted during the process of transportation of wind energy [94].

9.3 Contribution Toward Alternate Energy Source

Generation of wind energy decreases the dependency of countries on fossil fuels. It reduces geopolitical risk and enhances energy security, and encourages nation-wise sustainability as well as economic growth [61, 73].

9.4 Economic Viability and Job Creation

Wind energy installation increases a lot more job creation for the citizens of the nation. So it not only creates jobs but also helps in the economic growth of the country. Job creation at different steps of installation and production. So, it can be concluded that SDGs can be achieved by installing and operating wind power plants. Generation of wind energy contributes greatly to reducing poverty by giving employment, maintaining good health by providing clean and pure energy, contributing to economic growth of the nation, and development of infrastructure and industries [36].

10 Conclusion

Wind power can be utilized as safe, zero-emission, low-cost, and environmentally sustainable tool for energy generation. There are certain challenges related to the installation of WES, including noise pollution, impact on birds, failure prediction, and design optimization of turbines.

AI tools can further address these challenges and increase their utilization in different ways. AI models can monitor the movement of birds using various sensors and can reduce the harmful impact of wind farms on local wildlife. AI tools also have different applications in the wind energy industry, including forecasting, system performance prediction, failure prediction, design optimization of turbines, optimal placement, optimization of energy storage, optimization of energy trade, operation, and management. Different models like support vector machines, ANNs, back propagation algorithm, gradient boosting decision trees, and adaptive fuzzy techniques have been proposed. ANNs are efficient in modeling linear and nonlinear functions with arbitrary shapes. The prediction done by the ANNs has more accuracy as compared to other models. However, the type of model to be applied depends on different factors, including environmental conditions, site location, and data availability. ANNs can be applied for fault detection, wind speed prediction, and planning and management. Adaptive neuro-fuzzy techniques may be used for design optimization. However, for proper optimization of a wind plant, accurate data should be available. AI tools, with the availability of accurate location-wise data, can help in the placement and expansion of efficient, low-cost, sustainable, and reliable WES.

References

1. Abedinia O, Amjady N, Ghadimi N (2018) Solar energy forecasting based on hybrid neural network and improved metaheuristic algorithm. Comput Intell 34(1):241–260
2. Adeyeye K, Ijumba N, Colton J (2020) Exploring the environmental and economic impacts of wind energy: a cost-benefit perspective. Int J Sust Dev World 27(8):718–731
3. Ahmad T, Zhang D, Huang C, Zhang H, Dai N, Song Y, Chen H (2021) Artificial intelligence in sustainable energy industry: status quo, challenges and opportunities. J Clean Prod 289
4. Ahmadi A, Nabipour M, Mohammadi-Ivatloo B, Amani AM, Rho S, Piran MJ (2020) Long-term wind power forecasting using tree-based learning algorithms. IEEE Access 8:151511–151522
5. Ahmed SD, Al-Ismail FSM, Shafiullah M, Al-Sulaiman FA, El-Amin IM (2020) Grid integration challenges of wind energy: a review. IEEE Access 8:10857–10878
6. Albright R, Cooley S (2019) A review of interventions proposed to abate impacts of ocean acidification on coral reefs. Reg Stud Mar Sci 29:100612
7. Alencar DB, Affonso CM, Oliveira RC, Jose Filho CR (2018) Hybrid approach combining SARIMA and neural networks for multi-step ahead wind speed forecasting in Brazil. IEEE Access 6:55986–55994
8. Armin Razmjoo A, Sumper A, Davarpanah A (2020) Energy sustainability analysis based on SDGs for developing countries. Energy Sour Part A: Recov Utiliz Environ Effects 42(9):1041–1056
9. Band SS, Rabczuk T, Chau KH (2019) A survey of deep learning techniques: application in wind and solar energy resources. IEEE Access 7:164650–164666
10. Banik A, Behera C, Sarathkumar TV, Goswami AK (2020) Uncertain wind power forecasting using LSTM-based prediction interval. IET Renew Power Gener 14(14):2657–2667
11. Blenkinsopp TCSR, Coles SR, Kirwan K (2013) Renewable energy for rural communities in Maharashtra, India. Energy Policy 60:192–199
12. Carley M, Spapens P (2017) Sharing the world: sustainable living and global equity in the 21st century. Routledge
13. Chan CM, Bai HL, He DQ (2018) Blade shape optimization of the Savonius wind turbine using a genetic algorithm. Appl Energy 213:148–157
14. Charabi Y, Al Hinai A, Al-Yahyai S, Al Awadhi T, Choudri BS (2019) Offshore wind potential and wind atlas over the Oman Maritime Zone. Energy Ecol Environ 4:1–14
15. Charabi Y, Abdul-Wahab S (2020) Wind turbine performance analysis for energy cost minimization. Renew: Wind Water Sol 7(1):1–11
16. Chen Y, Xu P, Chu Y, Li W, Wu Y, Ni L, Bao Y, Wang K (2017) Short-term electrical load forecasting using the Support Vector Regression (SVR) model to calculate the demand response baseline for office buildings. Appl Energy 195:659–670
17. Danish MSS (2023) AI in energy: overcoming unforeseen obstacles. AI 4:406–425
18. Dawn S, Tiwari PK, Goswami AK, Singh AK, Panda R (2019) Wind power: existing status, achievements and government's initiative towards renewable power dominating India. Energ Strat Rev 23:178–199
19. Demirdelen T, Ozge Aksu I, Esenboga B, Aygul K, Ekinci F, Bilgili M (2019) A new method for generating short-term power forecasting based on artificial neural networks and optimization methods for solar photovoltaic power plants. Solar photovoltaic power plants: advanced control and optimization techniques, pp 165–189
20. Deng X, Shao H, Hu C, Jiang D, Jiang Y (2020) Wind power forecasting methods based on deep learning: a survey. Comput Model Eng Sci 122(1):273–301
21. Desai S, Sampat S, Vadodaria D, Pimpale M (2022) Wind energy forecasting using artificial intelligence. In: 2022 4th international conference on artificial intelligence and speech technology (AIST). IEEE, pp 1–6
22. Destek MA, Sarkodie SA, Asamoah EF (2021) Does biomass energy drive environmental sustainability? An SDG perspective for top five biomass consuming countries. Biomass Bioenergy 149:106076

23. Dhirani LL, Mukhtiar N, Chowdhry BS, Newe T (2023) Ethical dilemmas and privacy issues in emerging technologies: a review. Sensors 23(3)
24. Díaz D, Torres A, Dorronsoro JR (2015) Deep neural networks for wind energy prediction. In: Advances in computational intelligence: 13th international work-conference on artificial neural networks, IWANN 2015, Palma de Mallorca, Spain, 10–12 June 2015. Proceedings, Part I 13. Springer, pp 430–443
25. Dittner ME, Vasel A (2019) Advances in wind power forecasting. Advances in sustainable energy, pp 37–57
26. Do MH, Söffker D (2021) State-of-the-art in integrated prognostics and health management control for utility-scale wind turbines. Renew Sustain Energy Rev 145:111102
27. Dubey AK, Kumar A, Ramirez IS, Marquez FPG (2022) A review of intelligent systems for the prediction of wind energy using machine learning. In: Proceedings of the sixteenth international conference on management science and engineering management, vol 1, pp 476–491
28. Ekonomou L, Lazarou S, Chatzarakis GE, Vita V (2012) Estimation of wind turbines optimal number and produced power in a wind farm using an artificial neural network model. Simul Model Pract Theory 21(1):21–25
29. ElAlfy A, Palaschuk N, El-Bassiouny D, Wilson J, Weber O (2020) Scoping the evolution of corporate social responsibility (CSR) research in the sustainable development goals (SDGs) era. Sustainability 12(14):5544
30. Elkin C, Witherspoon S (2019) Machine learning can boost the value of wind energy. Deep mind field study
31. Elyasichamazkoti F, Teimourzadeh S (2021) Secure under frequency load shedding scheme with consideration of rate of change of frequency. In: 2021 IEEE green technologies conference (GreenTech). IEEE, pp 552–557
32. Fatimah YA, Govindan K, Murniningsih R, Setiawan A (2020) Industry 4.0 based sustainable circular economy approach for smart waste management system to achieve sustainable development goals: a case study of Indonesia. J Clean Prod 269:122263
33. Fei W, Opoku A, Agyekum K, Oppon JA, Ahmed V, Chen C, Lok KL (2021) The critical role of the construction industry in achieving the sustainable development goals (SDGs): Delivering projects for the common good. Sustainability 13(16):9112
34. Flores P, Tapia A, Tapia G (2005) Application of a control algorithm for wind speed prediction and active power generation. Renew Energy 30(4):523–536
35. Foley AM, Leahy PG, McKeogh EJ (2010) Wind power forecasting & prediction methods. In: 2010 9th international conference on environment and electrical engineering. IEEE, pp 61–64
36. Franco IB, Chatterji T, Derbyshire E, Tracey J (2019) Actioning the global goals for local impact. Springer, Berlin, Germany
37. Frangoul A (2017) China and US lead way with wind power installations, says global energy report [online]. CNBC. Available at: https://www.cnbc.com/2017/02/13/china-and-us-lead-way-with-wind-power-installations-says-global-energy-report.html. Accessed 27 Dec 2023
38. Fridleifsson IB (2001) Geothermal energy for the benefit of the people. Renew Sustain Energy Rev 5(3):299–312
39. Gao X, Yang H, Lu L (2016) Optimization of wind turbine layout position in a wind farm using a newly-developed two-dimensional wake model. Appl Energy 174:192–200
40. Ghadimi N, Akbarimajd A, Shayeghi H, Abedinia O (2018) Two stage forecast engine with feature selection technique and improved meta-heuristic algorithm for electricity load forecasting. Energy 161:130–142
41. Gomaa MR, Rezk H, Mustafa RJ, Al-Dhaifallah M (2019) Evaluating the environmental impacts and energy performance of a wind farm system utilizing the life-cycle assessment method: a practical case study. Energies 12(17):3263
42. Gu C, Li H (2022) Review on deep learning research and applications in wind and wave energy. Energies

43. Habibi S, Rahimi R, Shamsi P, Ferdowsi M (2021) Efficiency assessment of a residential DC nanogrid with low and high distribution voltages using realistic data. In: 2021 IEEE green technologies conference (GreenTech). IEEE, pp 574–579
44. Hansen LH, Madsen PH, Blaabjerg F, Christensen HC, Lindhard U, Eskildsen, KAEK (2001) Generators and power electronics technology for wind turbines. In: IECON'01. 27th annual conference of the IEEE Industrial Electronics Society (Cat. No. 37243), vol 3. IEEE, pp 2000–2005
45. Howard-Grenville J, Davis GF, Dyllick T, Miller CC, Thau S, Tsui AS (2019) Sustainable development for a better world: contributions of leadership, management, and organizations. Acad Manag Discov 5(4):355–366. https://www.saurenergy.com/solar-energy-news/engie-google-cloud-to-seek-more-efficient-wind-energy-through-ai-based-solutions. Retrived 29 Apr 2024
46. Hutchinson M, Zhao F (2023) Global wind report 2023 [Online]. Gwec.Net. Global Wind Energy Council, Belgium, pp 1–120. Available at: https://gwec.net/wp-content/uploads/2023/04/GWEC-2023_interactive.pdf. Accessed 27 Dec 2023
47. Ibrahim H, Ghandour M, Dimitrova M, Ilinca A, Perron J (2011) Integration of wind energy into electricity systems: technical challenges and actual solutions. Energy Procedia 6:815–824
48. Inhabitat 2014. http://inhabitat.com/breaking-the-new-sanya-skypump-is-a-wind-and-solar-powered-ev-charger/
49. Institute E (2023) Resources and data downloads [online]. Statistical review of world energy. Available at: https://www.energyinst.org/statistical-review/resources-and-data-downloads. 05 May 2024
50. Ju F (2019) Artificial Intelligence for Wind Energy (AI4Wind): a state of the art report. DTU Library
51. Jursa R, Rohrig K (2008) Short-term wind power forecasting using evolutionary algorithms for the automated specification of artificial intelligence models. Int J Forecast 24(4):694–709
52. Khosravi A, Machado L, Nunes RO (2018) Time-series prediction of wind speed using machine learning algorithms: a case study Osorio wind farm, Brazil. Appl Energy 224:550–566
53. Kim DH, Lim JH (2015) Methods for early fire detection and fire position determination inside the nacelle of wind turbine generator system. Trans Korean Soc Mech Eng B 39(12):935–943
54. Kumar S, Kumar N, Vivekadhish S (2016) Millennium development goals (MDGS) to sustainable development goals (SDGS): addressing unfinished agenda and strengthening sustainable development and partnership. Indian J Commun Med 41(1):1
55. Kumar Y, Ringenberg J, Depuru SS, Devabhaktuni VK, Lee JW, Nikolaidis E, Andersen B, Afjeh A (2016) Wind energy: Trends and enabling technologies. Renew Sustain Energy Rev 53:209–224
56. Kumar A (2024) Earth5R Research Article: assessing the impact of artificial intelligence on renewable energy in India. Earth5R Earth J
57. Kuşkaya S, Bilgili F (2020) The wind energy-greenhouse gas nexus: the wavelet-partial wavelet coherence model approach. J Clean Prod 245:118872
58. Lanke M, Porwal A, Sarolkar S, Palaskar G, Kakad S (2022) AI based student's assignments plagiarism detector. Int J Res Appl Sci Eng Technol 10(12): 991–999
59. Liu H, Mi XW, Li YF (2018) Wind speed forecasting method based on deep learning strategy using empirical wavelet transform, long short term memory neural network and Elman neural network. Energy Convers Manag 156:498–514
60. Lledó L, Torralba V, Soret A, Ramon J, Doblas-Reyes FJ (2019) Seasonal forecasts of wind power generation. Renew Energy 143:91–100
61. Majid M (2020) Renewable energy for sustainable development in India: current status, future prospects, challenges, employment, and investment opportunities. Energy Sustain Soc 10(1):1–36
62. Malz EC, Verendel V, Gros S (2020) Computing the power profiles for an Airborne Wind Energy system based on large-scale wind data. Renew Energy 162:766–778

63. Marugán AP, Márquez FPG, Perez JMP, Ruiz-Hernández D (2018) A survey of artificial neural network in wind energy systems. Appl Energy 228:1822–1836
64. Mitchell D, Blanche J, Harper S, Lim T, Gupta R, Zaki O, Tang W, Robu V, Watson S, Flynn D (2022) A review: challenges and opportunities for artificial intelligence and robotics in the offshore wind sector. Energy AI 8
65. Mokhtari S, Abbaspour A, Yen KK, Sargolzaei A (2021) A machine learning approach for anomaly detection in industrial control systems based on measurement data. Electronics 10(4):407
66. Mone C, Hand M, Bolinger M, Rand J, Heimiller D, Ho J (2017) 2015 cost of wind energy review (No. NREL/TP-6A20-66861). National Renewable Energy Lab.(NREL), Golden, CO (United States)
67. Moorthy M, Agrawal A, Deshmukh MK (2015) Artificial intelligence techniques for wind power prediction: a case study. Indian J Sci Technol 8(25)
68. Murthy KR, Raju MR (2009) Analysis on electrical energy consumption of agricultural sector in Indian context. ARPN J Eng Appl Sci 4(2):6–9
69. Nikolić V, Sajjadi S, Petković D, Shamshirband S, Ćojbašić Ž, Por LY (2016) Design and state of art of innovative wind turbine systems. Renew Sustain Energy Rev 61:258–265
70. Niu T, Wang J, Zhang K, Du P (2018) Multi-step-ahead wind speed forecasting based on optimal feature selection and a modified bat algorithm with the cognition strategy. Renew Energy 118:213–229
71. Olabi AG, Obaideen K, Abdelkareem MA, AlMallahi MN, Shehata N, Alami AH, Mdallal A, Hassan AAM, Sayed ET (2023) Wind energy contribution to the sustainable development goals: case study on London Array. Sustainability 15(5):4641
72. Olabi AG, Wilberforce T, Elsaid K, Salameh T, Sayed ET, Husain KS, Abdelkareem MA (2021) Selection guidelines for wind energy technologies. Energies 14(11):3244
73. Olabi AG, Wilberforce T, Elsaid K, Sayed ET, Maghrabie HM, Abdelkareem MA (2022) Large scale application of carbon capture to process industries—a review. J Clean Prod 362:132300
74. Pain S (2017) Power through the ages. Nature 551(7680):S134–S134
75. Pei S, Qin H, Zhang Z, Yao L, Wang Y, Wang C, Liu Y, Jiang Z, Zhou J, Yi T (2019) Wind speed prediction method based on empirical wavelet transform and new cell update long short-term memory network. Energy Convers Manag 196:779–792
76. Peiris AT, Jayasinghe J, Rathnayake U (2021) Forecasting wind power generation using artificial neural network: "PawanDanawi"—a case study from Sri Lanka. J Electr Comput Eng 2021:1–10
77. Peng Z, Peng S, Fu L, Lu B, Tang J, Wang K, Li W (2020) A novel deep learning ensemble model with data denoising for short-term wind speed forecasting. Energy Convers Manag 207:112524
78. Perera W (2023) Comparative analysis of wind energy towards sustainability. J Res Technol Eng [online] 4(2):195–209. Available at: https://www.jrte.org/wp-content/uploads/2023/04/Comparative-analysis-of-wind-energy-towards-sustainability.pdf
79. Qureshi S, Shaikh F, Kumar L, Ali F, Awais M, Gürel AE (2023) Short-term forecasting of wind power generation using artificial intelligence. Environ Chall 11:100722
80. Ragheb M (2017) History of harnessing wind power. In: Wind energy engineering. Academic, pp 127–143
81. Rahman MM, Khan I, Field DL, Techato K, Alameh K (2022) Powering agriculture: present status, future potential, and challenges of renewable energy applications. Renew Energy 188:731–749
82. Rashid H, Haider W, Batunlu C (2020) Forecasting of wind turbine output power using machine learning. In: 2020 10th international conference on advanced computer information technologies (ACIT). IEEE, pp 396–399
83. Robinson AL, Rhodes JS, Keith DW (2003) Assessment of potential carbon dioxide reductions due to biomass-coal cofiring in the United States
84. Rolik Y (2017) Risk management in implementing wind energy project. Procedia Eng 178:278–288

85. Sadorsky P (2021) Wind energy for sustainable development: driving factors and future outlook. J Clean Prod 289:125779
86. Sahu BK (2018) Wind energy developments and policies in China: a short review. Renew Sustain Energy Rev 81:1393–1405
87. Saidur R, Rahim NA, Islam MR, Solangi KH (2011) Environmental impact of wind energy. Renew Sustain Energy Rev 15(5):2423–2430
88. Salam A (2020) Internet of things for sustainable community development: introduction and overview. In: Internet of Things for sustainable community development. Springer, Cham, pp 1–31
89. Shafiullah GM, Oo AM, Ali AS, Wolfs P (2013) Potential challenges of integrating large-scale wind energy into the power grid—a review. Renew Sustain Energy Rev 20:306–321
90. Solarin SA, Bello MO (2022) Wind energy and sustainable electricity generation: evidence from Germany. Environ Dev Sustain 24(7):9185–9198
91. Stetco A, Dinmohammadi F, Zhao X, Robu V, Flynn D, Barnes M, Keane J, Nenadic G (2019) Machine learning methods for wind turbine condition monitoring: a review. Renew Energy 133:620–635
92. US Department of Energy (2014) Small wind electric system: an Ohio's consumer guide 6–9. http://www.greenenergyohio.org/page.cfm?pageID=654
93. Valdivia-Bautista S.M., Domínguez-Navarro JA, Pérez-Cisneros M, Vega-Gómez CJ, and Castillo-Téllez B (2023) Artificial intelligence in wind speed forecasting: a review. Energies 16(2457)
94. Van Vuuren DP, Stehfest E, Gernaat DE, Doelman JC, Van den Berg M, Harmsen M, de Boer HS, Bouwman LF, Daioglou V, Edelenbosch OY, Girod B (2017) Energy, land-use and greenhouse gas emissions trajectories under a green growth paradigm. Glob Environ Change 42:237–250
95. Walsh PP, Murphy E, Horan D (2020) The role of science, technology and innovation in the UN 2030 agenda. Technol Forecast Soc Chang 154:119957
96. Wang S (2021) An interview with Shouyang Wang: research frontier of big data-driven economic and financial forecasting. Data Sci Manag 1(1):10–12
97. World Bank (2015) Renewable energy consumption (% of total final energy consumption)
98. Wu Z, Luo G, Guo ZYY, Li K, Xue Y (2021) A comprehensive review on deep learning approaches in wind forecasting applications. CAAI Trans Intell Technol
99. Xiao L, Wang J, Dong Y, Wu J (2015) Combined forecasting models for wind energy forecasting: a case study in China. Renew Sustain Energy Rev 44:271–288
100. Yang M, Shi C, Liu H (2021) Day-ahead wind power forecasting based on the clustering of equivalent power curves. Energy 218:119515
101. Yang B, Yu T, Shu H, Dong J, Jiang L (2018) Robust sliding-mode control of wind energy conversion systems for optimal power extraction via nonlinear perturbation observers. Appl Energy 210:711–723
102. Yin M, Yang Z, Xu Y, Liu J, Zhou L, Zou Y (2018) Aerodynamic optimization for variable-speed wind turbines based on wind energy capture efficiency. Appl Energy 221:508–521
103. Zhang Y, McKechnie J, Cormier D, Lyng R, Mabee W, Ogino A, Maclean HL (2010) Life cycle emissions and cost of producing electricity from coal, natural gas, and wood pellets in Ontario, Canada. Environ Sci Technol 44(1):538–544
104. Zhao L, Nazir SM, Nazir HMJ, Abdalla NA (2022) A review on proliferation of artificial intelligence in wind energy forecasting and instrumentation management. Environ Sci Pollut Res 29(29):43690–43709
105. Zimmerling J, Pomeroy A, d'Entremont M, Francis C (2013) Canadian estimate of bird mortality due to collisions and direct habitat loss associated with wind turbine developments. Avian Conserv Ecol 8(2)

Transition from Fossil Fuel to Sustainability: Solar Energy—Prospects, Technology Advancement, Policies, Challenges, and Opportunities

Deepika Varshney, Rashmi Mishra, Shahid Pervez Ansari, Jamal Akhter Siddique, and Mohammad Omaish Ansari

Abstract This review underscores the increasing significance of solar energy, especially photovoltaics (PVs), as a clean and sustainable substitute for conventional energy. It details the historical development and underlying principles of PV technology, emphasizing progress in materials such as perovskites and organic cells. While recognizing solar energy's promise, the review also examines existing obstacles like efficiency limitations, elevated costs, and infrastructural constraints. Exploring future innovations like advanced energy storage and agrivoltaics, it emphasizes that resolving persistent issues related to efficiency, cost, scalability, and supportive policies is essential to fully harness solar energy's potential and create a sustainable energy future.

Keywords Solar energy · Renewable energy · Photovoltaic cell · Solar energy policies · Opportunities and barriers

---

D. Varshney (✉)
College of Engineering & Technology (CET), University of Technology and Applied Sciences, , Al Musanna, Oman
e-mail: deepika.varshney@utas.edu.om

R. Mishra
College of Economics & Business Administration (CEBA), University of Technology and Applied Sciences, , Al Musanna, Oman

S. P. Ansari
Department of Applied Chemistry, Faculty of Engineering and Technology, Aligarh Muslim University, Aligarh, India

J. A. Siddique
Department of Chemistry, Gandhi Faiz-E-Aam College (M.J.P Rohilkhand University), Shahjahanpur, India

M. O. Ansari
Center of Nanotechnology, King Abdulaziz University, Jeddah, Saudi Arabia

J. A. Siddique et al. (eds.), *Energy Sources*, Green Energy and Technology,
https://doi.org/10.1007/978-981-95-1209-6_13

1 Introduction

1.1 The Evolution of Energy

Energy has been a fundamental driver of human civilization and economic development throughout history. As societies have evolved, the demand for energy has steadily increased, reflecting humanity's progress and technological advancement [1–3].

From prehistoric times, humans have strategically utilized available energy sources in their environment. This journey began with fire approximately one million years ago, which provided heat, light, and cooking capabilities, then the harnessing of wind power around 8000 years ago for sailing vessels, followed by the widespread use of wood until the mid-nineteenth century [4] (Fig. 1).

The Agricultural Revolution marked a significant advancement as humans transitioned from hunter-gatherer lifestyles to settled communities [4]. A transformative period began with the Industrial Revolution in the eighteenth century, when coal emerged as the dominant energy source, powering steam engines and catalyzing unprecedented technological progress [6]. By the late nineteenth century, coal accounted for approximately half of the world's energy consumption, with biomass providing the remainder [7]. The discovery of oil in Pennsylvania in 1859 eventually led to petroleum becoming a major energy source, particularly after the mass

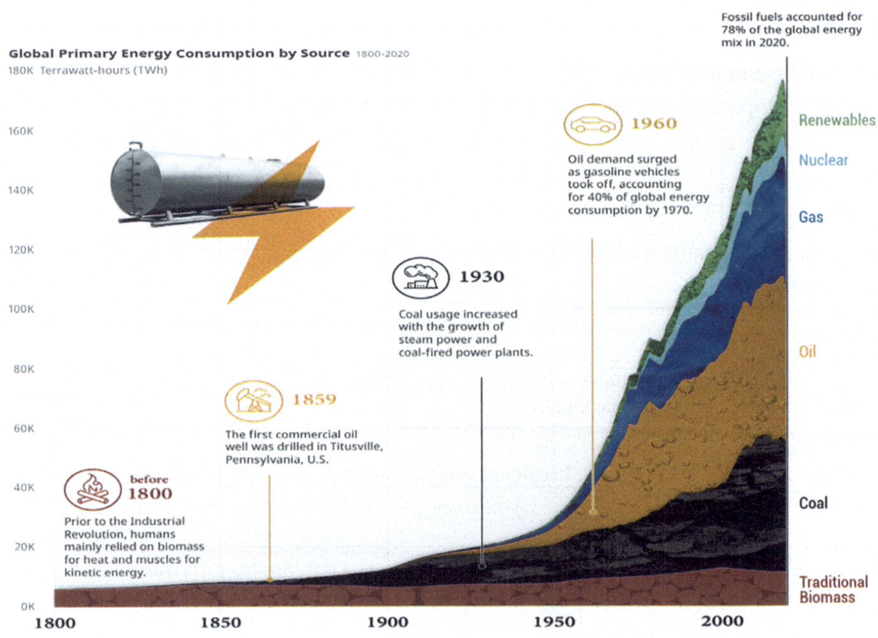

Fig. 1 Transition of energy. *Source* The world's energy transition [5]

production of automobiles with internal combustion engines. Natural gas similarly gained prominence for household and industrial applications [8]. The twentieth century witnessed the diversification of energy sources, with the addition of oil, natural gas, hydropower, and eventually nuclear energy in the 1960s. This period saw energy consumption grow at a compound annual rate of 2.43% from 1965 to 2020, demonstrating humanity's increasing energy demands [9].

1.2 Global Energy Demand

A century ago, electricity played a far smaller role in daily life compared to its indispensable status today, yet it now dominates as the fastest-growing form of energy consumption. 85% of global energy demand is projected to come from emerging and developing economies [10]. Under most scenarios, worldwide energy consumption is expected to increase by 50% between 2018 and 2050, driven by population growth, industrial development, and technological advancement [11]. With the global population anticipated to approach 10 billion by mid-century, and with China and India—comprising over one-third of the world's population—expected to increase their per capita energy usage three to five times over the next three decades, global energy demand will likely continue to grow throughout the twenty-first century [12].

Global electricity demand rose by more than 2.5% in 2023 and is projected to grow at an average annual rate of 3.5–4.5% by 2030 and till 2050, these growth rates range from 3 to 3.5%. Two-thirds of the increase in demand since 2013 was from China, driven by electrification of industrial processes and by growth in electricity demand for appliances, cooling, and data centers (which consumed an estimated 100 TWh of electricity in 2024). Around 1% of global electricity consumption only comes from data centers as of now, and while growth is expected to increase in different regions at different paces [10]. Worldwide electricity consumption is rapidly increasing, fueled by building construction and operation, especially in India, the Middle East, and Southeast Asia. China is the primary driver of this expansion, projected to account for roughly 45% of the global increase until 2030 and 25% from 2030 to 2050. India is forecasted to become the third-largest electricity consumer by 2050, with demand growing at over 4% annually under various projections. While other developing economies also exhibit strong growth, developed nations are expected to see more modest increases of 1.8–2.6% per year until 2050.

Until 2035, the construction industry in developing economies will be the largest contributor to rising electricity demand, driven by greater appliance adoption and increased air conditioning use. Although buildings are currently the largest electricity consumers, rapid industrial electrification, especially in China, will likely shift this balance to the industrial sector by the late 2020s. This industrial growth is fueled by the electrification of light industries, as well as the use of electricity for on-site hydrogen production in steel and chemical manufacturing. Looking ahead to 2050, the increasing adoption of electric vehicles (EVs) is projected to account for over 20% of the total increase in electricity demand. Additionally, hydrogen production via electrolysis will be a major factor, potentially requiring 7000–12,000 TWh, representing 25–30% of today's global electricity consumption [13].

1.3 Sources of Energy

Fossil fuels accounted for 60% of global electricity generation in 2023, marking a 50-year low. Coal was the leading fossil fuel source at 36%, driven by its cost-effectiveness, with prices averaging about $2 per MMBtu from 2012 to 2022. Natural gas provided 22% of electricity, with prices ranging from $2 to $5 per MMBtu. Petroleum, the most expensive and volatile fossil fuel, costs between $7 and $16 per MMBtu. Nuclear power's contribution decreased slightly to 9%, a 50% reduction compared to its share 30 years earlier [13] (Fig. 2).

The majority of energy technologies remain dependent on combustion-based mechanisms [14], where chemical energy stored in fossil fuels undergoes an exothermic oxidation reaction. Combustion of fossil fuels releases CO_2, constituting 60% of total greenhouse gas emissions that trap heat in the atmosphere and harmful by-products like nitrogen oxides (NO_x), sulfur dioxide (SO_2), and mercury [11]. This leads to adverse consequences such as droughts, rising sea levels, floods, extreme weather events, and loss of species. The severity of these impacts is directly related to the amount of carbon dioxide emitted.

Approximately 50% of all CO_2 emissions are absorbed by land and ocean "sinks," with the remainder staying in the atmosphere and contributing to climate change. In 2023, global CO_2 emissions from fires have exceeded the average due to an intense wildfire season in Canada, where emissions were six to eight times above normal levels. The current capacity for technology-based carbon dioxide removal (excluding

Fig. 2 Sources of energy. *Source* Compiled by Author

nature-based solutions like reforestation) is around 0.01 million tonnes of CO_2, which is over a million times smaller than the current emissions from fossil fuels [15].

The unprecedented reduction in emissions in 2020 amounted to 1.9 billion tonnes of CO_2 ($GtCO_2$), reflecting a decrease of 5.4% from 36.7 $GtCO_2$ in 2019 to 34.8 $GtCO_2$, primarily due to the COVID-19 pandemic. However, this was not the first instance in recent history where a significant global event led to a decline in emissions. For instance, the global recession in 2009 triggered a nearly two percent drop in CO_2 levels, and the recession in the early 1980s also notably affected emissions. The largest annual percentage reduction occurred at the end of World War II in 1945, with emissions falling by 17%. Projections indicate that emissions are expected to rise by 4.9% (ranging from 4.1 to 5.7%) in 2021, reaching 36.4 $GtCO_2$. Despite this increase, global emissions in 2021 are still approximately 0.8% lower than their 2019 levels. The growth of 1.6 $GtCO_2$ in 2021 is comparable to the 1.7 $GtCO_2$ increase (5.5% above 2009 levels) observed in 2010 after the global financial crisis of 2008–2009 [15].

China is the largest contributor to global GHG emissions, followed closely by the United States. China did not always hold the title of the world's top emitter; however, significant economic growth and industrialization over the past few decades have led to a dramatic rise in emissions. Since 1990, CO_2 emissions in China have surged by nearly 450%, while US CO_2 emissions have decreased by 6.1%. Despite this decline, the United States remains the largest cumulative carbon polluter in history [15] (Fig. 3).

"Emissions are moving in the wrong direction, and we must take action to limit global warming," stated Ben Poulter, a scientist at NASA's Goddard Space Flight Center. The concentration of carbon dioxide in the atmosphere has risen from roughly 278 parts per million in 1750, at the start of the industrial era, to 420 parts per million in 2023, surpassing pre-industrial levels (Emissions from Fossil Fuels Continue to Rise n.d.). This increase in heat-trapping carbon dioxide and other greenhouse gases is the main driver of the rising global temperatures. The global surface temperature in

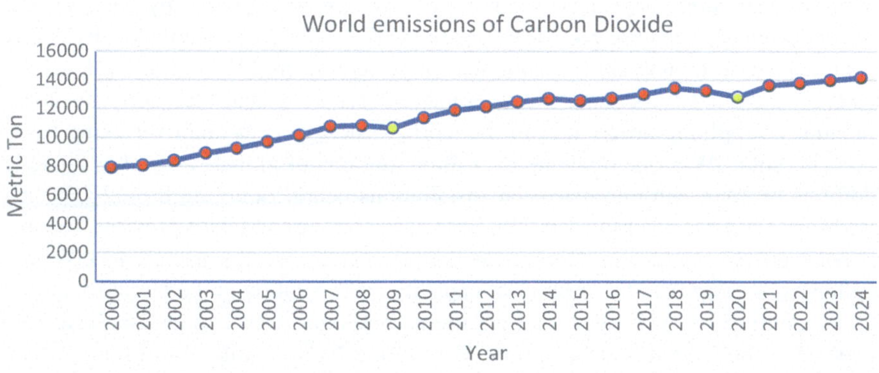

Fig. 3 Carbon dioxide emission. *Data Source* Ember (2024); Energy Institute—Statistical Review of World Energy (2024)

2023 was 1.2 °C (2.1 °F) above the average for NASA's baseline period (1951–1980), marking it as the hottest year on record [16].

1.4 Potential of Solar Power for the Global Energy

The Intergovernmental Panel on Climate Change (IPCC) cautions that without any intervention, global temperatures may increase by 6–7 °C by the end of the century, significantly surpassing the critical 2 °C threshold [14].

Recognizing this urgency, international climate efforts have intensified. At COP26 in Glasgow in 2021, 197 nations pledged to reduce reliance on fossil fuels while integrating health and equity into climate strategies. Building on the landmark Paris Agreement of 2015, signed by 196 countries, these commitments aim to mitigate greenhouse gas emissions, adapt to climate change impacts, and provide financial support to vulnerable communities. These initiatives align with the United Nations' Sustainable Development Goals (SDGs), which seek to eradicate poverty, protect the environment, and promote sustainable development by 2030 [1].

As countries commit to ambitious goals for reducing greenhouse gas emissions, the transition to renewable energy sources such as wind and solar power positions solar energy at the forefront of this shift. Beyond the advancements in capacity and efficiency, harnessing solar energy has the potential to play a pivotal role in reversing climate change and achieving energy independence. Solar energy is not just a critical resource; it represents a pathway toward a sustainable future. By embracing innovative strategies for solar power, we can promote environmental sustainability and significantly decrease our dependence on fossil fuels, marking a transformative shift toward a more sustainable lifestyle in energy consumption [17].

Solar energy and photovoltaic (PV) systems are rapidly gaining traction as leading renewable energy solutions. Solar panels convert sunlight into electricity, offering a cleaner and quieter alternative to fossil fuels. The global share of renewables in electricity generation is projected to rise from 29% in 2022 to 35% by 2025. Global renewable energy capacity grew at an accelerated pace in 2022, increasing by nearly 11%—a faster rate than the 9% annual average seen from 2017 to 2021 [18]. Solar power is expected to become the dominant source of global electricity by 2050, potentially supplying nearly half of the world's energy needs and overtaking fossil fuels [1]. Solar PV alone made up over three-quarters of global renewable capacity additions in 2023, with annual solar installations nearly doubling to 447 GW and total solar capacity reaching 1.6 TW. This surge far surpassed all previous records for solar energy expansion. China was the primary driver, installing 240 GW in a single year—more than the total solar capacity ever installed in the United States—and accounting for 57% of global additions, but significant expansions have also occurred in the United States, Europe, Brazil, and India. Unlike fossil fuels, which are geographically concentrated, solar and other renewables are widely accessible, enabling most countries to pursue greater energy self-sufficiency [14]. The International Energy Agency (IEA) projects that solar PV installed capacity will surpass

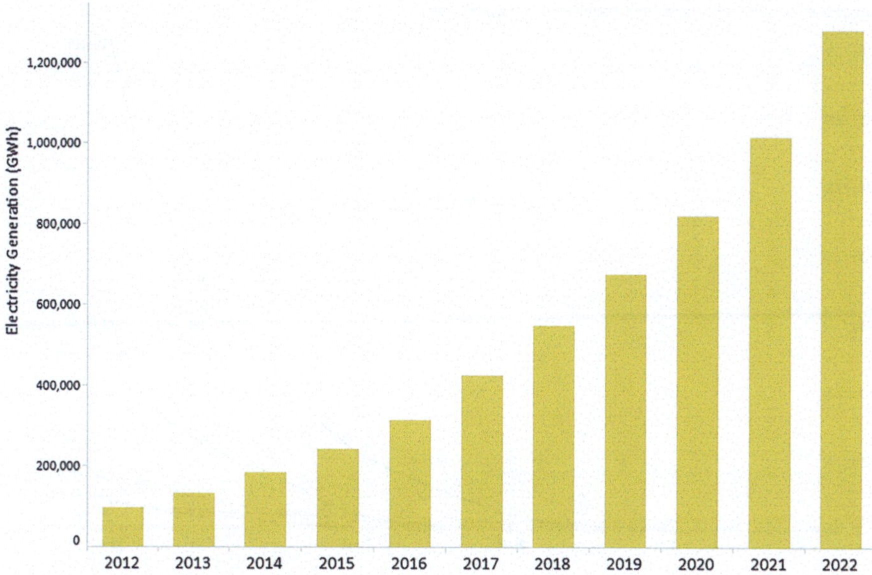

Fig. 4 Solar energy expansion. *Source* IRENA, http://www.irena.org/data

coal by 2027, marking a major shift in the global energy landscape. In 2022, solar PV generation grew by 26% to reach 1300 TWh, outpacing wind for the first time [19, 20] (Figs. 4 and 5).

2 Solar Cell

PV comes from "photo," meaning light, and "voltaic," meaning electricity [21]. A single PV device, known as a cell, is the basic building block of a solar energy system. Typically, an individual PV cell is quite small, producing about 1–2 W of power, and is made from semiconductor materials—most commonly silicon—that are often thinner than four human hairs. To increase power output, multiple PV cells are connected together to form a module, or solar panel. Several modules can be linked to create an array, and one or more arrays are integrated into a complete PV system that can be connected to the electrical grid. This modular approach allows PV systems to scale from small rooftop installations to large utility-scale solar farms [22].

PV cells are manufactured from a variety of semiconductor materials, with silicon being the most prevalent due to its high efficiency, low cost, and long lifespan. The core function of a PV cell is to convert sunlight into electricity through the PV effect. When sunlight, made up of photons, strikes the semiconductor material in a PV cell, it excites electrons, freeing them from their atomic bonds, leaving behind a vacant

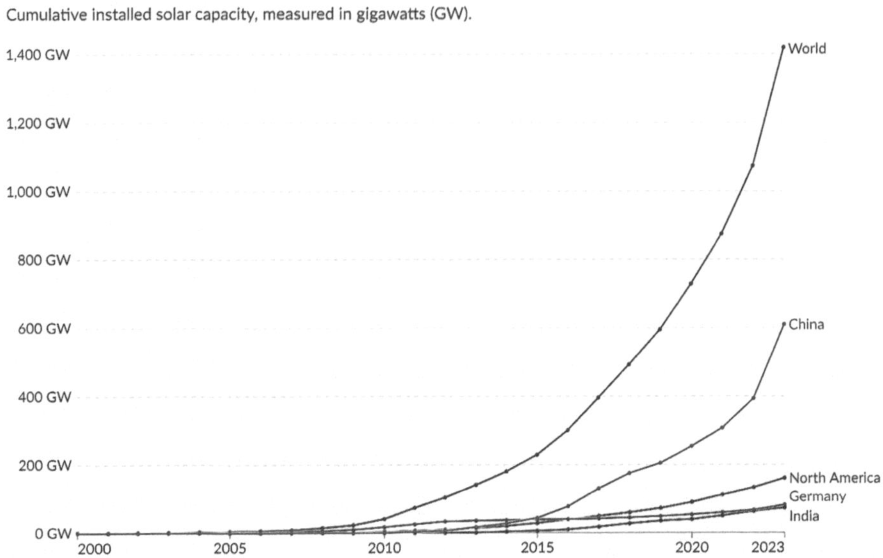

Fig. 5 Installed solar energy capacity. *Source* IRENA (2024), OurWorldinData.org/renewable-energy

space and creating a flow of electrical current. This process is the foundation of solar power generation.

Construction of Photovoltaic Cell

The construction of a PV cell involves several essential components and materials, each playing a critical role in converting sunlight into electricity. PV cells, largely made of silicon, a semiconductor, allow it to absorb photons and release electrons efficiently, incorporating a P–N junction. This junction is formed by doping silicon, creating regions with positive charge carriers ("holes") in P-type and negative charge carriers (electrons) in N-type. This junction is crucial for separating and directing the flow of electrons. Sunlight is absorbed in an absorption layer, releasing electrons, freeing them from their atomic bonds, and initiating the flow of electric current. These electrons are then channeled as electricity through metal contacts on the cell's top and bottom surfaces (Fig. 6).

An antireflecting coating enhances light capture and reduces the amount of sunlight reflected away, allowing more photons to enter the cell and increase its efficiency. The PV cell is usually protected with encapsulation such as glass or transparent plastics. Encapsulation shields the cells from moisture, dust, and mechanical damage while still allowing sunlight to reach the active layers. High-performance cells may also utilize a back-surface field to further enhance electron collection and overall efficiency [21].

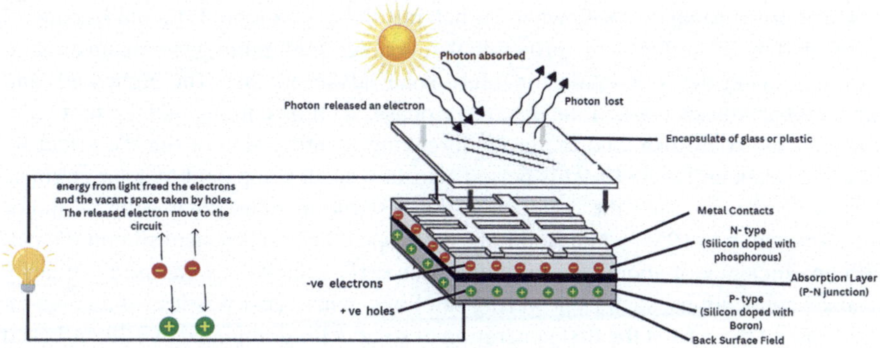

Fig. 6 Construction and working of solar cell. *Source* Compiled by author

Working of Photovoltaic Cell

A PV cell converts sunlight into electricity through the PV effect, a process driven by the interaction of light with semiconductor materials. Sunlight, composed of energy packets called photons, strikes the PV cell's semiconductor layer. When photons penetrate the material, their energy is absorbed by electrons in the semiconductor. The absorbed energy frees electrons from their atomic bonds in the semiconductor lattice, leaving behind positively charged "holes." This creates electron–hole pairs—the foundation for generating electric current. The PV cell's P–N junction creates an internal electric field. This field acts like a one-way valve, pushing freed electrons toward the N-type layer and holes toward the P-type layer, preventing recombination. Metal contacts on the cell's top and bottom surfaces collect the separated electrons. Electrons flow from the N-type side through an external circuit powering a lightbulb or charging a battery, while holes move toward the P-type side, completing the circuit [21, 22].

3 Evolution of Solar Technology

Solar energy is an inexhaustible and clean resource. The idea of a "solar revolution"—where we harness free, abundant energy from the sun to power our lives—has captivated imaginations for decades. On a clear day, the sun delivers about 1000 W of energy per square meter to Earth's surface [23], and the planet receives roughly 10,000 times more solar energy than humanity currently consumes [12].

The evolution of solar technology has been shaped by key breakthroughs in materials science, manufacturing, and a growing focus on renewable energy. The origins of harnessing solar energy date back to as early as the seventh century BC, when ancient civilizations used basic physics to concentrate sunlight for lighting fires, heating spaces, and, later, solar ovens. For centuries, solar applications were largely limited to heating and concentrated solar power (CSP) [11].

The journey began in 1839, when French physicist Alexandre Edmond Becquerel discovered the PV effect—the principle that enables sunlight to generate an electric current—laying the foundation for future solar advancements. The eighteenth and nineteenth centuries marked the discovery phase, with inventions like the first solar collector, the solar dish, and the groundbreaking identification of the PV effect by Edmond Becquerel in 1839. This period also saw the development of solar engines, solar water heaters, and the creation of the first rudimentary solar cells. A major milestone came in 1954, when Bell Labs developed the first practical silicon PV cell with an efficiency of about 6%. Initially, these early solar cells were used primarily in space applications, such as powering satellites, where cost was less of a concern. This marked the start of the first generation of solar cells, dominated by silicon-based technology. Despite their high cost and modest efficiency, these cells demonstrated the potential of solar energy and spurred further research and development [11].

The oil crisis of the 1970s heightened interest in alternative energy, leading to increased investment in solar research and the first commercial uses in remote telecommunications, navigation aids, and off-grid systems. The late 1970s also saw the introduction of thin-film solar cells, which reduced production costs and increased versatility. Second-generation solar cells, emerging in the 1980s, used materials like cadmium telluride (CdTe) and copper indium gallium selenide (CIGS), enabling flexible and lightweight panels for a broader range of applications. During this period, commercial efficiencies improved to around 10–12%, and advances in manufacturing made solar energy more accessible [24].

The 2000s ushered in third-generation solar technologies, including high-efficiency multi-junction cells, organic photovoltaics (OPVs), and perovskite solar cells. Multi-junction cells, which layer multiple semiconductors to capture a broader spectrum of sunlight, achieved laboratory efficiencies exceeding 40% and are mainly used in space and CSP applications. OPVs, using organic molecules or polymers, and perovskite cells, known for their rapid efficiency gains, represent the latest frontiers in solar innovation. OPV cells, which use organic molecules or polymers to capture sunlight, offer the advantages of being lightweight, flexible, and cost-effective, while their efficiency is currently lower than that of traditional silicon-based cells [24].

4 Photovoltaic Cells Generation

Solar cells are typically categorized into different generations, based on the materials and manufacturing techniques employed in their production [25] (Fig. 7).

- **First Generation**: Wafer-Based Technology

The first generation of solar cells is wafer-based, representing the oldest and most widely adopted PV technology. Its popularity is largely due to its relatively high-power efficiencies and the fact that its core fabrication techniques were adapted from established methods in integrated circuit manufacturing, particularly those used for

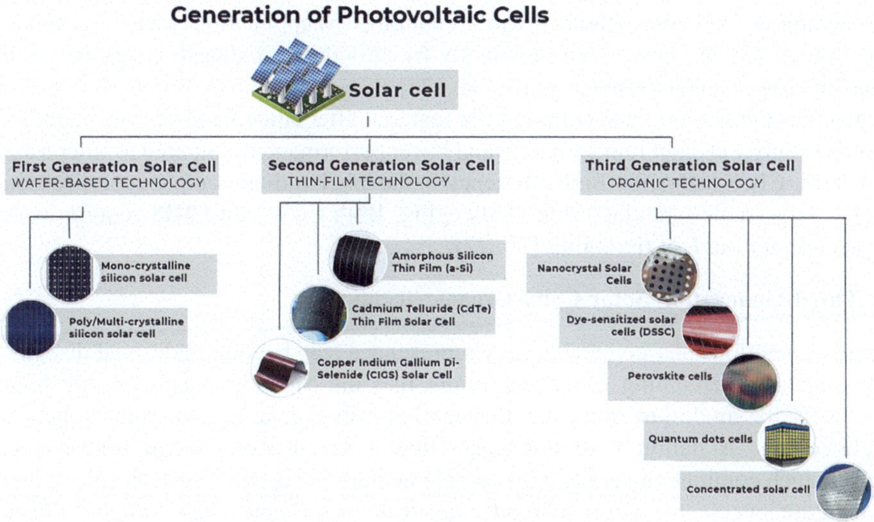

Fig. 7 Generation of photovoltaic cell. *Source* compiled by author

silicon wafer production. This allowed manufacturers to leverage existing expertise and infrastructure in the silicon industry. As a result, first-generation silicon-based solar cells—comprising monocrystalline and polycrystalline silicon—remain dominant in the PV market, accounting for over 80% of global installed capacity and producing about 95% of solar power as of 2020 [25].

Monocrystalline silicon solar cells are favored for their higher efficiencies, typically ranging from 15 to 24%, and long operational lifespans of up to 25 years. The production of pure silicon crystals for monocrystalline cells remains costly, which limits the scalability and cost-effectiveness of these high-efficiency technologies for mass production. Additionally, monocrystalline silicon cells tend to perform better at higher temperatures compared to other types, but their cost is offset by the use of multiple silicon crystals molded together in polycrystalline variants, making them less expensive but also less efficient [18].

- **Second PV Generation**: Thin-Film Technology

Second-generation solar cells were developed to reduce costs and improve certain characteristics compared to the first generation. These cells use thin-film technology, where the absorber layer is only a few microns thick (much thinner than the 100–200 μm layers of the previous generation). While this initially led to lower efficiencies, advances in manufacturing have now enabled efficiencies above 20%, with a record of 23.4% [25].

Thin-film solar cells can be made using materials like amorphous silicon, cadmium telluride (CdTe), and copper indium gallium diselenide (CIGS). These materials are deposited onto substrates such as glass, plastic, or metal using various cost-effective methods like spray coating and vacuum processes, which require lower production temperatures. Thin-film cells are flexible, enabling new applications such as wearable electronics, and are more environmentally friendly due to reduced energy payback time and lower greenhouse gas emissions [25].

CdTe is the second most common PV material after silicon and can be manufactured cheaply, but cadmium's toxicity and lower performance compared to silicon are concerns. CIGS cells offer high efficiency in the lab, but their complex composition makes large-scale manufacturing challenging. Both CdTe and CIGS require extra protection for outdoor durability [26, 27].

- **Third-Generation Solar Cells**: Organic Technology

Third-generation solar cell technology represents a leap beyond traditional designs, focusing on overcoming efficiency limits like the Shockley-Queisser threshold ($\sim 33.7\%$ for single-junction cells) through innovative materials and multi-junction architectures. A hallmark of this generation is the use of tandem (multilayer) cells, which combine materials with complementary bandgaps—such as silicon and perovskite layers—to capture a broader spectrum of sunlight. These structures have achieved laboratory efficiencies exceeding 44% for experimental four-layer gallium arsenide cells, while perovskite-silicon tandems now reach up to 33.9% in research settings [25].

Key achievements in technologies are:

Perovskite solar cells have rapidly gained attention for their high efficiency and low production costs. Laboratory efficiencies for single-junction perovskite cells now exceed 25%, and tandem perovskite-silicon cells have surpassed 33% efficiency. These cells use hybrid organic–inorganic materials and can be manufactured using low-temperature, solution-based processes, making them lightweight, flexible, and cost-effective to produce—often at costs lower than traditional silicon cells. The fabrication process is simpler and less energy-intensive, further reducing greenhouse gas emissions and enabling innovative applications.

Despite these advantages, commercialization faces significant hurdles. The main challenges include sensitivity to moisture, thermal instability, and concerns over lead toxicity in the most efficient perovskite formulations. Additionally, long-term stability and scalability for mass production remain critical issues that must be addressed before perovskite solar cells can achieve widespread adoption [18, 24, 26, 28].

Dye-sensitized solar cells (DSSCs), inspired by the natural process of photosynthesis, are attracting attention for their cost-effective production and design flexibility. These thin-film cells use a semiconductor—typically titanium dioxide—coated with a photosensitive dye to absorb sunlight and generate an electric current. DSSCs can be manufactured using simple, scalable techniques and can be integrated onto various surfaces, offering semi-flexibility and semitransparency.

Despite these advantages, DSSCs face notable challenges. Their power conversion efficiencies, though theoretically predicted to reach around 20%, currently peak at about 12% in the lab, which is lower than conventional silicon and other thin-film solar cells. Stability remains a significant issue, as the liquid electrolyte and some dye materials are prone to degradation under harsh environmental conditions, limiting long-term performance and large-scale commercialization. Ongoing research focuses on improving efficiency and developing more stable, environmentally friendly materials to address these limitations [18, 24, 28].

Organic solar cells (OSCs) use carbon-rich organic molecules or polymers that can be chemically tailored for properties like bandgap, transparency, or color. These cells are lightweight, flexible, and suitable for applications such as wearable electronics. OSCs can be produced using low-cost, energy-efficient printing methods, making them an eco-friendly alternative to traditional silicon cells.

However, their power conversion efficiencies, while recently exceeding 17% in the lab and even reaching over 18% with new designs, still lag behind silicon-based cells. OSCs also face significant challenges with stability and longevity, particularly due to degradation from prolonged exposure to ultraviolet (UV) light, which leads to a decline in performance over time. Despite ongoing improvements in efficiency and durability, short operational lifetimes and sensitivity to environmental factors remain barriers to widespread commercial adoption [18, 24, 26, 28].

Concentrated photovoltaic (CPV) technology also uses mirrors or lenses to concentrate sunlight, but instead of producing heat, it focuses the light onto highly efficient PV cells—often multi-junction cells. This concentration allows for higher efficiency and reduces the amount of expensive PV material needed. However, CPV systems require precise solar tracking and advanced cooling, making them more complex and costly to manufacture. While CPV can achieve higher efficiencies under direct sunlight, its adoption is limited due to the falling costs and widespread availability of conventional silicon PV modules [24].

Quantum dot solar cells leverage nanoscale semiconductor crystals (2–10 nm in diameter) that exhibit unique light-absorption and electricity-generation properties due to quantum confinement—a phenomenon where reducing particle size alters electronic behavior. By precisely controlling the dots' size and chemical composition, their bandgap can be tuned to capture specific wavelengths of light, enabling them to harvest energy across nearly the entire solar spectrum. Lab efficiencies now exceed 18% for quantum dot PV, with research focusing on mitigating surface defects and improving charge carrier mobility. While challenges like long-term stability persist, their tunability and compatibility still persist [24].

5 Technological Advancements

- **Bifacial solar panels** are designed to capture sunlight on both their front and back sides, allowing them to generate more electricity than conventional panels. They are especially effective in areas with highly reflective surfaces, like snow or sand, as they can utilize both direct and reflected sunlight. This dual-sided design leads to significantly higher energy output, making bifacial panels a valuable innovation for maximizing solar power generation [29].

- **Improved solar panel efficiency**: The efficiency of single P–N junction solar cells is fundamentally limited by the Shockley-Queisser limit, which sets a theoretical maximum efficiency of about 33% for converting sunlight into electricity. Most commercial silicon solar panels currently operate at efficiencies between 15 and 22%, with monocrystalline panels achieving up to 27% and polycrystalline panels typically ranging from 15 to 20%. Recent breakthroughs, such as those by LONGi, have pushed laboratory silicon cell efficiencies to 33.5%, approaching this theoretical ceiling. The advanced technologies have already demonstrated lab efficiencies above 33%, with perovskite-silicon tandem cells reaching over 33% and offering the potential for lower production costs and greater power output from smaller surface areas. As these innovations mature, they are expected to make solar energy more efficient, affordable, and widely accessible, further reducing reliance on fossil fuels and expanding the viability of solar installations in diverse environments [29].
- **Energy storage solutions** such as lithium-ion and flow batteries are essential for addressing the intermittent nature of solar energy by storing excess electricity generated during sunny periods for use at night or during cloudy weather. Lithium-ion batteries are the most popular choice for residential and commercial solar storage due to their high energy density, long lifespan (often up to 10,000 cycles), high efficiency, and minimal maintenance requirements [29]. Research is also accelerating on alternative technologies to address the limitations of lithium-ion batteries. Flow batteries, such as advanced vanadium redox flow batteries, offer scalable, long-duration storage with operational lifespans of up to 30 years, making them well-suited for large-scale renewable energy projects [30]. Solid-state batteries, which replace liquid electrolytes with solid materials, promise even higher energy density, faster charging, enhanced safety, and longer lifespans, with several automotive manufacturers planning demonstration vehicles and commercial deployment in the coming years [31]. Additionally, new chemistries like lithium-sulfur, lithium-air, and sodium-ion batteries are being explored for improved sustainability, cost, and performance.
- **Integration with smart grids**: Smart grids use digital communication and real-time data to monitor and respond to changes in electricity demand and supply. By connecting solar power systems to smart grids, energy distribution and consumption can be managed more efficiently, reducing waste and optimizing overall grid performance. This synergy enhances the reliability and effectiveness of renewable energy within the power network [24].
- **Building-Integrated Photovoltaics (BIPVs)** seamlessly incorporate solar panels into the structural elements of buildings—such as rooftops, façades, and windows—allowing these surfaces to function both as part of the building envelope and as sources of on-site renewable energy. This dual-purpose integration transforms buildings from mere energy consumers into active energy producers, supporting the transition to sustainable, low-carbon urban environments. BIPV systems not only deliver clean electricity but also enhance architectural aesthetics,

offering a wide range of materials, colors, and design options that can be tailored to complement or even elevate a building's visual appeal. As urbanization intensifies and available land for traditional ground-mounted solar installations becomes limited, BIPV unlocks the solar potential of vertical and horizontal surfaces in dense cityscapes, making solar adoption feasible in locations where space is at a premium. Beyond sustainability, BIPV contributes to improved energy efficiency, reduced greenhouse gas emissions, and lower long-term energy costs for building owners [24, 32].

6 Policies in Promoting Solar Energy Adoption

As the global community confronts the pressing necessity for a shift toward sustainable energy sources, government policies are vital in encouraging the adoption of solar energy. These measures encompass financial incentives, regulatory frameworks, and strategic initiatives, all of which are crucial for speeding up the expansion of the solar industry.

6.1 Investment Tax Credits (ITCs) and Grants

They are key financial incentives that support the adoption of solar PV systems by reducing the financial barriers.

Investment Tax Credits (ITCs): ITCs allow individuals, businesses, or organizations to claim a percentage of their solar installation costs as a tax credit, directly reducing their tax liability. This credit applies to both residential and commercial properties, covering expenses such as panels, batteries, labor, permits, and more, making it more affordable. The percentage of the credit and eligibility requirements varies by region. To qualify, the system must be owned (not leased), installed, and operational within the eligible year. Unused credits can roll over to future tax years.

Grants: Grants provide direct financial assistance from governments or organizations to help cover the costs of purchasing and installing solar PV systems. These are typically awarded through a competitive process and may have specific guidelines or priorities set by the grant provider [22, 33, 34].

6.2 Net Metering Policies

Net metering is a policy that enables consumers who generate their own electricity—typically from solar—to feed surplus energy back into the grid and receive credits on their utility bills, typically at the retail electricity rate. These credits can be used to offset future electricity consumption when the system is not producing enough power, such as at night or during cloudy periods. This mechanism provides a strong economic incentive for homeowners and businesses to invest in renewable energy systems, as it reduces their energy bills and shortens the payback period for their investments. By allowing consumers to offset their energy consumption with the electricity they produce, net metering makes renewable energy systems more financially attractive and accessible.

Net metering policies have played a crucial role in driving the adoption of renewable energy technologies globally, especially in residential and small commercial sectors. For example, California's Net Energy Metering (NEM) program has been instrumental in expanding the state's residential solar market by offering full retail rate credits for excess electricity generated. Similarly, Germany's Renewable Energy Sources Act (EEG) supports net metering and guarantees grid access, further incentivizing decentralized renewable installations.

These policies not only encourage investment in renewable energy but also contribute to grid stability and efficiency by integrating distributed energy resources. As a result, net metering helps diversify the energy supply, reduce transmission losses, and enhance grid resilience. Net metering policies vary by location, with differences in eligibility, system size limits, credit rates, and how long credits can be carried forward, depending on local regulations [22, 33, 34].

6.3 Feed-in Tariffs (FITs)

FiTs are policy instruments designed to accelerate the adoption of renewable energy by providing renewable energy producers with long-term contracts and guaranteed pricing for the electricity they generate and supply to the grid. These contracts typically offer a fixed or above-market price per kilowatt-hour over a period of 15–25 years, significantly reducing investment risk and providing economic certainty for developers and investors.

The structure of FiTs can vary, with some programs offering different rates based on technology type, project size, or location, and often incorporating a gradual reduction in tariffs to reflect declining technology costs and encourage continued innovation. The focus is on FIT programs that either provide a guaranteed payment per kWh (regardless of the market price) or a premium payment added to the market price. These payment structures can be categorized as:

- **Fixed price FITs**: Producers receive a predetermined, stable price per kWh for all electricity supplied to the grid. Germany exemplifies this approach, offering long-term contracts with fixed rates, which have been instrumental in driving the growth of its renewable energy sector.
- **Premium price FITs**: Producers receive the market price for electricity plus a fixed or sliding premium. Spain's system allows energy developers to choose between a fixed payment or a premium above the market price, with payments

evaluated annually and set within a defined range relative to consumer electricity prices.

Currently, the majority of EU countries have implemented feed-in tariffs, though the details—such as payment amounts, contract durations, and eligibility—vary significantly. For instance, Germany's Renewable Energy Act (EEG) offers fixed FITs for up to 20 years, while Spain's system provides flexibility between fixed and premium options, with annual adjustments based on market forecasts and within a price corridor of 80–90% of the pretax consumer price of electricity. FiTs have been widely implemented across the globe and have proven highly effective in driving the deployment of renewable energy technologies by ensuring stable and predictable revenue streams [22, 33, 35].

6.4 Renewable Portfolio Standards (RPS)

RPS are regulatory policies that require electricity suppliers to source a specified and increasing share of their power from renewable energy sources such as wind, solar, biomass, and geothermal. These standards are also known as quota systems, renewables obligations, or tradable green certificate programs and are widely implemented at the state level in the United States and in many countries worldwide, including Australia, Belgium, Canada, China, India, Italy, Japan, Poland, Sweden, and the United Kingdom. The key features of RPS are:

- **Mandated targets**: RPS policies set minimum requirements for the share of electricity that must come from renewable sources, often increasing over time.
- **Certificate trading**: Many RPS programs use tradable renewable energy certificates (RECs or TGCs), allowing flexibility and ease of compliance for electricity suppliers.
- **Market-based approach**: RPS mechanisms encourage competition and innovation, aiming to deliver renewable energy at the lowest possible cost while integrating renewables into the electricity supply.
- **Public benefits**: RPS policies help reduce greenhouse gas emissions, promote local economic development, conserve water, and can stabilize or lower wholesale power prices.

RPS programs have proven to be effective, flexible, and cost-efficient tools for increasing renewable energy deployment, especially when well-designed and supported by strong political and regulatory commitment. The design and implementation of RPS policies can vary widely, including differences in eligible technologies, target levels, certificate trading rules, and enforcement mechanisms [22, 33].

6.5 India's Integrated Solar Energy Policy 2023

India has established an ambitious goal of reaching 500 GW of renewable energy capacity by 2030, with solar energy serving as a central pillar of this goal.

India has set an ambitious target of achieving 500 GW of renewable energy capacity by 2030, with solar energy as a central pillar of this goal. To reach this target, the government has implemented a multipronged strategy:

- **Production-Linked Incentive (PLI) scheme**: This initiative encourages domestic manufacturing of high-efficiency solar PV cells and modules, reducing reliance on imports and boosting local industry. As a result, India's solar PV manufacturing capacity has surged from 2 GW to about 70 GW over the past decade.
- **Basic Customs Duty (BCD)**: Introduced in April 2022, the BCD on imported solar cells and modules aims to protect and promote domestic manufacturers. The duty rates were recently adjusted to balance local industry support with the need for affordable solar components.
- **Solar Parks and Ultra Mega Solar Power Projects**: Large-scale solar parks have been developed nationwide, benefiting from economies of scale and streamlined processes for installation and grid integration. These projects have significantly contributed to the rapid expansion of solar capacity.
- **Rooftop solar incentives**: Enhanced subsidies and net metering policies have made rooftop solar installations more attractive for residential, commercial, and industrial users. Programs like the PM Surya Ghar scheme offer substantial financial support for households adopting rooftop solar.
- **Solar in agriculture**: The PM-KUSUM scheme supports the installation of solar pumps and solarization of agricultural feeders, helping farmers reduce dependence on diesel and grid electricity while generating additional income by selling surplus power to the grid.
- **Renewable Purchase Obligations (RPOs)**: Utilities and large power consumers are now required to source a rising share of their electricity from renewables, with stricter enforcement and penalties for noncompliance. This policy ensures steady demand for renewable energy and accelerates the transition to clean power.

These integrated policies and incentives have enabled India to cross 200 GW of renewable energy capacity as of 2024, with solar alone contributing over 100 GW [36].

7 International Collaborations in Solar Energy

7.1 One Sun One World One Grid' (OSOWOG) Plan 2020

The OSOWOG initiative, launched by the Indian government, envisions a global network of interconnected solar power grids to enable the seamless transfer of renewable energy across international borders. The core idea is based on the principle that "the sun never sets," meaning solar energy is available somewhere on Earth at any given time. By connecting over 140 countries from the Far East like Myanmar, Vietnam, Thailand, and Cambodia to the Far West, which includes West Asia and the African region, OSOWOG aims to share solar power wherever it is needed, regardless of time zone or geography.

Its aim is to connect the Indian grid with those of the Middle East, South Asia, and Southeast Asia to establish a common regional grid for sharing solar and other renewable energy, and expand this network to include renewable resources from Africa.

Phase 1: Connects the Indian grid with those of the Middle East, South Asia, and Southeast Asia to establish a common regional grid for sharing solar and other renewable energy.

Phase 2: Expands this network to include renewable resources from Africa.

It is designed to maximize the use of renewable resources, reduce reliance on fossil fuels, and provide clean, reliable, and affordable energy worldwide. By enabling cross-border energy sharing, the initiative also aims to enhance grid stability, reduce storage needs, and support global climate and sustainability goals [36, 37].

7.2 International Solar Alliance

The International Solar Alliance (ISA) is a coalition of 121 solar-rich countries established to foster international cooperation and accelerate the global adoption of solar energy. Launched in 2015 by India's Prime Minister Narendra Modi and France's President Emmanuel Macron at the United Nations Climate Change Conference in Paris in November 2015, the ISA serves as a collaborative platform for governments, international organizations, corporations, and industry stakeholders to work together toward expanding solar energy deployment worldwide.

Headquartered at the National Institute of Solar Energy (NISE) in Gurugram, Haryana, India, the ISA is dedicated to advancing research, development, and implementation of solar technologies. In India, the ISA's targets include achieving 175 GW of installed renewable energy capacity by 2022, with 100 GW from solar power alone. On a global scale, the ISA aspires to facilitate the installation of 1000 GW of solar generation capacity, backed by an investment of $1 trillion, underscoring its commitment to sustainable energy development and climate action [38].

8 Opportunities in the Solar Energy Sector

- **Solar Energy in Transportation**

Solar energy is increasingly being integrated into the transportation sector, offering significant potential to reduce emissions and reliance on fossil fuels. Applications range from solar-powered EVs, buses, trains, and boats to solar-powered infrastructure like charging stations, bus stops, and even smart highways. Solar vehicles, including cars, buses, and aircraft, use PV panels to generate electricity for propulsion or auxiliary systems such as ventilation and cooling.

Innovations like solar-integrated cars from companies such as Aptera and Lightyear allow vehicles to charge directly from sunlight, while solar-powered buses and trains are being deployed in cities worldwide. Solar energy is also being used in public transport infrastructure, powering lighting, ticketing, and station operations, which lowers operational costs and carbon footprint.

Additionally, concepts like solar roadways and smart highways are being developed, where roads embedded with solar panels generate electricity, provide lighting, and support dynamic EV charging. Although current technology limits the full potential of solar-powered vehicles, ongoing advancements in lightweight, high-efficiency solar panels and infrastructure are expected to make solar-powered transportation more viable and widespread in the future [39, 40].

- **Solar Car Parking**

A solar carport is a covered parking structure elevated above the ground and fitted with solar panels, allowing vehicles to park in the shade beneath a canopy of PVs. This setup not only maximizes the use of parking areas by generating clean, renewable energy but also provides protection for vehicles from the sun, rain, and other weather conditions. The amount of power produced depends on the size of the solar array; for example, the solar carport at Rutgers University spans 28 acres and generates 8 MW of electricity at peak output.

By transforming underutilized parking areas into energy-producing assets, solar carports maximize space efficiency without requiring additional land. They help lower energy bills, offer the potential to sell surplus power back to the grid, and can be integrated with EV charging stations for sustainable transportation. Additional benefits include protection from weather elements, reduced vehicle cooling needs, and a decrease in carbon emissions. Solar carports are suitable for both residential and commercial use, making them a practical and sustainable solution for generating renewable energy while enhancing parking facilities [40].

- **Solar Kitchen Restaurant**

The Solar Kitchen Restaurant is a unique dining concept designed to promote the use of solar energy by preparing meals exclusively with the power of the sun, launched by Martí Guixé and Antto Melasniemi in 2011. Traveling across Europe and beyond, the restaurant uses parabolic mirrors to concentrate sunlight for outdoor cooking, creating unique flavors and textures distinct from traditional methods. The experience is weather-dependent, requiring flexibility from both chefs and diners, as menus and schedules change with sunlight availability. This project has inspired permanent solar-powered restaurants and highlights the potential of combining renewable energy, creative cuisine, and environmental responsibility to promote sustainable dining practices [40, 41].

- **Solar Canal**

The Canal Solar Power Project in Gujarat is India's first initiative to install solar panels over the Narmada canal network to generate clean electricity. The pilot project, launched near Chandrasan village in Mehsana district, covers a 750-m stretch of canal and produces 1 MW of power. This innovative approach eliminates the need for land acquisition, as panels are mounted above the canal, and significantly reduces water evaporation—saving millions of liters annually. The canal-top solar model not only generates renewable energy but also leverages existing infrastructure to address both energy and water conservation challenges, making it a pioneering solution for regions with extensive canal systems [40, 42].

- **Solar Space Power Station**

The idea behind a solar space power station is to harvest solar energy in space and send it back to Earth for utilization. It has been in research since the early 1970s. This idea addresses the significant loss—about 55–60%—of solar energy that occurs when sunlight passes through Earth's atmosphere due to reflection and absorption. By placing solar panels on satellites outside the atmosphere, these stations can convert sunlight directly into microwaves, which are then beamed down to ground-based receivers, bypassing atmospheric losses and providing continuous power regardless of weather or the day-night cycle. This technology provides continuous, uninterrupted power regardless of weather or time of day, offers higher energy output than ground-based solar, and produces clean energy that can be flexibly distributed worldwide. While challenges remain—such as high launch costs and the need for further technological development—space-based solar power is seen as a promising solution for providing reliable, large-scale clean energy in the future [29, 40].

- **Innovation in Solar Products**

Innovative solar products—such as solar windows, flexible panels, and solar tiles—are revolutionizing how solar energy is incorporated into modern architecture and daily life. Solar panels are now being incorporated directly into building materials, with products like Tesla's solar tiles blending seamlessly with traditional roofing for a more aesthetic and functional solution.

Emerging advancements include "solar skins"—thin, flexible layers of solar cells that can wrap around any surface, adapting to various architectural styles and making solar power generation a standard feature of future construction, as well as self-repairing solar cells that can heal minor damage and clean themselves, enhancing durability and reducing maintenance needs. These technologies not only provide clean energy but also integrate smoothly into the built environment, making solar power more practical, visually appealing, and accessible for everyday use. As material science and manufacturing techniques continue to advance, these solutions are expected to become even more efficient, durable, and widely adopted [17, 43].

- **Solar Paint**

Solar paint is an innovative technology that converts ordinary building walls into electricity-generating surfaces. It consists of nanoscale titanium dioxide particles coated with light-sensitive materials like cadmium sulfide or cadmium selenide, suspended in a water-alcohol mixture to form a paintable paste. When applied to walls or other structures, this paste acts as a PV layer, harvesting solar energy and producing electricity when exposed to light.

While current energy output per square meter remains modest, this breakthrough enables buildings to passively generate clean energy without the need for traditional solar panels. By integrating solar paint into construction materials, urban infrastructure could seamlessly contribute to renewable energy generation, blending functionality with aesthetics. This emerging application highlights the potential for solar technology to expand beyond conventional installations, offering a versatile and scalable solution for sustainable energy integration [40].

- **Solar Fabric**

Solar cell fabrics, or solar textiles, are innovative materials that integrate PV cells directly into textiles, enabling them to generate electricity when exposed to light. This integration can be achieved by embedding small-scale solar cells within textile fibers or by applying flexible PV films to fabrics, resulting in materials that maintain flexibility and can be used in a wide range of applications, from clothing and backpacks to canopies, building facades, and military tents [44].

Companies like Pvilion are leading the commercialization of solar fabrics, producing products that turn everyday surfaces into power sources for charging devices, lighting, or even supporting off-grid needs. Researchers have also developed cylindrical PV cells embedded in fibers, allowing solar energy collection to become silent and unobtrusive in everyday objects—not just rooftops or poles [45].

Solar fabrics are less efficient than traditional rigid silicon panels, but they offer greater versatility and can be produced quickly and cheaply using thin-film or organic polymer technologies. However, challenges remain regarding their lower efficiency, mechanical durability, and longevity, as repeated use and environmental exposure can degrade performance over time [40].

- **Solar Fountain**

Fountains have long been admired for their beauty and calming presence, but the pumps that power them typically consume electricity, contributing to environmental impact over time. By using solar panels to power fountain pumps, it's possible to create an eco-friendly and sustainable water feature. Solar fountains operate independently of the grid, harnessing sunlight to keep the water flowing, reducing energy costs and minimizing their carbon footprint while still enhancing the aesthetic appeal of gardens, parks, and public spaces [40].

- **Floating Solar Farms**

Floating solar farms—solar panels installed on bodies of water—are rapidly gaining popularity as an innovative solution to land and energy challenges. By utilizing lakes, reservoirs, and other water surfaces, these systems conserve valuable land, making them especially attractive for regions with limited space or where preserving agricultural land is a priority. The natural cooling effect of water increases the efficiency of floating solar panels, often boosting energy output by 5–16% compared to land-based systems. Additionally, these installations can significantly reduce water evaporation—studies have shown reductions of up to 70% in some reservoirs—helping conserve water resources, particularly in drought-prone areas [17, 29].

Floating solar farms also offer ecological benefits, such as reducing harmful algae blooms by limiting sunlight penetration and improving water quality. They are less prone to dust accumulation, which helps maintain higher efficiency over time. Many projects now combine floating solar with hydropower or wind, creating hybrid systems that further enhance grid stability, maximize renewable energy output, and optimize existing infrastructure [29].

The global capacity of floating solar farms has grown rapidly, with large-scale projects commissioned across Asia, Europe, and the Americas, and continued expansion expected in 2024 and beyond. Advances such as sun-tracking panels and integration with smart grids are further improving performance and scalability. As this technology matures, floating solar farms are set to play a crucial role in expanding renewable energy capacity, conserving water, and addressing both land and environmental challenges worldwide [43].

- **Agrivoltaics**

This innovative approach, known as agrivoltaics, enables solar panels and crops to coexist on the same land, providing shade for plants while simultaneously generating renewable energy. By allowing agriculture and solar power to share space, agrivoltaics addresses land use conflicts and can be transformative for regions where land availability is a critical concern. The partial shade from solar panels can improve crop yields by reducing heat and water stress, and also helps protect plants from extreme weather, leading to higher productivity and better crop quality [46–48].

Beyond agricultural and energy benefits, agrivoltaics strengthens community ties by preserving farmland, supporting local economies, and increasing returns for farmers. As a result, it enhances both food security and rural development [49]. The global agrivoltaics market reflects this growing interest, valued at $3.6 billion in 2021 and projected to reach $9.3 billion by 2031, with a compound annual growth rate of 10.1% [50].

- **Building Integrated Photovoltaics (BIPV)**

BIPV pertains to the integration of PV materials directly into the building's exterior, including windows, façades, roofs, skylights, and shading systems. Unlike conventional solar panels that are installed on top of existing structures, BIPV systems are crafted to function as both essential building elements and energy generators. BIPV technology enables buildings to generate electricity seamlessly without compromising architectural aesthetics, making solar energy generation an integral part of modern design. Examples include PV glass windows, solar shingles, curtain walls, and exterior wall panels, which can be installed during new construction or retrofitted when building components are replaced [24].

Notable projects like the Edge in Amsterdam and Kaneka Corporation's developments in Osaka showcase how BIPV can be successfully integrated into high-profile buildings, combining sustainability with visual appeal. As the technology advances and awareness grows, BIPV is expected to see broader adoption in the coming years, playing a key role in the transition to energy-efficient, zero-energy buildings [43].

- **Solar-Wind Hybrids**

Combining solar and wind energy in hybrid systems delivers a more consistent and reliable power supply, as the two resources often complement each other—solar power peaks during the day, while wind energy is typically stronger at night or during different seasons. This synergy maximizes energy production throughout the year and reduces reliance on any single source, resulting in greater grid stability and efficiency.

These systems optimize the use of existing grid infrastructure, lower transmission costs, and make better use of available land by colocating installations [51, 52].

- **Green hydrogen**

Harnessing solar energy to produce hydrogen—a clean and versatile fuel—has the potential to transform hard-to-abate sectors like steel, cement, and heavy transport. By coupling solar power with hydrogen generation, industries can significantly cut their carbon emissions, supporting the decarbonization of heavy manufacturing and contributing to broader climate goals. This approach not only offers a sustainable pathway for industrial processes but also paves the way for a cleaner, more resilient energy future [29].

9 Barriers in the Solar Energy Sector

Despite impressive progress and a promising outlook for solar technology, the industry still faces a range of persistent challenges that must be addressed to fully realize its potential. These obstacles span technical, economic, environmental, and social factors [24]. Key issues include policy uncertainty, high upfront costs, supply chain vulnerabilities, land and resource limitations, and the need for skilled labor

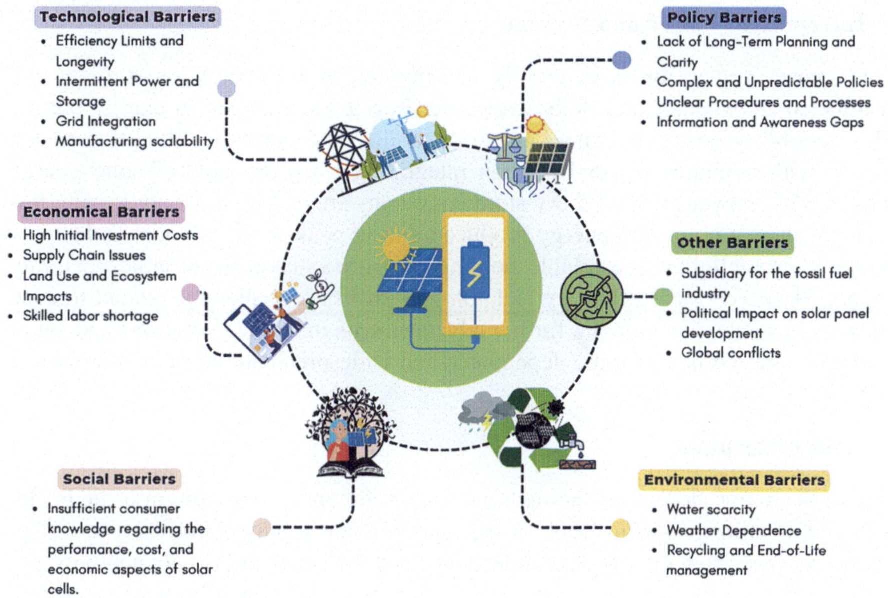

Fig. 8 Barriers to solar energy adoption. *Source* Compiled by author

and public awareness. Overcoming these barriers is crucial for the widespread adoption and long-term sustainability of solar energy, ensuring it can play a central role in meeting growing energy demands and supporting the global transition to cleaner [22] (Fig. 8).

9.1 Technological Barriers

- **Efficiency Limits and Longevity**

Laboratory tests have demonstrated high efficiencies for solar cells, particularly emerging technologies like perovskites, but maintaining these efficiencies in real-world conditions over long periods remains difficult. All solar panels experience gradual efficiency loss—known as degradation—due to factors such as UV exposure, temperature fluctuations, humidity, dust, corrosion, delamination, microcracks, and other environmental stresses. Regions with high temperatures and humidity see accelerated degradation, as heat and moisture can corrode components and reduce output more quickly. Over time, this leads to reduced energy production and shorter panel lifespans. Improving the durability and longevity of solar cells, especially for newer technologies, is essential to ensure consistent performance and reliability in real-world applications [22, 24].

- **Intermittent Power and Storage**

Solar power generation is inherently intermittent, as it depends on sunlight and is affected by weather and daily cycles, leading to fluctuations in energy output. This variability poses challenges for grid stability and makes it difficult to match supply with demand, especially when integrating large amounts of solar energy into existing power grids. Battery storage systems are essential for addressing this issue, as they store excess energy produced during peak sunlight hours and supply power when sunlight is unavailable, helping to ensure a steady and reliable electricity supply. However, current battery technologies still face challenges related to cost, capacity, and lifespan, making further advancements in storage solutions and smart grid integration crucial for the dependable and widespread adoption of solar power [22, 24, 53, 54].

- **Grid Integration**

One of the major challenges facing large-scale solar energy integration is the inadequacy of existing grid infrastructure, especially in developing countries where electrical grids are often outdated or underdeveloped. Insufficient grid connectivity can lead to energy losses and limit the ability to efficiently deliver solar power to where it is needed, ultimately restricting the growth of solar deployment.

Integrating a high share of solar power into current grids is complex and requires significant upgrades to transmission and distribution systems, as well as the adoption of advanced technologies like smart grids and real-time monitoring to manage the variability and intermittency of solar generation. Balancing supply and demand, ensuring voltage and frequency stability, and updating regulatory frameworks to support distributed generation are all essential steps for reliable integration [24, 33, 54].

- **Manufacturing Scalability**

Scaling production of emerging solar technologies like perovskite and quantum dot solar cells remains a critical hurdle, as it requires balancing high efficiency, cost-effectiveness, and quality control during mass production, as well as overcoming supply chain constraints and the need for skilled labor. Developing scalable manufacturing processes—such as roll-to-roll printing for perovskites or solution-based synthesis for quantum dots—is essential to reduce costs and meet global demand.

Additionally, large-scale solar farm deployment often competes with agriculture and urban development due to land requirements. Innovations like BIPVs, which embed solar cells into roofs and facades, and floating solar farms, which utilize reservoirs or lakes, are emerging to minimize land-use conflicts while expanding solar capacity without heavy reliance on traditional land resources [22].

9.2 Economical Barriers

- **High Initial Investment Costs**

Although solar energy costs have fallen sharply in recent years, further reductions are needed for it to compete with conventional energy sources without subsidies. High upfront installation costs remain a barrier, especially in developing countries. Distributed solar systems with net-metering, along with financial solutions, like leasing, power purchase agreements (PPAs), and green loans, are making solar more accessible. Advances in manufacturing and government incentives also help lower costs and support broader adoption [22, 24, 32].

- **Supply Chain Issues**

The sustainability of large-scale adoption of certain solar technologies faces challenges due to reliance on geochemically scarce or hazardous materials, creating supply chain bottlenecks and environmental risks. Many advanced solar panels rely on rare or hazardous materials, such as rare earth elements and specialized semiconductors, used in some thin-film technologies. For example, indium in copper indium gallium selenide (CIGS) and cadmium in cadmium telluride (CdTe) panels are either limited in supply or pose environmental and health risks [24] The extraction and processing of these materials can also lead to significant ecological impacts, including habitat destruction and pollution [17, 22, 54]. To ensure the long-term viability of solar technologies, it is essential to focus on sustainable sourcing and to develop effective recycling systems for these critical materials [24].

- **Land Use and Ecosystem Impacts**

Large-scale solar projects require substantial land, which can lead to the clearing of natural habitats, displacement of agriculture, and disruption of local ecosystems if not carefully managed. Poorly planned installations risk fragmenting habitats and reducing biodiversity, especially in regions where land is already under pressure from competing uses. In places like India, where half the land is devoted to agriculture, expanding solar infrastructure can create conflicts with food production and community needs. To address these challenges, thorough environmental assessments and the prioritization of brownfield or previously degraded sites are essential. For example, India's Bhadla Solar Park, developed on arid wasteland, demonstrates how solar generation can be scaled up with minimal ecological impact.

Innovative solutions such as agrivoltaics—where solar panels are integrated with crop cultivation or livestock grazing—offer a way to maximize land use efficiency and reduce competition between energy and food production. Floating solar farms, installed on reservoirs or lakes, further reduce land pressures and can provide additional benefits, such as lowering water evaporation rates.

With careful site selection, adaptive management, and the adoption of dual-use strategies, solar farms can balance renewable energy goals with the preservation of ecosystems and agricultural productivity, supporting both environmental sustainability and local economies [22, 24, 48, 54].

- **Skilled Labor Shortage**

The term "skilled labor shortage" in the solar industry highlights a significant gap between the demand for and supply of qualified professionals. This shortage is not just about the number of available workers, but also about the specific skills and expertise required across various roles in the sector. The need spans from trainers in solar installation to seasoned project managers, revealing deficiencies at multiple levels.

The issue is complex: the industry faces not only a lack of personnel but also a shortfall in critical skills and specialized training necessary to meet the sector's rapidly growing needs. High-demand positions include solar panel installers, electricians with solar expertise, and professionals skilled in solar business and asset management software. This shortage leads to delayed project completion, increased costs, and excessive workloads for existing staff. Such challenges pose significant obstacles to the future growth and efficiency of the solar industry.

Recruitment in the solar sector brings its own set of difficulties. Workers must possess a unique blend of technical knowledge and practical abilities specific to solar technologies—skills that are rarely covered in standard educational or vocational training programs. As a result, employees often require specialized training, further complicating efforts to quickly build a capable workforce. Addressing these workforce gaps is essential for the industry's continued expansion and success [55–57].

9.3 Environmental Barriers

- **Water Scarcity**

Certain PV technologies, particularly CSP, rely heavily on water for cooling—a significant challenge in arid regions where water scarcity is acute [22]. CSP plants can consume up to 3500 L of water per megawatt-hour (MWh) generated, surpassing even coal and gas plants in water intensity [58].

Solar panel maintenance further complicates water use, as traditional cleaning methods in water-stressed regions conflict with conservation goals. Advances in waterless technologies are addressing this dilemma. To mitigate this, adopting water-efficient cooling systems or exploring water-free alternatives is crucial for minimizing the water footprint of PV [22]. For example, MIT developed a no-contact, electrostatic system that repels dust without water, potentially saving 10 billion gallons annually, but it's too expensive, needs more improvement [59].

- **Weather Dependence**

Solar energy, thanks to falling costs and improved efficiency, presents a strong alternative to conventional fossil fuels. However, its performance is still influenced by atmospheric and weather-related factors. Air pollution, in particular, diminishes the amount of solar radiation reaching panels, thereby reducing their efficiency. Additionally, conditions such as cloud cover and rain can cause fluctuations in solar power output, making consistent energy generation more challenging [54, 60].

- **Recycling and End-of-Life Management**

Solar PV panels, with lifespans of 25–30 years or more, eventually become electronic waste (e-waste) requiring responsible disposal. Without proper recycling, hazardous materials such as lead, cadmium, and heavy metals can leach into ecosystems, posing environmental and health risks [22].

The production of solar panels relies on raw materials like silicon, silver, indium, and gallium, whose extraction can disrupt habitats, deplete water resources, and contaminate soil. While advancements in silicon processing and recycling have halved the industry's carbon footprint over the past decade, energy-intensive manufacturing and end-of-life waste management remain persistent challenges [22, 24].

The International Energy Agency (IEA) estimates that if end-of-life panels are systematically collected and recycled, recovered materials could supply over 20% of the solar PV industry's aluminum, copper, glass, and silicon needs, and nearly 70% of its silver requirements between 2040 and 2050. However, current recycling processes are often energy-intensive and may not fully recover high-value materials like silicon and silver, making it difficult to cover operational costs through material sales alone. This economic challenge hinders the widespread adoption of recycling practices [22].

By enhancing recycling technologies and adopting circular economy principles, the solar sector can further reduce its reliance on new raw materials, lower greenhouse gas emissions, and ensure that solar energy remains a truly green solution throughout its lifecycle [22, 24, 38].

9.4 Social Acceptance Barriers

Introducing new technologies to the public and overcoming deeply rooted mental barriers remains a significant challenge for any nation. For successful adoption, these innovations must be compelling and straightforward for governments, political leaders, investors, industry stakeholders, and end users alike. In the case of solar PV technology, widespread acceptance has been hampered primarily by a lack of social awareness and understanding, particularly in developing countries. Many local consumers remain hesitant to embrace solar PV due to limited knowledge about its benefits and operation, often exacerbated by low literacy rates, misinformation, and cultural resistance. Additionally, the implementation of large-scale solar PV projects is constrained by land availability, as substantial tracts are required for high-capacity plants, which can lead to competition with agriculture and urban development [32, 38]

To accelerate the adoption of solar PV, it is crucial to address these social and cultural barriers through targeted awareness campaigns, community engagement, and inclusive decision-making processes that consider local needs and contexts. Governments and stakeholders must also ensure that solar technologies are accessible, practical, and tailored to the realities of the communities they aim to serve, thereby fostering trust and encouraging widespread acceptance.

9.5 Policy Barriers

Frequent and unpredictable policy changes—such as "stop-and-go" approaches or retroactive adjustments to incentives—undermine investor confidence and disrupt long-term planning for solar projects. Consumers are also troubled by the lack of clear, long-term strategies and roadmaps for solar energy, making it challenging to make informed choices or trust the market's stability. Uncertainty about which entity—whether government, utilities, or the private sector—is responsible for leading solar initiatives adds to this confusion and hesitation. Additionally, consumers often struggle with navigating the complex procedures, policies, and requirements involved in solar installations, from contract terms and financing options to accessing incentives. This lack of transparency and standardized information further hinders widespread adoption [61] [62]. These issues, combined with financial and technical challenges, highlight the need for stable policies, clear leadership, streamlined procedures, and better consumer education to support broader solar adoption.

9.6 Other Barriers

- **Subsidiary for the Fossil Fuel Industry**

Fossil fuel subsidies disrupt the efficient use of resources, diverting funds that could otherwise support sustainable initiatives to enhance social and environmental well-being. According to the International Energy Agency (IEA), explicit global subsidies—where fossil fuels are sold below their supply cost—soared to $1.4 trillion in 2022, marking the highest annual total ever recorded. For comparison, fossil fuel subsidies in 2020 were just over $200 billion, following a three-year decline from nearly $600 billion in 2018. This dramatic increase in subsidies undermines the transition to cleaner energy and perpetuates reliance on fossil fuels [57].

- **Political Impact on Solar Panel Development**

The future of solar energy in the United Kingdom and the United States is uncertain, with the 2024 elections poised to reshape the political climate and potentially alter green policies under new leadership. A potential Labour government in the United Kingdom signals a possible intensification of environmental policies. Although the incumbent Conservative government has established ambitious goals, such as achieving net-zero emissions by 2050, its postponement of the ban on petrol and diesel cars indicates a flexibility that could undermine these targets. A Labour win, conversely, could revitalize green initiatives, perhaps leading to a faster expansion of solar energy [63].

The upcoming US presidential election features a rematch between President Biden and former President Trump, whose previous policies favored economic expansion and energy independence, particularly through support for fossil fuels, at the expense of environmental priorities. His 2018 solar panel tariffs, which added 30% to import costs, negatively impacted the US solar market's development. A potential return to the White House for Trump in 2024 could reverse the ongoing decline in solar installation expenses and revive the focus on fossil fuels, thereby substantially affecting the US solar power industry [64].

- **Global Conflicts**

The war between Russia and Ukraine in 2023 had a major effect on the global energy market. As a leading supplier of raw materials like nickel and palladium—key components for manufacturing solar panels and batteries—Russia's role in these supply chains became especially significant. The conflict caused disruptions in the availability and pricing of these materials, impacting both the production and cost of solar energy products.

Additionally, the crisis pushed Europe to reduce its dependence on Russian natural gas, accelerating investments in solar power and other renewables as part of a broader shift toward energy security and independence. If the conflict ends in 2024 and trade between Russia and Europe resumes, production of solar panels could increase, and gas prices might fall, potentially influencing the competitiveness and pricing of solar energy. However, any new global conflicts could again disrupt trade routes and supply chains, leading to higher costs for solar technologies [64, 65].

10 Conclusion

The future of the solar industry appears highly optimistic, even as it faces certain challenges. The industry is undergoing rapid growth and transformation, steadily advancing toward cleaner, renewable, and more independent energy solutions. Progress in technology, reduced costs, and strong governmental support are all contributing to a significant shift in how electricity is generated worldwide. Enhancements in solar panel efficiency, improved energy storage, and better integration with smart grids are reshaping the sector. At the same time, government incentives and international partnerships are accelerating the adoption of solar technologies. The development of advanced solar software is also expected to enhance system design and operational efficiency, further refining global energy systems.

As interest in alternative energy sources continues to rise, efforts to overcome hurdles within the solar industry are intensifying. With ongoing innovation and increasing investment, solar power is set to become a key element of the world's energy future. Still, it's wise to maintain a balanced perspective. While there are many reasons to be optimistic, not every solar project will succeed—failures often accompany advances. For example, the idea of solar roads, which proposed replacing traditional road surfaces with solar panels, proved impractical due to high costs, low energy production, and durability concerns. As the shift away from fossil fuels accelerates, it is expected that some new technologies will not make it to widespread adoption. These setbacks are a normal part of the transition to a more sustainable energy future.

References

1. Maka AOM, Alabid JM (2022) Solar energy technology and its roles in sustainable development. Clean Energy 6(3):476–483. https://doi.org/10.1093/ce/zkac023
2. Kumar A, Kumar K, Kaushik N, Sharma S, Mishra S (2010) Renewable energy in India: current status and future potentials. Elsevier. https://doi.org/10.1016/j.rser.2010.04.003
3. Zhang A, Yang J, Luo Y, Fan S (2023) Forecasting the progression of human civilization on the Kardashev Scale through 2060 with a machine learning approach. Sci Rep 13(1):11305. https://doi.org/10.1038/s41598-023-38351-y
4. History of energy_ the power through the Ages
5. The world's energy transitions_ a history told in infographics _ World Economic Forum
6. Human development is the history of energy development—Knowledge
7. How have the world's energy sources changed over the last two centuries_-Our World in Data
8. Alliant Energy Kids—The History of Energy
9. Evolution of Energy Sources _ The Geography of Transport Systems
10. Global electricity demand to grow 4% annually up to 2027
11. Tabassum S et al (2021) Solar Energy in the United States: development, challenges and future prospects. Energies (Basel) 14(23):8142. https://doi.org/10.3390/en14238142
12. Mousa OB, Taylor RA (2020) Global solar technology optimization for factory rooftop emissions mitigation. Environ Res Lett 15(4). https://doi.org/10.1088/1748-9326/ab702a
13. Internatinal Energy Agency (2024) World energy outlook 2024. [Online]. Available: www.iea.org/terms
14. Metayer M, Breyer C, Fell H-J (2015) The projections for the future and quality in the past of the World Energy Outlook for solar PV and other renewable energy technologies
15. Global CO_2 emissions
16. Emissions from fossil fuels continue to rise
17. Future of solar power energy in the world—Vishakha Renewables
18. Bouich A, Pradas IG, Khan MA, Khattak YH (2023) Opportunities, challenges, and future prospects of the solar cell market. Multidisciplinary Digital Publishing Institute (MDPI). https://doi.org/10.3390/su152115445.
19. Global PV Market Outlook, 4Q 2023 _ BloombergNEF
20. What is the future scope for the solar sector_ _ SolarCtrl

21. Photovoltaic cell _ GeeksforGeeks
22. Panagoda LPSS et al (2023) Advancements in photovoltaic (PV) technology for solar energy generation
23. Toothman J, Aldous S (2000) How solar cells work, vol 1
24. Verma V (2024) Advancement in solar technology: evolution, generation, future prospective, and challenges—a review. https://doi.org/10.20944/preprints202407.0003.v1
25. Marques Lameirinhas RA, Torres JPN, de Melo Cunha JP (2022) A photovoltaic technology review: history, fundamentals and applications. Energies (Basel) 15(5):1823. https://doi.org/10.3390/en15051823
26. Solar photovoltaic technology basics _ Department of Energy
27. Photovoltaic cells – solar cells, working principle, I_U characteristics, generations, material systems, multi-junction cells, organic photovoltaics, laser material process
28. Photovoltaic cell – definition and how it works _ Planète Énergies
29. Powering tomorrow_ The evolution and future of solar energy
30. Flow battery market report 2025—flow battery industry trends and growth
31. Solid-state batteries_ A new era for electric vehicles _ WhichEV.Net
32. Pandey AK et al (2022) Solar energy utilization techniques, policies, potentials, progresses, challenges and recommendations in ASEAN countries. Sustainability 14(18):11193. https://doi.org/10.3390/su141811193
33. Agupugo CP, Ajayi AO, Nwanevu C, Oladipo SS (2024) Policy and regulatory framework supporting renewable energy microgrids and energy storage systems. Eng Sci Technol J 5(8):2589–2615. https://doi.org/10.51594/estj.v5i8.1460
34. Government policies role in promoting solar energy
35. Rowlands IH (2005) Envisaging feed-in tariffs for solar photovoltaic electricity: European lessons for Canada. Renew Sustain Energy Rev 9(1):51–68. https://doi.org/10.1016/j.rser.2004.01.010
36. Upendra Kumar P, Patnaik RK, Kumar Gouda P, Kishore TS (2024) Policies for promotion of solar energy in India: a review
37. Patel H, One sun one world one grid. [Online]. Available: https://www.researchgate.net/publication/355779001
38. Raina G, Sinha S (2019) Outlook on the Indian scenario of solar energy strategies: policies and challenges. Energ Strat Rev 24:331–341. https://doi.org/10.1016/j.esr.2019.04.005
39. Intelligent Highways-692
40. Sadhu M, Chakraborty S, Das N, Sadhu K (2016) Role of solar power in sustainable development of India. Int J Appl Power Eng (IJAPE) 5(2):103–110
41. Solar kitchen restaurant by martí guixé and antto melasniemi returns to Lisbon, Portugal
42. Canal Solar Power Project—Wikipedia
43. How the future of solar energy will transform the power industry in 2025
44. Solar cell fabric—Wikipedia
45. Solar Power Fabric with Pvilion's Colin Touhey—Techstrong TV
46. Dinesh H, Pearce JM (2016) The potential of agrivoltaic systems. Renew Sustain Energy Rev 54:299–308. https://doi.org/10.1016/j.rser.2015.10.024
47. Tajima M, Iida T (2021) Evolution of agrivoltaic farms in Japan, p 030002. https://doi.org/10.1063/5.0054674
48. India's Solar Energy Sector_ Challenges, opportunities, and future prospects—PV magazine India
49. Navigating the trends of 2024_ Automation, AI, and the role of sheep in project success _ Utility Dive
50. Agrivoltaics Market Size, Share—2031 _ Analysis Report
51. Hassan Q, Algburi S, Sameen AZ, Salman HM, Jaszczur M (2023) A review of hybrid renewable energy systems: solar and wind-powered solutions: challenges, opportunities, and policy implications. Results Eng 20:101621. https://doi.org/10.1016/j.rineng.2023.101621
52. Nwagu CN, Ujah CO, Kallon DVV, Aigbodion VS (2025) Integrating solar and wind energy into the electricity grid for improved power accessibility. Unconventional Resour 5:100129. https://doi.org/10.1016/j.uncres.2024.100129

53. Albadi M (2019) Solar PV power intermittency and its impacts on power systems—an overview. J Eng Res [TJER] 16(2):142. https://doi.org/10.24200/tjer.vol16iss2pp142-150
54. Advancing solar panel efficiency—innovations, challenges and the way forward _ Industry EMEA
55. Briggs C, Atherton A, Gill J, Langdon R, Rutovitz J, Nagrath K (2022) Building a 'Fair and Fast' energy transition? Renewable energy employment, skill shortages and social licence in regional areas. Renew Sustain Energy Transition 2:100039. https://doi.org/10.1016/j.rset.2022.100039
56. Tackling the solar industry's skilled labor dilemma—cloud-based field service software
57. Mathur A, Foreword
58. Cutting water consumption in concentrated solar power plants
59. How to clean solar panels without water _ MIT News _ Massachusetts Institute of Technology
60. Kim MJ (2024) Air pollution and solar photovoltaic power generation: evidence from South Korea. Energy Econ 139. https://doi.org/10.1016/j.eneco.2024.107924
61. Report on barriers for solar power development in India 2010 South Asia Energy Unit Sustainable Development Department The World Bank
62. International Energy Agency, Solar energy policy in Uzbekistan: a roadmap. [Online]. Available: www.iea.org/t&c/
63. Scott E (2024) Library briefing king's speech 2024: energy security, net zero, environment and agriculture
64. What is the future of solar power in 2024 and beyond_
65. Zhang Z, Jiang C, Gao C, Tang B (2025) The impact of the Russia–Ukraine conflict on renewable energy trade in countries along the belt and road: a cascading failure model. J Renew Sustain Energy 17(1). https://doi.org/10.1063/5.0233853

The manufacturer's authorised representative in the EU is Springer Nature Customer Service Centre GmbH, Europaplatz 3, 69115 Heidelberg, Germany. If you have any concerns regarding our products, please contact ProductSafety@springernature.com

Printed and bound by CPI Group (UK) Ltd, Croydon, CR0 4YY
21/09/2026
02219990-0005